Enrico Fermi His Work and Legacy

SIF
Bologna

Springer
Berlin
Heidelberg
New York
Hong Kong
London
Milan
Paris
Tokyo

The translation of this work has been funded by SEPS
SEGRETARIATO EUROPEO PER LE PUBBLICAZIONI SCIENTIFICHE

Via Val d'Aposa 7 – 40123 Bologna – Italy
tel +39 051 271992 – fax +39 051 265983
seps@alma.unibo.it – www.seps.it

Translated from the Italian by: C. V. Pennison (*G. Salvini, T. Levi-Civita, C. Salvetti*), A. Olachea (*E. Amaldi, F. Rasetti, G. F. Bassani, G. Parisi, G. Gallavotti, B. Bertotti*), A. Crowe (*M. Cini, N. Cabibbo, A. Gandini, M. Cumo, U. Amaldi*), A. Casella (*E. Persico, M. Falcioni and A. Vulpiani, R. A. Ricci*).
L. Bonolis (author's original) revised by A. Casella. M. Jacob and L. Maiani (authors' original).

Original title:
Conoscere Fermi, nel centenario della nascita 29 settembre 1901-2001
edited by C. Bernardini and L. Bonolis
Editrice Compositori, Bologna, 2001 (1st edition), SIF, Bologna, 2002 (2nd edition)

English translation with authors' updating

Produced by the Editorial Staff of the Italian Physical Society, Bologna, Italy

Data conversion: K.H. Kuivalainen

Cover design: S. Oleandri

Printing and binding: Compositori Industrie Grafiche, srl, Bologna, Italy

Printed on acid-free paper

ENRICO FERMI
His Work and Legacy

edited by
Carlo Bernardini and Luisa Bonolis

Società Italiana di Fisica

Springer

Carlo Bernardini
Luisa Bonolis
Dipartimento di Fisica
Università di Roma "La Sapienza"
P.le A. Moro, 5
00185 Rome, Italy

ISBN 88-7438-015-1 SIF, Bologna, Italy
ISBN 3-540-22141-7 Springer-Verlag Berlin Heidelberg New York
Library of Congress Control Number: 2004108212

Jointly published by:

Società Italiana di Fisica, Bologna
http://www.sif.it

and

Springer-Verlag Berlin Heidelberg New York
Springer-Verlag is part of Springer Science+Business Media
springeronline.com

Printed in Italy

Enrico Fermi

Enrico Fermi's scientific work, noted for its originality and breadth, has had lasting consequences throughout modern science. Written by close colleagues as well as scientists whose fields were profoundly influenced by Fermi, the papers collected here constitute a tribute to him and his scientific legacy. They were commissioned on the occasion of his 100th birthday by the Italian Physical Society and confirm that Fermi was a rare combination of theorist, experimentalist, teacher, and inspiring colleague. The book is organized into three parts: three biographical overviews by close colleagues, replete with personal insights; fourteen analyses of Fermi's impact by specialists in their fields, spanning physics, chemistry, mathematics, and engineering; and a year-by-year chronology of Fermi's scientific endeavors. Written for a general scientific audience, "Enrico Fermi: His Work and Legacy" offers a highly readable source on the life of one of the 20th century's most distinguished scientists and a must for everybody interested in the history of modern science.

TABLE OF CONTENTS

Foreword

Without Enrico Fermi, 20th century physics would have been only a shadow of what it became. The essays in this book give evidence to his extraordinary role as superbly innovative scientist on a range of fields so large that it is unsurpassed by any other physicist in that century. He was an experimentalist and theorist —a very rare combination at that time— and an inspiring colleague and teacher.

To all this I would add that he was the very role model in other respects also. One was his willingness and ability to immerse himself in work that would change the course of history itself. This is what he and his team did in the 1920s and '30s in Rome, by raising the place of physics in Italy to world-class level, and thereby also laying the foundation for the continuation, after the war, of Italy's successes throughout the sciences.

A second example is of course his key contribution to the Allies' development of nuclear energy, during the race to preempt analogous efforts in totalitarian countries. As some of the essays in this book note, he also was careful to express his ethical concerns of social consequences and potential misuses of scientific and technical advances.

Those who were fortunate actually to meet him realized quickly that all his accomplishments and unchallenged authority issued from a person who seemed entirely unaware or uninterested in his own preeminence. To the support concerning this point, as presented in the essays in this book, I may be permitted to add a personal one. When Enrico Fermi came to give a series of invited lectures at Harvard University, we were astonished and delighted to find him, on the human level, to be so accessible, unprepossessing, in short to be a delightful colleague.

Some years later, I became interested in having a motion picture made for educational purposes, one that would concentrate on the personal and scientific biography of an outstanding scientist. Having asked around for suggestions of whose life to present, I found Enrico Fermi was constantly recommended as the most appealing and instructive choice. The interviews with his former students and colleagues that were eventually filmed for this biography showed how correct that choice had been.

Doing science is usually difficult and all-consuming. In the heat of the chase, we forget all too often to pause and to recall one of the giants on whose shoulders we stand. Luckily, here is a welcome chance for us to do just that.

GERALD HOLTON

Mallinckrodt Research Professor of Physics
and Professor of History of Science
Harvard University

Preface

I had the honor to chair the National Committee for the Celebrations of the centenary of the birth of Enrico Fermi (September 29, 2001). In this capacity I agreed with the Italian Minister of Public Instruction, Luigi Berlinguer, and with his successor Tullio De Mauro, that an important task of the Committee was to provide some contributions on relevant topics of Fermi's physics, suited for teaching in secondary schools (in Italy: "scuola secondaria superiore"). Both Ministers, accepting and financing our proposal, agreed in supporting the publication of a book in Italian to be distributed in every secondary school in the country. Our purpose was to stimulate physics teachers to initiate the students in the understanding of Fermi's activity. We were conscious of the difficulty of the task but were also urged on by the fact that Fermi was popularly known mainly, if not only, for his work on the atomic bomb rather than, more properly, for the extraordinary scientific achievements in many different fields. Therefore, a group of distinguished Italian physicists was selected to prepare simple texts suited for the goal of introducing Fermi's ideas in the terminal year of any such physics course. The authors were enthusiast about this opportunity and reacted very positively in a relatively short time (which is rather exceptional). Since when he was a student, Fermi was an excellent connoisseur of analytical mechanics, and the eminent Italian mathematician Tullio Levi-Civita greatly appreciated his work. For this reason I suggested to reproduce also the exhaustive contribution presented by this author at the Como Conference in 1927 on the so-called "adiabatic invariance", in order to implement the knowledge of the school teachers on this topic.

As far as the biographical notes are concerned, the Committee decided to utilize three talks delivered soon after Fermi's death by his close collaborators: Edoardo Amaldi, Enrico Persico and Franco Rasetti. These talks have the virtue of describing a "living exceptional colleague" rather than a "lamented exceptional leader". This is, I believe, an important quality of texts dedicated to young people. In addition to these, a very detailed and impressive chronology of the indefatigable scientific activity of Fermi was presented at the end of the book by Luisa Bonolis: it looks like a long tape full of extremely dense information covering nearly 40 years.

Giorgio Salvini prepared the introductory paper of the book, commenting on the various contributions it contains. Fermi was a multidisciplinary physicist, and this introduction was necessary to fix the coordinates of the more specific chapters touching problems scattered in fields quite far apart in physical sciences, from relativity to nuclear physics, from statistical mechanics to elementary particles, from reactor technology to computer simulations. Salvini's introduction is, therefore, the glue connecting the various topics and the pointer indicating the evolutions subsequent to Fermi's disappearance.

The Italian version of the book had a good success in the schools; many teachers consulted and utilized it following their didactic criteria. In any case, the volume was a reliable source of a large variety of issues in modern physics, even if referred to a single person's work. Therefore, we had the idea of promoting an English translation to allow a larger diffusion in the world. The Italian Physical Society (SIF) immediately endorsed the idea, and took care of the translation. Here is the result of this decision. All of us, the authors and I, hope to have done a useful job. Please let us know if some improvement would produce a better future issue.

CARLO BERNARDINI

Università " La Sapienza", Rome

Introduction

Enrico Fermi. His life and a comment on his work

Giorgio Salvini

With a series of papers written by a number of outstanding figures in Italian scientific research this book aims to illustrate Enrico Fermi's discoveries and their consequences for our present knowledge of physics, chemistry, mathematics, engineering, and all the technical progress that makes such a decisive contribution to all our lives.

Basically the book is divided into three parts. The first (sections **1-5** of this preface) is essentially biographical and it presents Enrico Fermi's life through three biographies written by physicists that lived and worked with him.

The second and most extensive part contains fourteen independent articles by specialists which illustrate Fermi's activity and discoveries in a specific field of physics and the consequences of his work up to our own day (section **6** of this preface).

The third part contains a chronology showing the evolution of Fermi's scientific work year by year.

Enrico Fermi's work was extremely wide ranging both in terms of its originality and of its range of interests. It is documented for the most part in Enrico Fermi's "Note e memorie"(Collected Papers) (hence forward FNM) published by the Accademia Nazionale dei Lincei and by the University of Chicago Press [1].

These fill two volumes of over 1000 pages each. They contain the scientific papers he wrote alone or in collaboration with others, with appropriate comments from his students and collaborators.

This book is one of a series of national initiatives organised in order to honour Fermi's memory in the centenary of his birth. He will be commemorated in various places, in Italy and abroad.

I wish to express our gratitude to the National Committee for the Fermi Celebrations and to its President Professor Carlo Bernardini for the conferences, exhibitions and museums prepared for this important occasion.

1. – Fermi's life, written by three outstanding physicists

The book, as we have said, contains three commemorations of Enrico Fermi: two by Enrico Persico (*) and by Edoardo Amaldi (*), read a few months after his death. One from 1968, written by Franco Rasetti, in the *Celebrazioni Lincee* of that year (*). Today Rasetti lives in Brussels and is a hundred years old.

These renowned scientists spent many intense years of study and work with Enrico Fermi, as they recall. Enrico Persico was also a friend and fellow student in their early youth in Rome. Amaldi and Rasetti took part in the fundamental nuclear research activity in Rome. They all maintained a continuous scientific relationship with him and they remain noble and unforgettable characters in our history of physics and of physics teaching, with different characters and different theoretical and experimental interests.

For someone like myself who had the privilege of knowing Fermi it is remarkable how his character emerges from these three accounts, with all his stature as an outstanding scientist, capable of flashes of intuition, but also an untiring, calm, tranquil worker, with great humanity towards his fellow scholars, students, his family and friends.

The three articles published here should therefore be read, because they introduce the figure of Fermi the man into world history and into teaching. They recall him in his younger years, with amusing episodes, and together they show a fraternal friendship and deep admiration for him without rivalry or envy.

This image would always accompany him, until his death. At the same time he is an intellectual giant, but also one of us, a human being who suffers and strives, who searches and works in the laboratory and at his desk, who wants to teach well and prepares for it thoroughly, who makes the knowledge he has acquired available to all.

I would also like to say with regard to Italy's evolution that Enrico Fermi's scientific line of research, from statistical mechanics to the understanding of the most complex atoms, from the flourishing development of the theory of solids to the settling of electrodynamics, to the outstanding theory of beta decay, opened up a line of thought that kept Italy in the front line theoretically and experimentally over the following decades. Consider for example, in the field of particles involved in the weak interactions, the famous discovery by Conversi, Pancini, Piccioni [2] on the fundamental properties of the lepton family, interpreted in its unexpected depth and extension by Fermi's collaborators (see the article by Jacob and Maiani); the historical clarification of field theory brought about by Cabibbo's angle; the discovery of intermediate bosons by Rubbia and his collaborators; the opening of new laboratories in Italy for research into the physics of aggregate states (something that Fermi paid great attention to, from his thesis at Pisa University and throughout his life).

As Ettore Fiorini observed in his introduction to the story of Enrico Fermi written by Pontecorvo [3], *"Rarely in the history of our country, and not only in the history of science, has there been a phenomenon as important and, at the same time, so remarkable and so long lasting, as the birth and development of the school of physics created by Enrico Fermi."*

2. – The papers written by various contemporary Italian authors

Thirteen of the papers in this book were written by Italian physicists now working in fields that he opened up and that have developed considerably over the century that has just ended. To help the historical setting a paper written for the Como Congress of 1927 by Levi-Civita has been added. He was a great mathematician who immediately took an interest in Fermi's work. Each of these authors has already left the mark of his personal results on the history of our science. But it is of great help to our understanding of history to see from the texts how their current results derive from, or are connected to, Enrico Fermi's initial seminal work in the twenties, thirties, forties, and fifties of the last century. In this preface we will try to run through them one by one. They will be commented on specifically in paragraphs **6**.1-**6**.13.

3. – A great teacher

This book is of general interest for everyone who loves science, but it is dedicated in particular to high school teachers of physics and mathematics who have to introduce the young to the study of the physical and natural sciences and mathematics in the age group from fifteen to eighteen. Of course you cannot take the high school student Enrico Fermi as an example for these students: at eighteen he knew much more about the physics and mathematics of the day than the average student then graduating after four years at university.

Nevertheless he was a quiet student, reserved, occasionally youthfully exuberant. I see something extraordinary and unique in this genius —who was well aware of it— as he undertook his years of study with his companions at the University of Pisa.

I would now like to make a few comments about him, not necessarily in order, on the basis of the three clear and coherent biographies presented here which offer a precise and ordered link to the life and figure of our Enrico, quoting the opinions of some famous physicists from our scientific world, on his work and style.

Inevitably he emerged early from this astounding, quiet youth. Many Italian physics and mathematics teachers appreciated his greatness, and in particular he was guided in his early scientific progress by our Orso Mario Corbino who offered him a scholarship and sent him abroad to environments more open to new ideas than the Italian scientific community. This is part of the motivation for the scholarship (ref. [4], p. 30) for the twenty-two year old Enrico Fermi.

"Only a few months after graduation he already shows a scientific maturity that will allow him to deal with problems of mathematical physics and experimental physics with confidence and clear intuition, as shown by: the interesting research into the dynamics of a rigid system of electrical charges; the explanation of the fundamental differences

between the expressions of the electromagnetic mass; the notes on pure general relativity concerning the phenomena approaching a time line... Faced with such massive and profitable activity at the very beginning of a career one can only admire and hope that, with a research post, this candidate will be able to extend the field of his knowledge of physics and to benefit to the utmost from his studies."

It should also be remembered that Enrico Fermi was a clear and excellent teacher who also taught his own professors, even during his student years at Pisa.

From then on Enrico Fermi managed to keep a constant unity of action, by which I mean that he was present at the same time in all the various physical sciences, calling on each in turn to interpret the other. He did this with the greatest simplicity, which made the theme under discussion fascinating, and that allowed him to reach efficiently and rapidly the central and most significant part of the problem under consideration. This characteristic may perhaps be unique in the history of our century By this I mean the ability to resolve the most complex aspects of a physical problem, theoretical and experimental, by going straight to its heart [4].

This ability soon became known around the world and aroused the admiration of other great scientists, as I shall soon relate. But first let me meditate on the early beginnings of the 20th century.

4. – Those extraordinary five years 1921-1926

Between 1900 and 1902 five great men who laid down the basis of our knowledge of physics were born. In order of birth they are: Pauli, Dirac, Heisenberg, Fermi, Jordan.

Between 1920 and 1933 these extremely young giants laid down the basis for our current knowledge. The way in which their different interpretations alternated and seemed to contradict each other, to then arrive at a new original picture of the universe, is fascinating and makes us thoughtful about mankind's adventure, or rather about the high and unexpected leaps men make to arrive at knowledge. We should remember the pages written by Abraham Pais [5] commenting on this period, which in Göttingen became known as the years of "Knabenphysik", the physics of youth.

Enrico Fermi did not participate directly in the early absolute foundation of these new ideas, but he contributed enormously to simplifying them and creating a unified vision. A difficult and delicate task, that came naturally to him (see for example Cini's comments in his article "Fermi and quantum electrodynamics".

Let me pause here for a moment. It has been said in this book of ours, by some authors, that it seems strange that immediately after his arrival at Göttingen, in the midst of the ferment surrounding the new matrix mechanics, Fermi did not immediately throw himself into the problem, and take his place amongst the founders. I am not a historian who tries to establish the reasons why something has happened in a certain way: maybe it was not easy for him to break into an already existing group which spoke a different language; maybe the matrix method did not seem to him to be the most appropriate to describe with certainty the atomic phenomena then known. Maybe he saw it as an arcane foreknowledge rather than a concrete scientific message. In which case he had not grasped the potential offered by the knowledge of matrices. We know that to this day there are those, including some amongst the writers of these papers, who prefer the wave path rather than the matrix path.

In any case Fermi made remarkable contributions in other fields in those years 1923-1924. Amongst other things he pointed out the almost insurmountable difficulties in the Bohr-Sommerfeld quantization (see the articles by Gallavotti, Falcioni and Vulpiani (*)). His ability to enter the quantum world however started in 1926 when he laid hands on Schrödinger's work. Fermi, as Segrè recalls in the history of his life, was dumbstruck, he set to work on the equation, and in accordance with his now well known characteristics, made a simplifying contribution to the new quantum mechanics, and to the analysis of the incompatibility between Heisenberg and Schrödinger [4, 6].

Thinking of that historical period, 1921-1926, it comes naturally to me to think of those great physicists as a team of mountaineers, roped together, setting out to conquer a new, high and difficult peak. There was a new way to be found, with no previous paths. It was an undertaking that could only be accomplished by geniuses with different characters, in a collaboration born from reciprocal esteem and from an enormous and swift critical capacity.

Fermi therefore only became important later. But in that same year of 1926 he became one of the greats with his development in statistical theory "Sulla quantizzazione del gas perfetto monoatomico" (On the quantization of the perfect monoatomic gas) (ref. [1] FNM, 30), first proposed by him, that was then named the Fermi-Dirac statistics. The story of this first international success is narrated in particular in the papers published here by G. Parisi and by F. Bassani (*).

Just let me say that everyone is made in their own way. You cannot blame a physicist who has done great things for not having done others. If I can express myself in a joke, you cannot blame Shakespeare for not having written "Doctor Faustus".

5. – Some judgements on Enrico Fermi by other great physicists

Returning to the theme of my story, I would like to recall some statements by some of the great scientists who recognized Enrico Fermi's simplifying genius: they are examples or tributes made in different situations and on different occasions.

In 1993, remembering Enrico Fermi, Victor Weisskopf [7], one of the great world physicists who helped present clearly the scientific results in nuclear and electrodynamic physics, wrote:

"Fermi was unique in his way of doing physics. He had a very special way of attacking problems. He always managed to find the simplest and most direct approach, with the minimum of complication and sophistication. In the early 1930's, when I tried in vain to understand the new quantum electrodynamics (see M. Cini's article in this book) *I was lucky enough to find in "Review of Modern Physics" Fermi's article called "Quantization of radiation in the Coulomb gauge". I studied it, and from then on I understood field theory. I know I am not the only one who reached this result and who has this opinion."*

We find a similar idea in a speech by Bethe on Enrico Fermi, which recalls difficulties with field theory (ref. [4] p. 54: BETHE H. A., *Memorial Symposium in honor of Enrico Fermi at the Washington Meeting of the American Physical Society, 1955, Rev. Mod. Phys.*, **27** (1955) 253): *"... It is an unsurpassed example of simplicity in a difficult subject. It appeared after a group of extremely complicated papers on the subject, and preceded another group of papers that were equally complicated. Without Fermi's luminous simplicity I think that many of us would have been unable to explore field theory in depth. I am certainly one of them."*

This is the great theoretician Hans Bethe speaking, the Nobel prize winner for stellar evolution and for the origin of heavy nuclei.

And here is a thought from C. N. Yang on Enrico Fermi's teaching. It is to be found in the "Note e memorie" (Collected Papers) already mentioned (FNM p. 673), as an introduction to the classic article "Are mesons elementary particles?".

Yang, a future Nobel prize winner along with T. D. Lee for the discovery of the non conservation of parity, said amongst other things: *"On every subject, Fermi had the habit of always starting from the beginning, examining simple examples, avoiding formalisms as much as possible (he used to joke that complicated formalism was reserved for High Priests). The very simplicity of his way of thought gave the impression of effortlessness, but it was a false impression: simplicity was also the outcome of careful preparation and of a careful weighing of the various possible alternatives... Finding that Fermi had kept detailed notes on various physical subjects for years was an important lesson for all of us. We learnt that this was physics: it has to be built up from the foundations, brick by brick and layer by layer. We learnt that abstractions come after profound preparatory work"* (ref. [4], p. 172, and ref. [1], Vol. 2, p. 373, which also reports: FERMI E. and YANG C. N., *Phys. Rev.*, **76** (1949) 1739).

A final statement comes from an experimental physicist who worked with Fermi for fifteen years, from the construction of the pile to the final research on elementary particles. This is how Herbert Anderson remembers Fermi's style [8]:

"I was immensely attracted by Enrico Fermi: he had an exceptional grasp of physics, and he kept it between his fingers, always ready for use. When a problem was born he was ready to go to the blackboard, and simply solve it. Physics flowed freely from his chalk."

Here Anderson refers to Fermi the theoretician and experimentalist working on the construction of the nuclear reactor in the forties.

6. – Some comments on the fourteen papers by the individual authors

But now I would like to go through the fine articles in this book for a brief presentation, to help in the continuity between one theme and the next, for us all to feel Enrico Fermi as an inspirer and as one of us, and to arrive at a conclusion that may help us in our work.

Moving rapidly through the papers presented by the various authors, we will begin with three of them that show Enrico Fermi's definite entrance into the Olympus of the great physicists, in 1926-1933. They are the papers by Giorgio Parisi on statistical physics, by Franco Bassani on solid state physics, by Nicola Cabibbo on weak interactions (*).

6·1. *Giorgio Parisi:* "Fermi's statistics" (*). – This essay begins with a clear presentation of the general concepts of probability and statistics for elementary particles. In some aspects it is appropriate for a lesson for final year High school students, given its clarity and method of explanation. Later he develops a subtle critical analysis of Bose's statistics for photons and of Einstein's interpretation. Parisi recognizes Fermi's quickness of mind and priority in the statistics that bears the name Fermi-Dirac. He immediately applied the Exclusion Principle formulated by Pauli (a principle, Parisi points out, to which Fermi had come very close) in his work on the quantization of perfect monatomic gas (February 1926). But Parisi also recognizes, with a precise analysis, the clarifying contribution made by Dirac (August 1926), who re-examined Fermi's statistics and distinguished the symmetrical and antisymmetrical wave functions.

I would like to quote Franco Rasetti's comments in this paper of Fermi in "Note e memorie" (Collected Papers) (FNM) on page 178, Vol. I.

"As soon as Fermi read Pauli's article on the exclusion principle, he realized that he possessed all the elements for a theory of an ideal gas that could satisfy Nernst's principle at absolute zero and give the correct Sackur-Tetrode formulae for the absolute value of the entropy in the low density and high temperature limit." This rather long comment by Rasetti also quotes Fermi's contribution at the Como conference.

Here too I would like to emphasize again, as Parisi points out, the strict reciprocal critical attention between great scientists of differing qualities to clarify finally, fundamentally and perhaps for ever, the statistics of elementary particles and to lay the essential bases for the new quantum mechanics.

For Enrico Fermi the years from 1926 to 1930 were particularly intense and productive. Parisi says that *"Fermi did not take a direct interest in these applications of his theory, on which generations of physicists will work, but turned instead to what was in that moment the new frontier of theoretical physics: quantum electrodynamics, with all its problems linked to the emission and absorption of photons, and he wrote a series of magistral works, widely admired for their extreme clarity."*

6·2. *Marcello Cini* (*). – This brings us to Marcello Cini's paper "Fermi and quantum electrodynamics". The first pages of the article explain Fermi's position at Göttingen, slightly in the shadow with regard to the revolution Heisenberg, Pauli, Jordan were achieving. He would then come out of those shadows in 1926-1927 with those "lion's strokes" of his statistics and the Thomas-Fermi atom, not to mention his major theoretical work "Tentativo di una teoria dell'emissione dei raggi beta" (A tentative theory of beta ray emission). But the intense work of enormous synthesis and clarification of quantum electrodynamics that Cini tells us about is what allowed Fermi to arrive with astounding lucidity at the explanation of beta radioactivity and to the opening up of a new field theory.

The successes of matrix mechanics, and hence of Schrödinger's wave mechanics, soon aroused great enthusiasm, but as Cini points out, *"The situation was very different as far as the solution of problems involving the electromagnetic field and its interaction with material charged particles was concerned. The extension of the quantistic paradigm to the relativistic systems, with infinite degrees of freedom, in fact presented difficulties which were much greater than those which had to be dealt with to apply the rules of quantum mechanics to resolve the "brain-teasers" of "normal science"* [9].

Fermi's contribution to electrodynamics in these years was therefore of great benefit to the whole advanced scientific community, theoreticians and experimentalists.

I have already mentioned the comments of physicists like Weisskopf and Bethe on Fermi's work. Marcello Cini stresses the theoretical and conceptual difficulties of electrodynamics from its birth to the seventies. It is a clear account which should be read and which brings us to the current synthesis: the electroweak theory. From this essay too Fermi's other great historical function emerges: the synthesis and coordination between theories that seemed to be in disagreement. To enormous gain in the advancement of research and the history of science.

In the second part of his paper (points 7, 8, 9) Cini makes some observations *"which result from an examination through the eyes of an historian"*. They are interesting and profound pages, with which one may or may not agree, but which deserve to be thought about because they lead us to consider the position of the theoretical physicist isolated in his creative genius compared to the scientific and cultural environment in which he lives. The question that Cini asks is *"the approaches chosen by the physicists —Jordan, Dirac, Heisenberg, Pauli and Fermi— . . . have been characterized by substantial methodological and epistemological differences. A spontaneous question comes to mind: Are these differences due only to characterial and psychological factors or philosophical prejudice which is purely individual, or is it possible to try and trace the origin, at least in part, in the different cultural and social context in which they were working?"*

The answer is not easy and in his paper Cini provides a valid analysis of the problem.

As for any "philosophy" of Fermi, we must say that it is difficult to find Enrico Fermi's attitude towards these problems. He did not talk about them readily, almost as if he did not have time in his short life. In reality nobody knows to what extent they were alive or cogent within him.

On this subject —the extent to which Fermi kept general and abstract questions to himself and remained in the immediate concrete world— I will quote a subtle observation by Eugene Wigner on Fermi's famous work on beta rays (ref. [4], p. 75 and ref. [25]: WIGNER E. P., *Yearbook of the American Philosophical Society*, quoted pp. 435-439).

"The work seems to be pervaded by an apparent naivety, that invites criticism and generalisations, and a more learned presentation. This apparent naivety, amongst the various possibilities, was after all correct and characteristic of Fermi's tastes, and did not represent the state of his knowledge when he wrote the article on beta rays. He could certainly have added even then a quantity of abstract ideas that others would have considered important and highly significant."

This consideration by Wigner is quoted in the biography of Fermi written by Emilio Segrè, and he adds:

"Fermi always sought simplicity, and his choice of vector interaction, of all the various possibilities, was after all correct: and so, instead of discussing all the possibilities, by intuition or by luck, he chose the right one."

Leading on from Cini's contribution, I would like to say that historical meditations such as he has written are important, and should be of interest to future historians as well, because what happened in physics between 1920 and 1930 is logical, new, astonishing. And at the same time it seems to spring unstoppable from a meditation which matured over many centuries.

The end of Cini's essay should be noted, where he quotes Fermi's conclusion to his article on thermodynamics.

"We can conclude that practically all the problems of radiation theory which do not involve the structure of the electron have a satisfactory explanation; while the problems connected with the intrinsic properties of the electron are still very far from being resolved."

6·3. *Nicola Cabibbo.* – With Nicola Cabibbo's article, "Weak interactions" (*), the scientific history of Fermi opens out towards immortality. In particular we refer here to Fermi's essay in 1933 "Tentativo di una teoria dell'emissione dei raggi beta" (A tentative theory of beta ray emission) (FNM 559; 575). This is a work that Parisi (*) considered Fermi's greatest theoretical contribution which opened up a new field of physics, which was born from a nuclear experiment but which soon dominated the whole Universe with its presence.

Cabibbo's article starts in "the Fermi style", starting with the early experimental facts, and explaining the immediate interest of these things. It is thus a model of educational presentation, and there remains nothing to do but read his masterly account.

Fermi's theory fits into the general picture of quantum fields outlined by Jordan and Klein. Indeed Cabibbo says *"the language of fields allowed the description of phenomena in which particles are created or destroyed, but Fermi's work on beta radioactivity is the first in which this possibility was used outside the photon theory."*

In his essay Cabibbo presents the history of weak interactions, not only up to the fundamental discoveries Fermi was able to make in the time available to him but also after 1954 and up to the present day. He can and must do so because in the field of weak interactions the theoretical contribution and the discoveries of Italians played an essential part.

Cabibbo divides the history of weak interactions into two periods. The first starts with Fermi's work in 1933 and ends with the discovery of the violation of the symmetry of parity (1956). It includes the experiments of Conversi, Pancini, Piccioni, (1946) [2], the discovery of π-mesons, the discovery of new particles, K-mesons and hyperons.

A new phase of research into weak interactions opens up around 1960, when the decay of the new particles seemed to violate, in its rates, the universality of weak constants. We refer you to Cabibbo's essay for this fundamental story. But we wish to emphasize that the solution to these problems was actually found by him, in 1963, in an essay [10] which is now a classic in the scientific literature, and with the introduction of the famous "Cabibbo angle", which in the author's essay is called mixing angle, which is what it actually is.

In his essay Cabibbo brings us up to date with current results, to the mixing of the mass of neutrinos, to the role of weak interactions in the theory of the Universe, to the problem of solar energy. It is a field of research which is far from having been fully explored. Here too I would to quote the conclusion of his essay.

"Fermi's theory of weak interactions has become an essential component of the more general theory of elementary particles which goes under the name of "Standard Model". These developments are described in the contribution by M. Jacob and L. Maiani in this volume. It is important to remember though that Fermi's theory still maintains its value today, both for the validity of the solutions proposed and as a stimulus for research which has kept physicists busy for almost seventy years, and that will still do so for decades to come. In this theory Fermi's greatness is reflected, the signature of a great physicist."

6·4. *Franco Bassani.* – Franco Bassani's paper has the title "Enrico Fermi and Solid State Physics" (*). Bassani starts with an illustration of Fermi's statistics and shows with great clarity the awareness Fermi immediately had of the importance of his discovery. Of particular importance is Fermi's "prophetic" speech at the Como conference in 1927 (ref. [5] in Bassani's essay).

"One can try to construct a theory of metals capable of accounting for the forces that hold together the structure of the metal. It would be sufficient to consider the positive ions at the vertices of the metal's crystalline lattice and then calculate the valence electron distribution under the effect of electrostatic forces ... applying of course the new statistics instead of classic statistics. The calculations necessary for this theory are however rather long and are not yet complete."

Bassani, one of the founders of solid state physics in Italy, then observes: *"This expression has been universally used in solids throughout the second half of the century."*

In his comprehensive and effective article Bassani shows how the understanding of thermal properties, transport properties, and the optical properties of all crystals derive from Fermi's theory and intuition. But his statistics gave birth consequences of great intellectual value, and finally to a first explanation of superconductivity and the fundamental concept of electron "hole".

Bassani recalls the success of the Thomas-Fermi model for solid state physics. He reminds us that Thomas fully deserved to have his name associated with this model, which he arrived at before Fermi, if in a partial form, based only on the exclusion principle.

In the second part of his article Bassani recalls a new field of physics opened up by Fermi: neutron physics, as it would later be called, which he defines as "neutron diffraction".

It is right to remember that Fermi always had a lively interest in solid state physics and that he made a fundamental contribution to the birth of this science. Perhaps this is also the right time to recall Corbino's faith in and inspirational force towards Fermi [4].

Before we pass on to the second part of Fermi's scientific life —in the fields of nuclear research both fundamental and applied— we would like to recall, through the essays of B. Bertotti, G. Gallavotti and M. Falcioni and A. Vulpiani, Enrico Fermi's first research in his very early twenties. They precede the overwhelming period of quantum mechanics, they date to the years 1921-1922 and they will be returned to in an extremely interesting way, as Falcioni and Vulpiani recall (*), in the early fifties.

Enrico Fermi in his twenties had a period of intense work, with a joy and desire to understand, to calculate, to arrive at new applications and new original discoveries. As I have already mentioned, Italy's mathematical-physicists soon realized this.

6˙5. *Bruno Bertotti* (*). – Amongst Fermi's early work, all published in "Note e memorie" (Collected Papers) already mentioned [1], let us begin by remembering his third paper "Sopra i fenomeni che avvengono in vicinanza di una linea oraria" (On the phenomena that occur in the vicinity of a time line), presented to the Accademia dei Lincei in January 1922 (FNM 3).

This is the paper clearly commented on by Bertotti in his essay "Fermi's coordinates and the Principle of Equivalence". Fermi's value was not immediately fully understood. Perhaps this is the place to remember how these precise and important papers came to life again fifty years later, when scientific interest returned to the point of departure outlined fifty years before, albeit by unforeseeable routes and enriched by new developments.

Bruno Bertotti, who has worked in this field with great competence, presents us with the reasons for confirming our interest, even today, in Fermi's coordinates within a laboratory in arbitrary motion. I quote Bertotti's final considerations at the end of his essay:

"Thus, if we want to describe gravitational phenomena in a region that is small relative to the characteristic scale, Fermi's generalized coordinates are essential. ... This conceptual and computational tool is important nowadays, for example, for the design and operation of gravitational wave detectors: instruments whose sizes are, in general, much smaller than the wavelength of interest. They are basically used to determine the time-dependent curvature of spacetime through the geodetic deviation experienced by two neighbouring material points."

6˙6. *Giovanni Gallavotti.* – This is a very careful and critical analysis of Fermi's work around 1921-1922: "Classic mechanics and the quantum revolution in Fermi's early works".

The author recognises the value of Fermi's early papers, in a clear synthesis. *"They were carried out during a period when Physics was undergoing radical changes of which the young Fermi was well aware in spite of the isolation due to the almost non-existent involvement of Italian science."*

In the rest of the essay he concerns himself with another group of papers, on mechanics, related to matters generated by the new-born quantum mechanics.

This was a general problem which interested all European physicists, and it is presented well by Gallavotti, who is a theoretical physicist, indeed of international fame in the field of mechanics, in his section "Adiabatic invariants and the quasi-ergodic hypothesis".

It is a matter of examining the validity limits of the Bohr-Sommerfeld theorem. Fermi concluded, in an important article, written while he was still at Göttingen, with a fundamental and, in a sense, final criticism of the idea of adiabatic invariants, developing the doubts already raised by Einstein in 1917. Gallavotti's section is challenging to read, but the explanation that Gallavotti himself included in item [4] of his bibliography provides excellent support.

The author clearly indicates the young Fermi's limits and errors in evaluation, and this too is interesting and worth studying. I remember that in Fermi's "Note e memorie" (Collected Papers) (FNM) Emilio Segrè was of the opinion that *"the ergodic theorem given by Fermi is not considered rigorous from the mathematical point of view, and it is difficult to make it so."*

It seems to me that Gallavotti gives us a picture of the young Fermi as an eagle sweeping in wide circles around its nest before setting off decisively towards goals that are unique and of worldwide recognized value.

In the last section, "Theoretical physics in Italy and Fermi's initial uneasiness about Matrix Mechanics", Gallavotti returns to Fermi's lack of interest towards the work of the young scientists ("Knabenphysik"), that I have mentioned. I have already commented on this interesting period in Fermi's life. Here I quote Gallavotti's final conclusion on his note.

"For the sake of clarity, we should point out that in other countries too, especially in its native land, Matrix Mechanics has always been and still is essentially ignored in basic courses, in favor of the ondulatory approach. Why this should be, poses an interesting historical problem."

6'7. *Tullio Levi-Civita.* – The editors of this book thought it appropriate, to give a living image *"delle cose occorenti ne' tempi suoi"* ([1]), to add a paper by the great mathematician Levi-Civita who took an interest in the problem of adiabatic invariants and realized its clear limits.

"On the adiabatic invariants" (*). This paper by Levi-Civita is clearly linked to Fermi's work on the same subject. It is published in the proceedings of the Como conference of 1927 [11]. In paragraph 1 of Levi-Civita's paper "Recent atomic theories and their formulations" there is a clear summary of the problem treated by himself and by Fermi. Levi-Civita writes *"Many physicists, and not only traditionalists, disliked the combination mentioned above of Newtonian mechanics with a selective principle of quantum discontinuity."* In this Como report of 1927 there is Sommerfeld's wide ranging article devoted amongst other things to Fermi's statistics, which confirmed its fundamental value in solid state physics, as shown in Bassani's paper published here (*).

6'8. *Massimo Falcioni and Angelo Vulpiani.* – Falcioni and Vulpiani's paper on "Enrico Fermi's contribution to non-linear systems: The influence of an unpublished article" (*) is of great interest. It too, as already observed, takes us to an historical consideration. Fermi, Pasta and Ulam opened up a new line of thought and research in the years 1954-1955. They stressed the importance of a systematic study of non linear systems,

---

([1]) "Of the events occurring in his times". D. Compagni, Florentine writer, politician, and author of a famous chronicle of political events in his lifetime (1255-1324) [T.N.].

and of the need to use numerical simulations alongside theoretical studies and laboratory experiments. Fermi had probably felt the need for a clarification of non linear systems for decades, considering the analyses and criticisms of his papers between 1922 and 1924. His contribution to world physics between 1930 and 1950 did not allow him to research further the strictly mathematical-physics sector. The clarifying paper by Falcioni and Vulpiani puts us in contact with an aspect that "front line physicists", both theoretical and experimental, sometimes ignore, the ponderous, sometimes slow progress of mathematical physics which may lead to important new themes of fundamental research, such as those linked to the study of integral systems and dynamic chaos, which over the last few years have opened up new paths for scientific thought.

Falcioni and Vulpiani begin with Fermi's analysis of 1923. These authors explain how the young Fermi argued that Hamiltonian systems should generally be considered ergodic, in other words approximately capable of touching all points of assigned space with their solutions, as soon as an epsilon perturbation, however small, is added to the unperturbed Hamiltonian. This conclusion was accepted by all physicists.

Well, as these physicists explain with precise references, this affirmation is not generally true. It was the theoretical research of Kolmogorov, and the computer analyses of Fermi, Pasta and Ulam, that lead to unexpected results. The consequences of this research, told in a fascinating way by Falcioni and Vulpiani, have increasingly developed over the last few years in the field of theoretical physics and of mathematical physics. In particular they recall the discovery of solitons in the sixties. I would like to conclude by emphasizing the importance of Falcioni and Vulpiani's final affirmations in their paper:

"Most physicists were convinced of the importance of non-linear systems not so much because of the significance of the Kolmogorov, Arnold and Moser theorems, but rather because of the suggestive power of the drawings obtained from the computer simulations of the pioneers of chaos: Fermi, Pasta, Ulam, Chirikov, Lorenz, Hénon. The systematic use of the computer has favoured the rebirth of entire research sectors: the most significant examples are perhaps turbulence and chaotic dynamical systems, which had been marginal and relegated among the engineering applications (turbulence) or were considered as being more esoteric branches of mathematics (dynamical systems)."

This paper by Falcioni and Vulpiani throws more light on Fermi's advice on electronic calculators that I quote in my comment on "Fermi's last lessons" described by Renato Angelo Ricci.

Let us now turn to the nuclear period in Fermi's creative activity, which begins in the thirties and continues until his premature death.

6˙9. *Ugo Amaldi.* – This paper (*), "Nuclear physics from the nineteen thirties to the present day" rapidly sums up developments in nuclear physics from its birth, stops to comment on the great results achieved by Fermi's school in the thirties, but then extends further up to the results and problems of today. He quotes the most relevant passages from the writings of Fermi's students, particularly Edoardo Amaldi and Emilio Segrè, concentrating on the glorious years of nuclear physics in Rome, 1934-1938.

In my opinion this is an excellent method of presentation: we glimpse Enrico Fermi at work with his students, we feel the thrill of discovery, the anxiety, we see Amaldi, Segrè, Rasetti and Fermi himself running along the corridors of Via Panisperna. I would like to point out that much of their account could be presented and be understood in a high school, indeed it could become the inspiration for a further thirst for knowledge.

Certainly the discovery of the properties of slow neutrons and the speed of its interpretation give us a new example of Fermi's power when faced with an unexpected natural phenomenon.

From 1935 to 1938 there were years of intense consolidation, if somewhat confused, partly because of the uncertain Italian political situation. And then Ugo Amaldi's account brings us to the discovery, unexpected and unforeseen, of nuclear fission. In the final part of Amaldi's paper the results are given from the 1960's onwards in particle physics and nuclear physics. Two fields that Fermi himself never chose to differentiate clearly.

The account of the enormous consequences for society of the properties of neutrons and of fission comes from Carlo Salvetti's paper, that I will now comment on. We will then return to the tale of fundamental research into nuclei and into subatomic particles.

6˙10. *Carlo Salvetti.* – Salvetti's account begins with the discovery of fission (1938) and presents us with the "The birth of nuclear energy: Fermi's pile" (*).

Salvetti gives a detailed and rapid account of the whole historical and scientific debate in which the question of the "pile" evolved, in a time of war and anguish. We cannot be other than thoughtful and moved when we consider the fact that Enrico Fermi and the other great physicists who had revealed weak interactions and nuclear forces to the world for purely scientific research had to dedicate themselves to arriving first at the conquest of a great new resource with civilian and military applications such as nuclear energy. Salvetti's account, with Fermi inevitably in the main role, stretches from 1939 to that 2nd of December 1942 when the Pile came on line and shows the world how a new form of energy became available. Note in this intense dramatic account the "peak" of our human adventure, when the pile became critical and was left to work for 28 minutes, with a $k_{\text{eff}} = 1.0006$ and a maximum power of 1/2 watt. Let us remember that monument in Chicago to Fermi and the pile in the very place where the first pile worked.

This success, and the long hard road to arrive at it, are well described by Carlo Salvetti, who was one of the main figures in the study and application of nuclear energy for peaceful purposes in Italy. His tale opens up the path to a variety of considerations that lead to admiration as well as to tormented meditation on what men, well coordinated by an outstanding mind, can achieve.

After Salvetti's account, which stops on that 2nd of December 1942, which shows us Fermi's completeness as an experimental physicist, and his coherence and determination to reach a goal, one spontaneously questions the consequences of Enrico Fermi's first pile, and modern developments in new energies.

6˙11. This is done in two following papers (*). One by *Augusto Gandini* "From the Chicago Pile 1 to the next-generation reactors", the other by *Maurizio Cumo* "Reactors and nuclear technology: Development in the world" These papers emphasize the impetuous development of nuclear energy after the first pile and current thinking on safe development of reactors and of all nuclear sources, on land, at sea and soon in space. The problem of safety with regard to the risks of radioactivity and accidents is dealt with here by two authors who have devoted many years of intense study and research to this subject.

These are two articles that introduce us to the current situation in the world of nuclear energy, such as the grave problems of a fair distribution of energy between countries, something mankind has a growing need for. Perhaps we can say that the two paths, the practical road of nuclear energy and the road of fundamental research, that both flow together by historical fate into the hands of Enrico Fermi, are today rather distinct.

To return to the field of fundamental research, Ugo Amaldi, in the final part of his paper, talks of the discoveries that descend from the work of the preceding generation, down to the "Focal points of present-day research". This is a useful up-to-date summary that helps us to understand that we are still far from a final picture in the field of particles and nuclei. We have reached new plateaux, climbed high mountains, the new discoveries whet our appetite for knowledge. We are gluttonous for it (if you will allow a term from Dante) but perhaps the final goal is still far off. Part of the paper by Jacob and Maiani (*) is also dedicated to these questions and problems.

Here the nuclear part, fundamental and applied, ends and I shall move on to Enrico Fermi's last lessons, commented on by Renato Angelo Ricci.

6·12. *Renato Angelo Ricci.* – "Fermi's last lessons" (*). This paper does not take its place in line, ordered by subject matter, as the others naturally did. It is instead a memoir of Enrico Fermi's final years in Italy and his wide ranging activity of lectures and conferences on various subjects in contemporary physics.

In his paper Ricci reminds us also of Enrico Fermi's teaching activity in his final publications, such as "Nuclear Physics", 1949-1950, and his brief and prophetic work on "Elementary Particles" in 1951. They are books that still deserve a place on the shelves today and not in the archives of physics. As Ricci recalls, in many places, the general ideas they contain are a valid guide even today for those who wish to set physics students off on the path of research. They are lessons that I too have indicated to physics students in the eighties and nineties, for Fermi's universal way of inserting a specific problem into the general physics picture.

The reproductions Renato Ricci gives us of Fermi's hand written notes and jottings in those final years are direct documents of his constant search for concision and clarity.

In section 3 of his paper Ricci talks of Fermi's last lectures at Varenna, in 1954, at Villa Monastero, on Lake Como. Renato Ricci gives us Fermi's fine page in his lecture on isotopic spin, invented by Heisenberg. Heisenberg's lecture on that same course is devoted to the "Production of mesons in very high energy collisions." This is a problem Fermi and Heisenberg tackled in the fifties with great style and clarity. This made me, and makes me, consider the great contribution made to physics by these two great men, so different and so subtle in their approach to fundamental problems. I was in part a witness to these events and would like therefore to add some of my own direct memories to Ricci's paper, also, as you will see, as a debt of gratitude to Fermi.

The last lessons of the Varenna school in Italy that Fermi attended, in August 1954, were those devoted to European projects for accelerating machines, with contributions by E. Amaldi, J. Adams, E. Persico and myself [12]. In particular the last two lessons were devoted to the "Italian Project for an accelerating machine". Enrico Persico presented his theory on the capture of electrons in a high energy electrosynchrotron, which came on line in Frascati in 1958.

Fermi listened and was very interested in the project, particularly in the elegant theory presented by E. Persico and his collaborators (the synchrotron group was represented at Varenna by Persico, the young Carlo Bernardini and Ferdinando Amman and myself). But I must recall here that two pieces of advice sprang from Fermi's interest, which duly attentive we accepted, and which in fact guided our efforts over the years that followed.

After listening to our lectures Fermi said —and he confirmed it with clear reasoning— that in our situation, and with that opening of the ring, the final intensity of the gamma beams from the machine was more than proportional to the injection energy E of the electrons. It would be best to aim at raising E, within the limits set by the finance and time available. (I must note that this was Fermi's last contribution to Italian physics, at least in Italy, along with another one that I shall describe later.)

We remembered this when we had to chose a final electron injector for our project [13], and we opted decisively for the new Van de Graaf machine, now available on the market. If our synchrotron held the record for intensity for over two years compared to similar machines in the United States and elsewhere, this is due at least in part to Fermi's advice, which we did not hesitate to emphasize to the Italian Scientific Community which had to decide on our undertaking.

Enrico Fermi also gave us another decisive piece of advice in those magic days. Gilberto Bernardini, the great physicist who was our master for muons, mesons and cosmic rays, was in practice the source of inspiration for the Frascati synchrotron in 1952-1953. When the location —Frascati— was finally decided on, some funds remained available, which had by now been assigned to fundamental research, in particular offered by Pisa and Lucca. Fermi, walking along the paths of Villa Monastero, immediately advised us *"make a computer with it"*. G. Bernardini and M. Conversi immediately followed this advice, and so an avant-garde electronic computation project was born in Pisa which greatly assisted the evolution of Italian scientific research.

In other words I am one of many who have received important advice from Enrico Fermi. He was always able to get to the fundamental point, even on the basis of a variety of facts he had only just been presented with. While working on the preface to these papers, from which I have learned so much, I discovered while reading Falcioni and Vulpiani's article (*), the source of Fermi's swift recommendation of a computer. He was still surprised, on the basis of his recent work with Pasta and Ulam, by the decisive contribution made by a computer appropriately applied to a fundamental problem such as the behaviour of the solutions to a perturbed Hamiltonian equation.

I now come to the last paper in this fine series which ranges from weak interactions to their origins, to the problems of elementary particles in our own day.

6·13. *Maurice Jacob and Luciano Maiani.* – The article "The scientific legacy of Fermi in particle physics" (*) provides results of great interest, that remain part of the history of physics, and that have been the launchpad for our knowledge today.

Let us say at once that it is a great merit of this article, and not only of this one in this book, that it shows that the theoretical and experimental path from the death of Fermi to today is certainly immense, but perhaps we are aware as never before that we are far from a coherent and complete theory of our physical Universe, and many problems will have to wait for many years for a solution.

These authors give an organised list of the results obtained by Enrico Fermi in his lifetime and recall some relatively unknown but impressive episodes (the Flerov intuition). They chose three of the nine subjects that they indicated in the article as those which made Fermi famous worldwide: the theory of beta decay, the theory according to which the π-meson (pion) is bound state of a barion and of an antibarion, the inelastic pion-nucleon collision.

Beta decay, and the discovery of weak interactions, is, as we have already mentioned, Enrico Fermi's most significant theoretical contribution to the history of physics. It has already been told, in another paper devoted to it, Nicola Cabibbo's (*). But allow me to defend the presence of two presentations of beta decays in the same book. The facts and results agree but are seen from a different prospective by authors that have contributed to this branch of physics with original discoveries. For example, from these papers, and from Marcello Cini's paper devoted essentially to electrodynamic research, we find the close link between these theories, a link which in the seventies led to the electroweak theory. After sixty years and more we see with great historical interest the noble and intense competition that had Enrico Fermi and Wolfgang Pauli as main protagonists.

The second argument dealt with "Is the pion a fundamental particle or a barion bound state?" brings us to a central problem in the fifties and sixties. It starts, as Jacob and Maiani say, from a paper by Fermi and Yang in 1949. It is a hypothesis that at the time may have struck many physicists as grotesque, that the pion was not elementary, but rather a bound state between a nucleon and an antinucleon. Jacob and Maiani show us how the world of true elementary particles (or so we believe), quarks and leptons, springs from these beginnings.

The third subject (the pion-nucleon collision) brings us to those lectures in Varenna already recalled by Ricci. Here Fermi's years —the last years of his life— are narrated, when he worked with enthusiasm on experimental research on pions using the new Chicago 450 MeV synchrocyclotron, and he opened up the new analysis of barion resonances. He maintained a strictly phenomenological attitude to this analysis, although he was well aware of the theoretical forecasts of a possible 3-3 resonance.

From those years Jacob and Maiani take us to the birth of quantum chromodynamics, to chiral dynamics, to the Standard Model and to the possible extrapolations from it.

It is a remarkable summary, which expands our vision of the world, and at the same time shows us how long a road physics still has ahead. In the final part they remind us that the hunt is on for the Higgs boson and say that *"The Higgs boson is actively searched at the present accelerators. As of today, no convincing signal of a Higgs boson has been found with LEP, which implies that its mass has to be larger than about 114 GeV."*

Perhaps it is legitimate to observe that, despite all the achievements reached, today we realize that there are facts, theories and phenomena in the universe, and in our minds, that we do not know how to conceive and that we do not even know that we do not know. A future which has barely begun.

7. – Some final comments

Enrico Fermi's intense life is contained in the three opening biographies (*). Let me add here a few comments and recollections.

7·1. *How much we know.* – Reading these fourteen papers, written for the most part by people who knew Fermi directly and benefited from his advice, shows us the extent of his intellect. He and the men who accompanied him on his adventure have showed us what man can understand and perhaps will be able to do in the future: a power destined to expand his presence on the planet. It does not tell us what the outcome of all this may be. Indeed Fermi was always extremely cautious about such matters, while well aware of the good and evil contained in human destiny.

Please allow me to emphasize man's capacity to think, to help and to harm. Progress has been continuous in the past millennia, from the use of tools to language, to art in the past thirty or forty thousand years, to the invention of writing some six thousand years ago, and then of geometry, of printing, of optical instruments, of electricity and of nuclear knowledge. It is an immense process, carried forward by a few men gifted, like him, with extraordinary qualities. But alongside them, and just as necessary, you find the students and the following generations that understand them, follow their example, take up and broadcast their intuitions and discoveries. The joy of the new, the joy of understanding.

In other words, when I think of this run through time, I cannot be other than astonished and admiring of our progress over these centuries. Where will we be in three thousand years time?

7·2. *A meditation, with Enrico Fermi, on our future responsibilities.* – But here, on this wave of hope, we must slow down, stop and think. Will mankind be able to control the new scientific and technological knowledge, from physics to biology, with an agreement or rule for universal behaviour, arriving at peace between nations? Are we equally aware of the risk of an immense catastrophe, limited of course to our small planet? Or in the future will we able to enjoy the pleasure, the curiosity of understanding the Universe and ourselves, which we still know so little about?

This ambiguity of ours towards the future brings us back to Enrico Fermi again. With the discovery of nuclear energy he, and not he alone, brought us to the narrow divide between two worlds, between liberation and catastrophe.

Many have asked what Fermi thought of the new forces he brought forth, of the consequences for the future of humanity.

It is a major problem, because it leads us to evaluate Fermi the man, the scientist, and with him the responsibility of scientists in future human history. Much has been thought and written about this problem, and about what Fermi's views really were (see the Fermi Celebration of 1992 [7].) One thing should be remembered: in his writings and statements he did not like to go beyond the field of physics.

I will mention in partial illustration two cases in which he engaged his moral conscience, and also a thought quoted by Edoardo Amaldi and Gilberto Bernardini.

One is the end of the message Fermi sent to Hutchinson, president of the University of Chicago, on the 6th of September 1945 (ref. [4], p. 163; ref [3], p. 144). He commented on the development of the hydrogen bomb after Hiroshima: the powerful bomb which now too hangs over our heads like a sword of Damocles. In this letter he wrote, amongst other things:

"The new means of destruction has such a power that in the event of war between two powers both equipped with these weapons, both would see their cities destroyed... The possibility of an international agreement should be explored immediately with energy and hope. The most fervid hope of the men who contributed to these developments is that such an agreement is possible."

The other case is to be found in a letter he wrote along with I. Rabi in 1947 (ref. [7], p. 231; [4], p. 169; [3], p. 144):

"The fact that this weapon's capacity for destruction is without limit means that its very existence and the knowledge of how to build it represent a danger for the whole of humanity. It is inevitably an evil from any point of view. For this very reason it seems important to us that the President of the United States should declare to the Americans and to the whole world that on the basis of fundamental ethical principles we judge that it would be a grave error to undertake the development of this weapon." This was a declaration of Rabi and Fermi as members of GAP (General Advisory Committee), on request of AEC (Atomic Energy Commission), October 29, 1949.

I hope finally to report correctly a thought of his which I have only heard verbally from Amaldi and Bernardini. In Varenna in 1954, a few months before he died, he said that the coming century might be decisive for the history of mankind, wavering as it is between a possible absolute tragedy, and the beginning of an age of new serenity.

Here I would like to conclude by addressing in particular science teachers in all levels of schooling, to comment on our great responsibility. We have said that this human adventure may have a magnificent future but could end within a few centuries in a disaster for civilisation.

Well this prospect cannot be excluded, but it depends on us. It is our responsibility to show clearly to the new generations the terms of the problem, both scientific and human. I know from my own experience teaching in junior high schools that young people can become passionately interested in, and understand these problems, and that even an objective scientific comment can become Ariadne's guiding thread that will lead them out of the labyrinth: We have to explain the seriousness of our future, but at the same time give hope, in which I believe, that the world can be improved by our efforts. We must free ourselves of every idea or superstition that makes men enemies to each other. We must persuade ourselves, that scientific progress may now force us, almost violently, to accept the need for peace and unselfishness.

With this hope we bow today, on his centenary, to Enrico Fermi and to the great men who have opened up a new world to us.

(*) These articles and papers are presented in this book with their bibliographical references and in full.

REFERENCES

[1] Works by Enrico Fermi: *Enrico Fermi. Note e Memorie (Collected Papers)* (Accademia Nazionale dei Lincei and University of Chicago Press), Vol. I, 1961 and Vol. II, 1965. Each paper is indicated by FNM followed by the number according to the order in which it is presented.
[2] CONVERSI M., PANCINI E. and PICCIONI O., *Phys. Rev.*, **71** (209) 1947.
[3] PONTECORVO B., *Enrico Fermi* (Edizioni Studio Tesi, Pordenone) 1993.
[4] The information is given in a fine book by Emilio Segrè: *Enrico Fermi Physicist* (University of Chicago Press) 1970.
[5] PAIS A., *Inward bound* (Oxford, Clarendon Press) 1986, p. 250 and following pages.
[6] DE MARIA M., *Un fisico da Via Panisperna all'America, Le Scienze*, collection *"I grandi della Scienza"*, **8**, II, 1999.
[7] WEISSKOPF V., *Proceedings of the Accademia dei Lincei. Atti dei Convegni Lincei,* Vol. **104**, *Symposium dedicated to Enrico Fermi, Rome, 10 December 1992*, p. 231.
[8] ANDERSON H. E., in *"All in our times. The reminiscences of twelve nuclear pioneers", The Bulletin of Atomic Scientist*, 1975, p. 66: *Assisting Fermi.*
[9] FERMI E., *Rev. Mod. Phys.*, **132** (1932) 87.
[10] CABIBBO N., *Phys. Rev. Lett.*, **10** (531) 1963.
[11] LEVI-CIVITA T., *"Sugli invarianti adiabatici", Congresso Internazionale dei fisici, 11-20 Sett. 1927* (Zanichelli, Bologna) p. 475. (In honour of Alessandro Volta, in the first centenary of his death.)
[12] Supplement to Vol. II, Series X of *Nuovo Cimento*, **1** (1955). Devoted to the Memory of Enrico Fermi.
[13] Proceedings of the International Conference on *"The restructuring of Physical Sciences in Europe and the United States", 1945-1960* (World Scientific, Singapore) 1989.

About the Author

GIORGIO SALVINI, physicist, was born in Milan on the 24th April 1920. Professor of Physics at the Universities of Pisa (1952-1955) and Rome (from 1955; Emeritus Professor since 1995). From 1952 to 1960 he directed the work on the construction of the Italian 1100 MeV Electrosynchrotron which came on line in Frascati in 1958. He has carried out research on extended swarms of cosmic rays and on the photoproduction of mesons. In cooperation with CERN (European Centre for Nuclear Research) he established the existence of W and Z intermediate bosons (1978-1983). He is honorary president of the Accademia dei Lincei. He has been Minister for Universities and Scientific Research.

Commemoration talks

Edoardo Amaldi (1908-1989), *Enrico Persico* (1900-1969), *Franco Rasetti* (1901-2001) are the Italian physicists who were nearer to Enrico Fermi, for the sincere friendship that charcterized their scientific professional collaboration.

We publish here the three commemoration talks presented by Amaldi and Persico a few months after Fermi's death, and by Rasetti on occasion of the "Celebrazioni Lincee" of 1968 in honour of Enrico Fermi.

Commemoration of the Academy Fellow Enrico Fermi(*)

Edoardo Amaldi

When in the distant future the history of science in our times will be written, the first half of the twentieth century will emerge as a period of particular importance, not only for the discovery of many new facts and the development of new concepts, but for their influence, both direct and indirect, on the organization of human life as well.

It was precisely between the end of the nineteenth century and the beginning of the twentieth that certain experimental observations threw the classical concepts of the physical world into crisis: on the one hand the behavior of light with respect to various frames of reference in motion among themselves, and on the other the first evidence regarding the granular structure of energy emitted or absorbed by bodies in the form of radiation.

It was in the twentieth century that these early queries, and many others deriving from them, found their solution, some in the theory of relativity, others in the quantum theory of matter and radiation.

In the twentieth century theories of the atom and its molecular and crystalline aggregates originated and evolved to the point of accounting for practically all known phenomena in this field, which quickly increase in number and variety thanks to the simultaneous development of new experimental techniques.

In the twentieth century the existence of the atomic nucleus came to light, and a vast new chapter of physics —nuclear physics— was dedicated to its study. In recent decades a new applied science —nuclear engineering— branched off from nuclear physics in the strict sense; it has already created economic and social repercussions and is destined to do so even more in the future.

(*) Commemoration held in the joint Sections meeting of March 12, 1955. Uncut text from the Accademia Nazionale dei Lincei, notebook no. 35, Rome (1955).

In the last twenty years the existence of many new unstable particles was demonstrated; their study is just commencing, and is the latest, most recent chapter opened to man's exploration. When future historians will have to note in their books that our country was not absent from this movement but participated in it and contributed to it in a fundamental way, they will have to acknowledge that this is mainly due to Enrico Fermi. In this half-century of extraordinary development of the physical sciences, the life of Enrico Fermi was played out and drew to a close; in the last thirty years he contributed with numerous theoretical and experimental discoveries to the formation and indeed the creation of many new chapters in physics.

But if this is what will be written in the history of science, the history of our country's culture will also portray other aspects of Enrico Fermi beyond that of consummate scientist and ingenious researcher. His role as "maestro" which he dispensed throughout his life wherever he happened to be, was surely of particular importance in Italy. Without Enrico Fermi, the new physical concepts evolving in other parts of Europe and throughout the world would have arrived in our country much later and with greater difficulty.

One might say that he set out on his path without "masters", often in the face of opposition, and if he managed to achieve success it was through his own wits and tenacity; if others followed him it was thanks to the power of his example.

Enrico Fermi was born in Rome on September 29, 1901, the son of Alberto and Ida De Gattis, third of three children of whom the first, Maria, became and still is a professor of literature in a high school in Rome, while the second, Giulio, died as an adolescent in 1915.

His father, originally from Piacenza, was employed by the Ministry of Transport, and through him Enrico Fermi, while still very young, came into contact with various railroad engineers. One of them, struck by the boy's unusual aptitude for mathematics, loaned the fifteen-year-old Fermi university-level science books, which he quickly read and understood. At the same time he became friends with a boy of about his own age, Enrico Persico, who like himself had a passion for scientific problems. The two met to talk and exchange ideas on their studies, which they fueled by searching in book stalls and at second-hand book dealers for treatises on mathematics and physics.

Fermi finished high school in Rome and then, following a competitive examination, entered the "Scuola Normale Superiore" of Pisa as a resident student. He took his degree in 1922 with an experimental thesis on the reflection of X-rays by curved crystalline surfaces. In the meantime, however, he had already begun to publish various works of theoretical nature on questions of electrodynamics and relativity. The calculation of the electrodynamic mass of an electric charge in motion dates back to that period; until then it had been deduced using an incorrect procedure.

After graduating he went to Göttingen on a scholarship from the "*Ministero della Pubblica Istruzione*" (Ministry of Public Instruction), which allowed him to continue his studies as well as broaden his culture; he stayed for about six months at the Max Born Institute at the very time when ideas were maturing which after a few years would evolve into quantum mechanics. During this time at Göttingen Fermi met young physicists like Heisenberg, Pauli, Jordan, who a few years later were to become the most immediate shapers of the new concepts.

Even before going to Germany, Enrico Fermi had met Orso Mario Corbino, professor of experimental physics and director of the Physics Institute of the University of Rome. Corbino was impressed by Fermi's maturity and discernment in discussing difficult problems of the most varied kind, whether theoretical or experimental. He did his utmost to assure that when the young Fermi returned from Göttingen, the Science Faculty of the University of Rome would appoint him to teach the course in Mathematical Institutions. From that first encounter until Corbino's death —which occurred prematurely in 1937— there was a relationship of mutual esteem and deep friendship between the two men. For his part, Corbino admired Fermi for his brilliance and he realized, as few others did, the exceptional abilities of the young physicist, while Fermi felt the influence of Corbino's authority that asserted itself through his lively intellect, his experience and culture as a scientist, his quality of broad-mindedness and his fervent spirit as animator of physics studies in Italy.

In the autumn of 1924, with a grant from the Rockefeller Foundation, Fermi spent three months at Leiden with Ehrenfest, whose statistical works had greatly attracted his attention. At Leiden, for the first time in his life, Fermi became fully aware of his own possibilities, and was encouraged to prepare himself to work with problems of a fundamental nature.

During the biennium 1924-25 and 1925-26 Fermi was again appointed to teach mathematical physics and rational mechanics at the University of Florence, where he found an old friend and fellow-student from the days of Pisa, Franco Rasetti. His friendship with Rasetti helped keep his interest in experimental problems alive. But his activity during that period culminated in a theoretical work on the quantization of the perfect monoatomic gas, in which he developed a new statistics for antisymmetrical particles, wich he called antisymmetrical statistics. This was universally known as the statistics of Fermi or Fermi-Dirac, since Dirac had achieved the same results —though using a different approach— shortly after Fermi.

At the end of 1926 Fermi, having won the competitive examination, was named to the chair of theoretical physics that had been instituted at the Science Faculty of the University of Rome, mainly through the efforts of Corbino. It was in Rome, in 1928, that he married Laura Capon. They subsequently had two children, Nella in 1931 and Giulio in 1936.

In the Physics Institute, directed by Orso Mario Corbino, Enrico Fermi threw himself into his teaching, as well as his unceasing research activity.

But before passing on to that aspect of Fermi, I would like to view his scientific work as a whole during the ten-years from 1922 to 1932, which can be considered his first period as a researcher. There can be no doubt that Fermi's most prominent work is his statistics, which sets up a general law followed by a broad category of particles called "fermions". Electrons, protons, neutrons, μ mesons and certainly some of the recently discovered unstable particles that are still being studied, are all fermions.

In nature, in addition to fermions, there are "bosons" —particles as photons and π mesons that follow the Bose-Einstein statistics. This distinction between fermions and bosons is fundamental: it refers to the value of the corresponding spin or intrinsic moment and is seen in various properties of the corresponding eigenfunctions. Ferretti has underscored that the discovery of antisymmetric statistics does not mean merely to have introduced Pauli's principle into statistics, but rather to have given this principle new significance as a general law and not as a simple property of the atom, as its discoverer had originally thought.

Fermi's statistics may be applied to numerous problems, from electric and thermal conduction in metals to a model of the atomic nucleus which, schematic though it may be, is currently used for its simplicity. Among these applications we must not forget the model of an atom developed by Fermi himself, and independently by Thomas, in England, in 1927. The Thomas-Fermi model, in which the electrons are presented as a totally degenerate gas by Fermi, held around the nucleus by its Coulombian attraction, was used by Fermi, by various collaborators and students, and by numerous other scholars, to calculate all those properties of the atom that vary regularly with the varying of the atomic number. Fermi himself, or his direct collaborators, applied his statistics to the theory of the periodic system, to calculating the value of optical and Röntgen terms, the intervals of optical and Röntgen multiplets, the relationship of the intensities of the first lines of the main series of alcalies, the theory of rare earths with calculation of the corresponding $4f$ orbits, the theory of the electronic affinities of halogens, calculating ionic spectra and that of the eigenfunctions of the ∞s orbits of the elements.

In addition to Fermi's statistics, the Thomas-Fermi model and all the applications just mentioned, other contributions of the 1922-1932 period concern the theory of phenomenona which until then had eluded every attempt at quantitative interpretation. The ability to grasp immediately the general law hidden behind a table of rough experimental data, or recognize at once the mechanism for which the results of certain experimental observations, which seemed strange or insignificant at first, were instead natural or of considerable physical significance when compared with other phenomena or general theories, was, throughout his life, a characteristic that contributed to making Enrico Fermi one of our century's most remarkable figures in the field of the physical sciences.

The relationship of the intensities of alkaline doublets, the Raman effect in molecules and the Raman effect in crystals, the oscillation and rotation bands in ammonia, the effect of pressure on spectral lines and the theory of hyperfine structures are examples of this type; in each can be found some essential aspect of the physical mechanism of the phenomenon that had, until then, escaped detection.

A new formulation of Dirac's theory of radiation also dates back to this early period; from then it became the one usually followed in subsequent presentations of this subject. This theory, accompanied by many new applications to different phenomena, lends itself especially well to illustrating Enrico Fermi's varied "faces" as scholar, researcher and teacher.

As I mentioned earlier, Fermi arrived in Rome in 1926 and, with Corbino's support and the help of Franco Rasetti, who had followed him from Florence, he founded a school. He had assembled a small group of young people who were enthusiastic about physics and the new horizons that were opening in the field, and he dedicated himself to their preparation. This involved, on the one hand, lessons in theoretical physics and, for a few years, in geophysics, which he taught with diligence and exemplary simplicity, presenting only what was essential, stripping the subject of any useless adjuncts; and, on the other hand, the personal and characteristic method of gathering some of his collaborators and students around his desk, usually toward the end of a long afternoon spent in the institute or in the laboratories, discussing and trying to resolve "in public" an unresolved problem suggested by a question from someone present or that he himself brought up in connection with a topic that had captured his attention. The treatises that he developed using this method were written directly in a notebook with very few changes, ready for publication, with the addition of his comments and criticisms which he voiced but did not write at the moment for fear of slowing down the steady, calm and continuous process of reasoning.

The new formulation of Dirac's electrodynamics theory was born in this way. In one of these meetings, someone in the group asked Fermi to explain the just-published theory. Enrico Fermi, turning to us experimentalists with his usual slightly ironic but good-natured smile, observed that if he presented this beautiful theory in the way Dirac had, we would not understand it; he could try however to help us understand it by explaining it in his own way. He then began, and after a dozen meetings there was a thick notebook on his table containing the whole general treatise and the applications of the theory. This notebook, complete with his comments, appeared several years later in the "Review of Modern Physics", and was the subject of courses that he taught at the Institut Poincaré in Paris and at the University of Ann Arbor, Michigan, in 1930.

To give an idea of his influence on the development of physics in Italy while he was professor of theoretical physics at the University of Rome, I would like to mention that in addition to Enrico Persico, his collaborators or students in the field of theoretical physics were: Ettore Maiorana, Gian Carlo Wick, Giulio Racah, Giovanni Gentile junior, Ugo Fano, Bruno Ferretti and Piero Caldirola. The experimentalists, along with Franco Rasetti, were: Emilio Segrè, Oscar D'Agostino, Bruno Pontecorvo, Eugenio Fubini Ghiron, Mario Ageno and myself.

But his influence was felt not only by those who were fortunate enough to be in Rome or who were able to transfer themselves there for a while. Enrico Persico came to Rome from time to time, first from Florence and then from Torino; Antonio Carrelli came from Naples, with his problems in spectroscopy, and Bruno Rossi from Florence and later from Padua, with problems on cosmic radiation. And each time discussions were sparked that often provided the clue for a new work, or even concluded in a finished work. Thus, for example, during a Sunday walk with Bruno Rossi, visiting Rome for a few days, a well-known treatise concerning the action of the earth's magnetic field on penetrating radiation, published by Enrico Fermi and Bruno Rossi in the "Rendiconti dei Lincei" (1) of 1933, was conceived and completed.

I remember another Sunday walk, around the same time, in the company of the Florentines, as we called them, Gilberto Bernardini and Giuseppe Occhialini who had come to Rome with some of their collaborators. I remember Fermi's exhaustive discussion of

---

(1) "Reports of the Lincei" [T.N.].

the various problems that interested the two young physicists, and his own attention to the technical details regarding counters and proportional counters and Wilson's chambers. At that time Fermi was already internationally famous both as a researcher and as a "maestro", so that many established physicists came to Rome to spend their sabbatical year or a few semesters with grants from the Rockefeller Foundation. H. A. Bethe, H. J. Bhabha, F. Block, E. Feenberg, H. S. Goudsmit, F. London, C. Møller, R. F. Peierls, G. Placzeck, E. Teller and G. E. Uhlenbeck were among the many who visited Rome during those years or shortly afterward. With 1932 the first period of Fermi's scientific activity, mainly directed to the theoretical treatment of problems of atomic and molecular physics, came to a close.

The second period, lasting from 1933 until 1949, includes the activity Fermi dedicated to nuclear physics, and differs from the previous one for the predominance of experimental over theoretical research, not in terms of quality, since both are outstanding, but for the relevance of the results obtained.

A few months after the discovery of the neutron, Fermi developed a fundamental theory which is certainly not second to his statistics for importance and renown. The first work, entitled "Tentativo di una teoria della emissione dei raggi beta" (Attempt at a Theory on the Emission of Beta rays), appeared in "Ricerca Scientifica" early in 1933, and was followed by more extended papers appearing less than a year later under the same title, in "Nuovo Cimento" and "Zeitschrift für Physik". In this theory, using recently conceived field theory methods that until then had been employed only in the radiation theory by Dirac, Fermi and others, he traced, quantitatively, the process of beta disintegration of a nucleus back to an elementary process that may be described as a transition of the nucleon from the "neutron" state to the "proton" state, with the emission of an electron and a neutrino. Other authors subsequently introduced variations to Fermi's theory in its original form. It remains, however, the basic scheme not only for processes transforming one nucleus into another with the emission of an electron, but for others discovered many years later as well, such as the disintegration of a μ meson or its capture by a nucleon. Today, twenty years after the work first appeared, the tendency is to believe that all processes in which there are four fermions obey the same law with the same coupling constant, which in fact is called "Fermi's universal interaction". The neutron that is transformed into a proton with the emission of an electron and a neutrino, the negative meson that is absorbed by a proton with the emission of a neutron and a neutrino would be, according to this recent view, nothing more than specific examples of a universal law.

The first report on beta disintegration processes had not yet appeared in "Ricerca Scientifica" when news arrrived that the wife-husband team of Curie-Joliot in Paris had successfully produced radioactivity in certain light elements by subjecting them to the action of α particles. Fermi immediately decided to try to provoke artificial radioactivity by substituting α particles with neutrons, which, for lack of an electric charge, should have been more effective. For this purpose he had his friend Giulio Cesare Trabacchi, director of the Physics Laboratory of the Institute of Health, prepare a neutron source by mixing beryllium powder with radium emanation. After a few fruitless attempts, the experiment was successful, and Fermi could then announce, in a letter to the editor of

"Ricerca Scientifica" in March 1934, the discovery of radioactivity provoked by neutrons.

In the Physics Institute at the University of Rome, feverish activity immediately got underway. Investigations were organized on a broad base so as to try many approaches and let the fewest phenomena possible slip through the nets skilfully spread in all directions. Fermi directed the work of others but at the same time took part himself in all types of physical measurements and chemical manipulations, lending himself also to actually making pieces in the machine shop and glassworks.

Within a few months more than forty new radioactive bodies were produced; many of these were chemically identified, and the corresponding nuclear process for their production was demonstrated. In this way it was decided which cases used (n,α) processes and which (n,p), and was shown that a (n,γ) process was often produced, that is, a process of radioactive capture unknown until then and destined to be of great importance in the future. Until 1934 the known nuclear reactions were so few that a systematics could not be determined. Following Fermi's discovery of radioactivity provoked by neutrons, this suddenly became a simple problem of data-gathering.

In that period it was also found that bombarding uranium —the last of the elements of the periodic system— with neutrons, caused it to produce many new radioactive bodies, some of which were wrongly interpreted as transuranic elements. This interpretation, confirmed shortly afterward by different researchers, was generally considered valid until 1939, when Hahn and Strassmann discovered that uranium acted upon by neutrons underwent the phenomenon of fission, *i.e.* the breaking into two lighter nuclei. The number of radioactive products that resulted masked the transuranic elements, which were also produced but which, as was later demonstrated, were different bodies from those separated by Fermi and his colleagues originally.

In October of 1934 Fermi and his collaborators made another significant discovery. In all cases where radioactivity provoked by neutrons was due to a process of radioactive capture, the intensity of the phenomenon could be increased by a very high factor, in some circumstances of the order of 50 or even 100, simply by surrounding the neutron source and the irradiated body with a hydrogenated substance like water or paraffin.

A few hours after discovering this phenomenon, Fermi had already given a clear explanation of it and written some of the basic formulas that govern it. The hydrogenated substance effect, as it was then called, involved the discovery of two phenomena: one was the fact that, because of consecutive impacts with hydrogen atoms, neutrons can lose their energy to the point of becoming "slow neutrons", that is, neutrons whose energy spectrum extends as far as the energies corresponding to the velocity of thermal agitation. The other phenomenon, more profoundly significant, in a sense, for the development of our ideas on nuclear structures, is that these slow neutrons are far more efficient than the fast ones in producing certain processes, such as radioactive capture, for example. This discovery paved the way to a study regarding the properties of slow neutrons, carried out by Fermi and his collaborators and lasting from the end of 1934 to the middle of 1936. The most striking results obtained, and published like the preceeding ones in "Ricerca Scientifica" in the form of short letters to the editor, then in final form in "Nuovo Cimento", "Proceedings of the Royal Society" of London, "Physical Review" and "Ricerca

Scientifica" were the following. Various new reactions produced by slow neutrons were discovered and completely explained, among them the reactions of boron and cadmium. The γ-rays emitted following the capture of the neutrons were observed, and it was found that slow neutrons were absorbed by certain nuclei having exceptionally large collision cross-sections, of the order of thousands of times greater than the corresponding geometrical cross-sections. It was shown that this fact was usually linked to the existence of resonances, that is, very limited energy intervals within which, as for the light-absorption lines typical of atoms, the neutrons were absorbed to an exceptional degree.

Furthermore, the mechanism that decelerates the neutrons and their diffusion through the decelerating substance —the moderator, to use what later became a widespread expression— was explained quantitatively. Among the results obtained on this topic were the determining of the spatial distribution of neutrons in the moderator and the dependence on the neutrons' initial and final energy, of the mean square distance they travel during the deceleration process; the distribution of the neutrons near a surface that bounds a moderator; the discovery of the effect of the chemical bond on the neutrons' elastic collision cross-section; determining the parameters that play a role in the diffusion of thermal neutrons, such as the mean free path, the diffusion length and the mean lifetime.

As soon as an experimental discovery, be it large or small, was made, Enrico Fermi was able to give it within a few hours' time a corresponding theoretical interpretation, and to suggest new experiments to determine which among the various possibilities still open should be considered the definite one.

The weighty note which appeared in the August 1936 issue of "Ricerca Scientifica" is a work of particular importance, containing many results published only in Italian. In this note Fermi expounded the theory of the deceleration and diffusion of neutrons and set in perspective the experimental results he and his collaborators had obtained over the preceding two years. This work, presented simply and easily, yet with great elegance, is the starting point for all the far more complex treatises subsequently written by other authors, and gives the basis for calculations regarding the utilization of moderators in constructing nuclear reactors.

The results of this work were so many, and they followed each other in such rapid succession, that it seemed, especially to Fermi's younger colleagues, as if this state of things should continue indefinitely. Fermi was of course satisfied with the work and excited by the quantity and quality of the results, but not to the point of losing his calm, detached judgement; every so often he would say: remember boys, these are the years of the fat cows; the lean ones are sure to follow.

The body of scientific work carried out at the Physics Institute of the University of Rome in those years made an extraordinary contribution to the development of nuclear physics; for this accomplishment Enrico Fermi was awarded the Nobel Prize in 1938.

In the years preceding 1938, following his first visit to the United States in 1930, Fermi was often invited to the new continent to teach courses during the summer sessions of different universities, and on many occasions he had been offered important chairs on a permanent basis. He had felt uncertain about these offers, torn between the desire to remain in Italy, holding onto and reinforcing the strong ties of his life and work, and the wish to live, and have his family live, in a less troublesome climate than the one that was evolving in Italy. The decision was precipitated in 1938 by the introduction of the racial laws, which struck his family, since his wife Laura was of Jewish origin.

In the autumn of 1938 Enrico Fermi accepted an offer from Columbia University and with a regular leave of absence from the Ministero della Pubblica Istruzione (Ministry of Public Instruction), he moved to the United States. After two years the Ministry refused to renew the leave, and Fermi's transfer to the U.S.A. formally assumed the character of a permanent departure. At Columbia University, Fermi took up his work as "maestro" and researcher with the same calm aggressiveness toward unsolved scientific problems that had been typical of his period in Rome.

Hahn and Strassmann's discovery, in 1939, of uranium fission caused by neutrons, opened new possibilities just a few months after Fermi's arrival in the United States. If along with the high energy, a sufficient number of neutrons were liberated during this process, they could have produced other fissions in their turn, and thus triggered what is usually identified as a chain reaction. Fermi was not the only one, but certainly one of the first to realize this, and begin exploring the field with the help of various collaborators, among them L. Szilard, W. H. Zinn and H. L. Anderson; the latter would remain with him from then until his death.

Several important works on this subject were published by Fermi and his colleagues in the "Physical Review", but it was soon obvious that would it have been possible to provoke a chain reaction, this could have had highly important practical consequences. The war was already underway in Europe and was reaching out to involve the rest of the world; as a result, the expession "practical consequences" signified, in such a climate, not only a new energy source, but the faint possibility of new explosive devices as well. Research results in this field were therefore considered of military importance, and kept secret. At that time Fermi and his group began to work for the American government in the laboratories at Columbia University, remaining there until 1942, when they were transferred to Chicago. There, on December 2, 1942, the first nuclear reactor went into operation, planned and constructed by Enrico Fermi with his collaborators H. L. Anderson and W. H. Zinn. L. Woods and G. L. Weil also participated in the final phase of construction.

This accomplishment and this date made Enrico Fermi's name known throughout the world, in every sphere. A very natural reaction, since it was this accomplishment that suddenly transformed neutron physics from being a complex and refined chapter of nuclear physics, cultivated by a limited number of specialists, to a body of knowledge destined to affect even the life of the common man.

But even if Fermi had never built the nuclear reactor, his name would still be among those of the greatest scientists of our century, since his fame as a scientist is based on all the theoretical and experimental discoveries that he contributed in more than thirty years of tenacious and ingenious work.

The construction of the first nuclear reactor leads us however to make two observations. The first is that this success belonged to Enrico Fermi by right, since no other person in the world was better prepared to resolve the numerous theoretical and experimental problems implicated in such an undertaking. The second is that Fermi should also be considered the founder of nuclear engineering, which came into being with the construction of the nuclear reactor. He was, in fact, not only a consummate theoretician and an ingenious experimentalist, but an expert engineer as well, able to project complicated machines with that attention to the least detail which, while not essential in an early draft or even in a laboratory experiment, is of the greatest importance in a machine meant to last and perform with regularity.

Enrico Fermi devoted himself to the development and improvement of the reactor until the winter of 1944, when he moved to Los Alamos where, at the end of 1942, laboratories for developing wartime applications of nuclear energy had been organized. At Los Alamos he did not disappear into the large, complex organization already existing, but was given a position of particular importance as consultant. The production of the first atomic bombs in the Los Alamos laboratories, and the use made of them in the summer of 1945, are universally known facts that mark a moment of extreme gravity in man's history.

The possibility of using nuclear energy for military purposes posed very serious problems at the human level. It is not a question of new problems however, but problems as old as man, that arise again today in a more dramatic way, and on a far greater scale than in the past. Obviously it is not possible, either from a practical point of view or from that of what is to man's advantage, to consider interrupting the normal evolution of the study of natural laws whose results, in themselves, are neither good nor bad. It is only the use made of these discoveries that is good or bad, and man must seek to direct this use towards beneficial rather than destructive ends.

This natural conclusion, which we all agree to in times of peace, is abandoned in war-time, since each of the warring parties, fearing that the enemy may utilize all sorts of weapons, is induced to equip himself with every means of destruction. Thus it was that the vast majority of physicists working at Los Alamos, Enrico Fermi among them, while hoping that it would not be possible to liberate nuclear energy rapidly enough to create an explosion, felt that if this possibility did exist, they must manage to discover it before the enemy did.

These historical circumstances have brought into sharper focus a problem that no longer concerns only the scientist, but mankind in general: within what limits should one contribute to the defense of one's own group, one's own society and to attacking its enemies, even if this contribution is made knowing that the most onerous decisions will not be made by oneself, but by others. No definitive solution to this problem has yet been found. In the past humanity has already managed to overcome, not without suffering, crises no less critical than this. It is to be hoped that, once the initial period of disorientation has passed,we will adjust to the idea of having an exceptional font of energy at our disposal, and will direct its development exclusively to peaceful ends. At that point, even more than today, December 2, 1942 will become a date of fundamental importance in the history of mankind.

At the end of the war, Enrico Fermi returned to Chicago where in January, 1946, he was named to the Charles H. Swift chair of the University as professor of physics, and a member of the Institute of Nuclear Studies. Here, using the nuclear reactor as an intense source of neutrons, he devoted himself to a study, more subtle than any that had yet been done, of slow-neutron properties for which he developed the optics, stressing and studying quantitatively the corresponding phenomena of reflection, refraction and diffraction.

Among the many interesting results obtained in this field was the determining of the phase difference with which a neutron wave is scattered by different nuclei of the same crystal. The refined investigative methods of neutron optics, applied to the study of crystals, marks the beginning of a new chapter in experimental crystallography. These works, carried out in collaboration with H. L. Anderson, J. Marshall, L. Marshall, A. Wattenberg, G. L. Weil and W. H. Zinn, close the second period of Fermi's scientific activity, the phase dedicated to nuclear physics. Possibilities in the field were not yet exhausted and obviously he could have continued in it for many years, but the work no longer had the character of ground-breaking exploration of new, fundamental knowledge that it did in the past.

For this reason Enrico Fermi turned to another field, the physics of π mesons, which were being produced with the big new cyclotron in Chicago. The third, and unfortunately the last period of his research activity, began in this way; among his many collaborators were E. F. Alei, H. L. Anderson, M. Glicksman, E. A. Long, A. Lundby, F. Martin, N. Metropolis, D. E. Nagle and G. B. Yodh. Together with his colleagues, Fermi achieved a number of important results in this field as well, including the measurement —in terms of energy— of the collision cross-section of the π mesons, of both signs, against protons and of the corresponding angular distributions, whose analysis may lead to determining the phases of diffused waves with a given angular momentum and isotopic spin.

But his activity in this field was just developing, or rather was just beginning, when his exceptionally strong fibre, as a researcher and as a man, was undermined by an incurable illness. Regarding the various aspects of π meson physics, two other significant works from 1947 should be considered part of this final phase of Fermi's activity, dedicated to studying the properties of subatomic particles. The first, theoretical, was carried out in collaboration with Teller and Weisskopf and concerns absorption by the nuclei of negative μ mesons; the second, experimental, with L. Marshall, has to do with the interaction between neutrons and electrons.

This schematic division of Fermi's work into three periods —the first devoted to molecular and atomic physics, the second to neutron physics and its application to nuclei, the third to the physics of subatomic particles— does not give the whole picture.

His work in cosmic radiation, in particular, cannot be ignored; he turned his attention to these problems only at intervals, but in a way that led to substantial contributions.

In addition to the work of 1933 already discussed, in 1940 he focused on the influence of the dielectric constant of the medium on the energy loss of a fast particle, a phenomenon often referred to in the literature as the Fermi effect. In 1949 he conceived a theory on the origin of cosmic rays, and in 1950 a theory on particle production in the collision between two high-energy nucleons. The latter, based on purely statistical and thermodynamic considerations, is a highly useful scheme for the comparison with experimental data and for determining the corresponding laws, which are still very unclear today.

The theory on the origin of cosmic rays, as he himself pointed out from the beginning of his work in this field, cannot be based only on the mechanism through which a part of the ionized material in the universe is accelerated, creating the very high-energy particles that form the primary products of cosmic radiation. It is however an existing mechanism, whose relative importance with respect to those suggested contemporaneously or shortly after by other authors, is not yet established, and will be decided only in the future on the basis of a more exhaustive experimental study of primary radiation and its variations. The theory, which Fermi presented at the International Congress on cosmic radiation of Como in 1949, is based on the principle of a collision mechanism between ionized hydrogen atoms and the clouds of ionized matter that drift around interstellar space. The elegance, grand scale and simplicity of the concept are typical of Enrico Fermi's genius. This problem led to his interest in the magnetic fields existing in the galaxy and in the arms of the spiral nebulae in general, subjects to which he dedicated various papers in his last years, some in collaboration with the astrophysicist S. Chandrasekhar.

All of this, and much more that time does not permit me to mention, was accomplished by Enrico Fermi in a little over thirty years of intense, tenacious work. His activity was still so lively, so youthful, without the least sign of fatigue, as to give the impression that he would have gone on producing works of fundamental importance for many years to come.

His scientific accomplishment is so vast and ingenious, the practical consequences of some of his work of such significance and gravity, that one who never had the good luck to know him could easily be led to form an idea of him that is quite different from the reality. Only his relatives and friends, only those who knew him, are aware that if on the one hand it was hard to separate the various aspects of Fermi as scientist, researcher, maestro and man, since they were profoundly interwoven with each other; on the other, his simple tastes and life style, his serenity in the face of life's difficulties, the lack of pretentiousness or eccentricity in his character were human qualities that were even more remarkable for their contrast with his extraordinary qualities as a scientist.

A tireless worker, calm and confident, he spent most of his day at the university and in the laboratory concentrating his efforts on a well-defined problem which he faced with exceptional resolve, keeping his eye on the main target without worrying about unnecessary details. The mathematical theories or experimental techniques that he used with ease according to his need, were for him only means to an end, that of clearly understanding a particular phemonenon.

It was always the natural phenomena that caught his interest, and their discovery always took on the character of a final conquest, while the theories were seen as useful, even necessary schemes linking the phenomena to each other, schemes that are created, fit the purpose or are substituted, according to experimental data.

Although his scientific culture was extensive, he never studied in the normal sense of the word. For example, when he came upon an important theoretical work by someone else, instead of reading it he tried to get an idea of its content by running rapidly through the introduction and the conclusions and then he reformulated it himself, most of the time taking an original approach.

Fermi loved to work with collaborators, who, though they might already be mature scientists, usually ended up learning so much from him as to become, in a sense, his students. And his students, though they might be very young, were immediately confronted with the problems that interested him in that moment, thus becoming his collaborators. Work was done efficiently but calmly, in a low-key atmosphere as might have been that of a demanding but enjoyable game.

His collaborators and students were also his friends, who visited his house and spent their vacations or Sundays with him, taking walks or playing tennis, or going skiing together.

His remarkable sense of duty and respect for the law; his calm temperament, foreign to every act of violence; his respect for his fellow-men and their opinions; his lively critical spirit, that led him to set, without prejudice, a scale of scientific and human values; his sense of friendship and his generosity in recognizing the merits of others: for these qualities he will be unforgettable.

Not until the beginning of October 1954, following exploratory surgery, was the illness that had long been consuming him identified, but by then it was too late. Even under these circumstances, of which he was fully aware, he remained calm and at peace. He died in his home, near the University of Chicago, on November 28, 1954.

Fermi had been in Italy during his last summer, and at the invitation of the Italian Physical Society had held a course on the physics of π mesons at the International School of Varenna. On August 6th he gave his last lecture —the last of the course and of his life— on nucleon polarization, the theory to which his final work was dedicated, published in the April 1954 issue of "Nuovo Cimento". About forty students, half Italian and half from other countries throughout the world, followed this lecture, given with that simplicity of form, incisive clarity, compelling logic and stimulating critical spirit typical of all his lessons. At the end, there was a moment of emotion-charged expectancy, followed by unforgettable applause, full of gratitude and admiration. Everyone who was present remembers him: small, thin, his face hollowed by the illness that had not yet been diagnosed but was already destroying his robust constitution, his eyes bright from the pleasure of teaching, of communicating to others his simple, profound and elegant analysis of the most recent experimental results.

In commemorating Enrico Fermi today, the Academia Nazionale dei Lincei, of which he was a correspondent since 1932 and a member since 1935, feels the painful weight of sorrow for such a premature loss. All physicists, scientists from all over the world, participate in this sorrow. But we are not remembering just the scientist. Most of us knew him, many were his friends, some recall him as a young man or little more than adolescent, when his name was only an undoubted promise.

It is an unfillable void that he has left, but also a legacy of inestimable cultural and human value, which we must do our best to accept and preserve.

Enrico Fermi was born and educated in this country, he studied in our schools, he taught and worked in our universities, he conceived theories and discovered new phenomena in our laboratories. The events of life drove him away from our country, and where he went, he was outstanding both as a man and as a scientist; but he never broke away from the tree of our culture, to which his contributions were unceasing and essential. In the history of this culture he remains, before posterity, among the great men of the past.

Commemoration of Enrico Fermi(*)

ENRICO PERSICO

Excellency, Ladies and Gentlemen,

Today the University of Pisa and the Scuola Normale di Pisa would like to honour one of their most distinguished former students. This man, who died prematurely about two months ago, has contributed to our understanding of the laws of nature and to our mastering of them as few others have done.

My only claim to having this honour of commemorating him is to have been one of his closest and oldest friends, so please excuse me if in this talk I make frequent reference to our personal relationship, which for me was a rare opportunity of intimately knowing someone who had such a lucid and excellent mind as well as such a strong character and well balanced personality.

Enrico Fermi was born in Rome on September 29, 1901. His exceptional aptitude for the exact sciences was evident at a precocious age. When I met him I found, to my surprise that I had a fourteen-year-old school mate who was not only (as they say in scholastic terms) "very good", but who was endowed with a type of intelligence which was completely different from the "very good" students I knew.

We made the habit of taking long walks from one end of Rome to the other, during which we discussed, with youthful presumption, philosophy, politics and science. Indeed, during these discussions Enrico had very clear, original and definite ideas which never ceased to amaze me. Furthermore, when mathematics and physics were discussed he was much more knowledgeable about these subjects than most of us, and it was evident that this learning had been acquired over and above that learned in school. His was not merely a scholastic knowledge, he could readily discuss these subjects with absolute confidence since, already at that time, knowing a theorem or a law was tantamount to

(*) Commemoration held at the University of Pisa, January 1955.

knowing how to use it.

Thinking about that sense of amazement and admiration which Enrico's intelligence generated in me, I wonder if I had ever spoken of him, who was almost my own age, as a "genius".

I probably did not because for a normal boy and also for many adults this word is associated with an elderly, famous and aloof person rather than with someone's mental capabilities. What struck me about my new friend's mind was something too new for me to be able to define.

After finishing lyceum Fermi applied to the Scuola Normale di Pisa and brilliantly won a place there. In the autumn of 1918 he went to Pisa and registered at the University. Even though he diligently attended the courses and sessions, his studies focused on subjects which he had personally chosen independently of his obligatory curriculum. For example, in February 1919, that is during his first year at the University, he wrote to me:

"... since I have almost nothing to do for school and since there are lots of books available here, I'm trying to learn more about mathematical physics and I'll try to do the same for mathematics".

One year later he wrote to me:

"My studies are going very well because I was able to get rid of inorganic chemistry and I've decided to learn organic chemistry in the classroom". To learn it in the classroom was the only thing he could do for a subject which he was not particularly fond of. (As a matter of fact, this exam together with that of chemistry preparations were the only ones in which he got only 30 *sine laude* (1).) In this same letter he also added that he was gaining prestige within the physics department, and he had been asked to give a talk for various professors on the subject of the *quantum theory* (almost totally unknown in Italy at that time) and about which he was making a lot of propaganda.

Luigi Puccianti, the Director of the Institute of Physics in Pisa at that time, was a very distinct person but modest enough to immediately recognize the exceptional qualities of Fermi and to deserve him a special treatment. He considered him more as a friend than as a student, and never was he too proud to ask Fermi to teach him or to give him the latest update on what was going on in physics.

While a student at Pisa, Fermi published his first papers. They were theoretical notes on the problems of electromagnetism and relativity which were fairly current subjects at that time. You could say that these publications were an "overflow" of his deep interest in broadening his knowledge of physics. His way of studying a book has essentially always been to write down only the data of the problem or the results of the experiment found in the book; he would than re-elaborate everything himself and then compare his results with those of the Author. He, at times thus, discovered and resolved new problems and he even corrected errors in what were generally accepted solutions. This is how he came to produce his first publications.

---

(1) In Italian Universities, 30 actually is the highest mark, but for especially distinguished students the *cum laude* notation can be added to mean perfect knowledge of the subject.

These scientific investigations, so atypical of a student, did not keep him from taking part in the goliardic activity at Pisa, on occasion playing innocent boyish pranks. He physically exercised by hiking in the "Alpi Apuane Mountains" and he continued this beloved sport throughout his life.

In July, 1922 Fermi received his degree in physics *summa cum laude*. His thesis was an experimental one on X-ray diffraction. Since there were no available positions in Pisa, he went back to Rome where the Director of the Physics Department, Corbino, who already knew him and greatly esteemed him, offered him a position as assistant professor, and he had the Faculty assign him the Introductory Mathematics course. Corbino's initial intention of getting such an extraordinary teacher and researcher at the University of Rome on a permanent basis was since then perfectly clear.

During the two year period between 1924 and 1926 Fermi went to the University of Florence as an associate professor of Mathematical Physics and Rational Mechanics.

It was at this time, that is in 1926, that he published his first internationally recognized paper, "Sulla quantizzazione del gas perfetto monoatomico" (On quantization of the monoatomic perfect gas), which would influence the future development of physics. The idea which he elaborated was an apparently simple one, but it actually was really rather subtle. The subject was how to statistically pose problems involving a large number of identical particles which follow Pauli's principle. Today this method is known as "*Fermi's statistics*" as well as Fermi-Dirac, since Dirac, using another method, developed it a short time later. We now know that this is used for electrons, protons, neutrons and other types of particles, collectively called *fermions* to distinguish them from other particles which obey a different statistical law.

In 1927 a chair in Theoretical Physics was established at the University of Rome and Fermi, having applied for and won the post, returned to Rome and was assigned this position.

This was the beginning of what was to become a long, fervent activity in which Fermi was the fulcrum of a small group of brilliant students and collaborators.

Fermi was a born teacher and to him teaching came naturally. Besides teaching his regular courses, he devoted a great deal of time to his own personalized type of teaching which was intimately linked to his research activity. This method consisted in joining four or five of his students around a table and in solving a problem in their presence, "thinking out loud", so to say. Often these problems were related to his research at that moment. Nothing could have been more useful for his students who followed, with admiration, the meanders of this unique mind as it moved between the limits of the known and unknown. The students were about his age or slightly younger and one of his most admirable traits was that he was able to spontaneously create a very congenial working atmosphere while at the same time these students were very respectful of their great teacher.

Fermi's teaching method, either direct or indirect and his very personalized working style, raised the level, in a little over 10 years, of the School of Italian Physics to a point which had previously been deemed impossible. Deep and lasting traces of his presence were to be felt years after he had left Italy.

During his first six years, Fermi's work at the University of Rome was mainly theoretical and dealt with some applications of his statistical method to the structure of the atom, the Raman effect and other spectroscopic phenomena.

During this period of time, in 1928, he married Laura Capon and they had two children, Nella and Giulio.

The year 1932 was a banner year for physics since it was the year in which a series of fundamental discoveries about the atomic nucleus were made and it is considered as being the year in which nuclear physics was "born". Fermi immediately turned his attention to this new field. In 1933 he contributed to this research with a note entitled "Tentativo di una teoria dei raggi beta" (Tentative theory of beta rays). Although it was presented as a tentative approach, this extremely original theory is still a basic tool for interpreting beta ray emission phenomena from nuclei.

In 1934, that is the following year, Fermi would make an even more memorable contribution to nuclear physics.

Until 1933 it had been believed that radioactivity was spontaneous, and found in only a few rare elements. It was also thought that it could neither be artificially produced nor modified by man, just as the course of the stars cannot be modified. Instead at the end of 1933, Frédéric and Iréne Joliot-Curie discovered that certain elements, such as aluminium, which are not radioactive spontaneously, if bombarded by polonium alpha rays would become radioactive: these rays are extremely fast charged particles which by penetrating in the aluminium nucleus can modify its structure and make it unstable and thereby radioactive.

Thus, an artificial radioactivity was created for the first time. This, however, could be produced in only a few of the lightest elements and only in an extremely small quantity.

The news of the Joliot-Curie extraordinary discovery triggered off a very simple, but ingenious idea in Fermi's head. He understood that the reason the alpha rays were so ineffective in producing the artificial radioactivity was that since both the alpha rays and the atomic nucleus are positively charged they repel each other, thereby very few alpha particles can get into the aluminium nucleus, a necessary condition for making it radioactive. Fermi hypothesized that particles without a charge, the *neutrons*, which two years earlier had been discovered, would be much more effective. On the basis of this simple line of reasoning Fermi decided to prove that if a material is exposed to a neutron source it would become radioactive.

Even though in theory the experiment was a simple one, it did necessitate a completely new set-up which did not exist at the University of Rome, and in addition to this, Fermi was, up to that moment, primarily a theoretical physicist. Nevertheless, Fermi working alone prepared the experiments, and in March 1934 for the first time produced artificial radioactivity in aluminium and fluorine by using a neutron source. He soon found that, as expected, the phenomenon was reproducible in many other elements even if their atomic weight was higher, thus making them less penetrable to the alpha rays but still penetrable to neutrons.

Fermi immediately formed a small group of collaborators and they began to systematically study the phenomena in almost all the known elements, and they found a very interesting variety of behaviour patterns. The various publications and related theories based on this work, which was carried out between 1934 and 1936, placed the Department of Physics of the University of Rome in the international scientific limelight.

Once the phenomena and the methods for studying it had been discovered it seemed that investigations of the more or less ninety known elements would just be a routine job. In the course of this work instead, a new phenomenon was casually discovered. Even today the impact of this discovery, scientifically, socially and economically, cannot be fully appreciated.

During the morning of October 22, 1934 the group noticed that the radioactivity in silver, artificially produced by neutrons, was enormously greater if a piece of paraffin was placed close to it. No logical explanation for this could be found. That afternoon, Fermi did come up with an explanation which initially appeared to be a paradox. He hypothesised that the hydrogen in the paraffin had the effect of slowing down the neutrons and that the slow ones were more effective in producing artificial radioactivity. This hypothesis is so far from what a logical type of reasoning could come up with, that only someone like Fermi, with his very acute and perceptive scientific intuition, would dare to suggest.

Further experiments confirmed Fermi's explanation. This slowing-down of the neutrons would be in the future the building block for nuclear piles and the basis for all the peaceful uses of atomic energy.

It is worth mentioning at this point that Fermi was adept at doing both good experimental as well as fine theoretical work, a rare quality in physicists. He was not an experimental physicist who could build complicated apparatuses or carry out experiments which required great precision, his competence lay in his knowing which was the right experiment, what was the right moment for planning and carrying it out and which were the simplest and most efficient ways of doing this task. He was a very hard and patient worker and he did not waste energy or time on non-essentials. His theoretical and experimental works were always closely linked and both were calmly and methodically performed with an iron will which was sustained by an incredible physical and mental force.

In 1938 Fermi was awarded the Noble Prize for Physics "for having demonstrated the existence of new radioactive elements using neutrons and for the discovery of nuclear reactions by slow neutrons".

During this same year, the fascist government in Italy began a policy of racial persecutions. Although it did not directly affect Fermi, it did pose a threat to his wife and threatened his daughter's and son's future. It was at this point that he decided to write to four universities in the United States saying that the reasons for which he had declined their previous offers now ceased to exist. He selected Columbia University, in New York, from among the five chairs which had been offered to him.

Upon his arrival in New York in January 1939 he immediately began working on the current "hot" topic which was exciting nuclear physicists —the *fission* of uranium. A few months before, in fact, the German physicists Hahn and Strassmann, had discovered that, among the many nuclear reactions produced by slow neutrons, using the process discovered by the group in Rome five years earlier, a peculiar phenomenon appeared only in uranium. It was called "fission" because it was interpreted as being the breaking-up or splitting of the uranium nucleus into two almost equal parts. At that time, this newly discovered phenomenon appeared to have no practical applications whatsoever, because this, like all nuclear reactions, necessitated an enormous amount of material. However, for a physicist the importance of a phenomenon is not measured by the quantity of the matter involved but rather in how much of the unknown can be revealed through it: thus many theoretical and experimental physicists, both in Europe and America, were studying the fission problem. And then, in the most unexpected way, the possibility that nuclear reactions were not only pure science speculation, but that they could be utilized in the field of applied physics became a reality.

Fermi was one of the first, if not the first, to theoretically figure out this remote possibility. In fact, in a meeting of physicists held in January, 1939 he mentioned that there were good theoretical arguments supporting the hypothesis that when an atom of uranium splits there is an emission of one or more neutrons. And it was clear that, if more than one neutron were emitted, they could in turn set off other reactions in an increasing number and thereby the reaction would propagate through the entire uranium mass, instead of involving just an infinitesimal fraction. A similar "chain reaction", if it could be produced, would create energy and radioactive substances of an order of magnitude unimaginable up to those times.

Experiments performed in various laboratories in Europe and America immediately confirmed this assumption. But after the publication of these first results, possible military applications were foreseen and thus both in Germany and America research continued under a veil of secrecy.

In New York, Fermi was the leader of a small group of hard working collaborators who had the task of creating the "chain reaction" which the theory had predicted as probably possible, and which would have made nuclear energy useful for practical applications. However, between the theoretical possibility and the concrete realization there was an enormous gap because the substance capable of producing the reaction, uranium-235, is always present in nature mixed with a much larger quantity of its isotope uranium-238. This isotope prevents the reaction by absorbing the neutrons before they can produce it. To separate the isotopes was an enormous problem on the technical level and only later, with the advent of colossal instruments was it possible to overcome this problem. Fermi and Szilard however, got around this difficulty using a simple and elegant system: they placed the pieces of uranium, at chosen intervals, inside a huge mass of high-purity graphite. This is how the first *nuclear pile* was invented.

The construction ran into many problems since the available uranium was very limited, its properties were not well known, and besides this, the smallest traces of impurities in the metal or in the graphite might absorb the neutrons thus posing an obstacle to the reaction. Work on the pile was carried out first at Columbia University and then at the University of Chicago, where Fermi transferred in April, 1942. On December 2 of that year, a memorable historical date, the first chain reaction on earth was created using Fermi's pile: this event opened up an unlimited horizon bursting with new potentialities. Even today we are only minimally aware of this potentiality.

As known, the pile is a great source of energy and radioactive substances, however, together with these beneficial products, the pile produces the *plutonium* used for making atomic bombs. It is an obvious but distressful fact that without this aim in mind no government would have made the financial and industrial sacrifices and efforts necessary for creating the nuclear pile. But when, in 1939, the world was divided between two adversaries, the fundamental facts about fission were already public domain so that during the war each of the two could be certain that the other was working in secret to exploit them to the adversary's damage. And no one could predict if it would take a long or short time (today we know that both sides had underestimated the difficulties). It was, therefore a vital and inevitable necessity, but still not without grief, as is all war effort, that every effort be made to create the new weapon before the adversary.

During the summer of 1944 Fermi together with almost all of the best American physicists, transferred their residence to Los Alamos. In this "city laboratory", built for finding ways of using energy for military purposes, Fermi was not assigned a specific task, he was rather a consultant and someone who put the final touches on things.

Once the war had ended, at the beginning of 1946, Fermi returned to the University of Chicago where the *Institute for Nuclear Studies* had been established. This Institute, essentially based on Fermi's project, was devoted exclusively to nuclear physics research. He did not wish to be the director since this would imply administrative as well as social obligations. Nevertheless, that he was the real head and intellectual leader there was recognized by everybody.

In Chicago Fermi could fruitfully use the new great research facilities that this powerful source of neutrons, the reactor, offered and he was able to carry out fundamental research on diffraction and neutron scattering as well as on a variety of other subjects. In addition to this, he published various theoretical papers, among which was a publication in which he developed a rather sophisticated and original theory about the origin of cosmic rays.

In 1951, about the time of his 50th birthday new research opportunities became available at the Institute of Nuclear Studies at the University of Chicago when the big cyclotron, which could produce *mesons*, was inaugurated. Thus, the mesons, about which little was still known, and which in nature are found in cosmic rays in a very limited quantity, were now readily available in very large quantities. With youthful enthusiasm Fermi launched into this new field. The physics of mesons was the subject of the lessons which he taught last summer at the International School of Physics at Varenna. Students, both young and old, from several different countries, paid close attention to Fermi's lessons.

In the midst of this fervent research and teaching activity death struck Fermi at the age of 53 on November 28, 1954.

If I were to sum up in a single sentence Enrico Fermi's very complex mental physiognomy, I would say that his prodigious capacity of immediately spotting the essential element in everything and getting to the heart of it in the simplest way was his principle natural gift. Connected with this fundamental capacity are many other qualities which distinguished him as an extraordinary man and scientist.

Above all, his intuition and foresight, already evident in youth, in the selection of his research topics, is worthy of note. If we were to leaf through the Nuovo Cimento dating around 1920, we would see that most of the physics published in Italy dealt with intricate and obscure matters which could only be studied empirically without a clear theoretical guide, for example, the properties of the arc and the various forms of electrical charges in gas, the electrification by rubbing, the strange electrical properties of bismuth and selenium. Many of these phenomena in the light of today's experience can be more or less labouriously interpreted, but they appear to be a snarl of many different elementary phenomena and thus, particularly unsuitable as the key to the really fundamental questions. At this same time, the student Fermi in his efforts to interpret the hydrogen spectrum and other spectroscopic phenomena concerning the motion of electrons in the atom, turned his attention to Denmark, Holland and Germany to see what research was being carried out there. It was this type of research, aimed at establishing the laws that govern the basic elements of matter (beginning with the simplest atom, hydrogen) which was then the key to learning about more complex phenomena, as for example, discharges in gases and the electrical properties of bismuth and selenium.

Another feature of Fermi's concrete approach to things can be evidenced in his lessons and writings. The most complex problems are reduced to the essential and resolved, for the most part, using original approximate mathematical methods which were exactly the most suited ones for finding the solution to the problem. The researcher and student are often fooled by the apparent simplicity of Fermi's reasoning and only when they try to abandon his guide to move alone following a similar train of thought do they realize that they have been masterly led through an impervious maze.

Good examples of these didactic qualities are "Introduzione alla Fisica Atomica" (Introduction to Atomic Physics) published in 1928, "Molecole e cristalli" (Molecules and crystals) published in 1934 and "Thermodynamics" published in 1937. I remember that he wrote the first of these while we were on vacation together in Valtellina: he wrote while in bed, on a school notepad, without hesitation, afterthoughts or erasures as if someone had been dictating to him.

Not only in scientific and didactic matters was this innate love for simplicity apparent, simplicity was his whole style of life.

He accepted, with grace, acknowledgement of his merits, however, any form of adulation disgusted him. You could not say that he was modest, he realized that he was better than others, and he was honest enough to admit rather than deny it. But vanity and search for glory, influent positions or places were not in the least a part of his character. Nevertheless, many unsought honours were bestowed upon him. He was a member of 18 Italian and foreign Academies, 8 foreign universities granted him a honorary degree in physics. In addition to the Nobel prize he received many other prizes from Scientific Institutions of several countries. He was president of the American Physical Society in 1953. But, as a rule he usually declined non-scientific positions.

He had extremely simple tastes, he enjoyed a peaceful family life and considered money as a basic commodity for essentials and for giving him the necessary tranquillity for pursuing his studies. Any ostentation and display of luxury was a useless complication for him.

He loved to physical exercise: tennis, skiing, mountain trekking were all done with youthful enthusiasm. Even last summer I had the pleasure of his company during vacation in Tuscany and in the Alps. Even though he was suffering from the illness which would shortly put an end to his life, he was still the cherished and simple comrade of our youthful outings. In one of our excursions, in the Isle of Elba, an old habit which I think few knew about and which would surprise those who knew him superficially, surfaced. Often when he was relaxed, admiring a beautiful landscape or hiking, I would hear him reciting, to himself, long verses of classical poetry. This rich treasury of poetry, which from youth he had cherished, he enjoyed more than music which he was less inclined to appreciate.

The name Fermi for most of us is linked to the reactor and atomic energy, but for physicists it is directly or indirectly associated with much of the progress which has been made in this science over the past thirty years. However, for those who knew him well and cherished him, he will be remembered as an unpretentious, wise and good person, endowed with the serene goodness of the strong.

Enrico Fermi and Italian Physics (*)

Franco Rasetti

In the brief time at my disposal it would not be possible, even touching just on the main points, to describe Enrico Fermi's multifaceted work as a scientist and as a maestro. I will therefore focus above all on those aspects of his character and his activity that exerted a profound influence on Italian physics, an influence that, far from disappearing with the death of the distinguished scientist, continues to make itself felt today through his students, with a mentality, an environment, a tradition that he has created.

Enrico Fermi was born in Rome on September 29, 1901, and completed his high school studies there. Very early, when he was just ten or eleven years old, he showed lively interest and exceptional aptitude for the mathematical and physical sciences. We know that at thirteen and fourteen he was reading and assimilating, without anyone's help, mathematical treatises at the level of third- or fourth-year university courses. At that age he already revealed the capacity to concentrate his interests on a specific topic, an aptitude that he maintained throughout his life, and made him consider mathematics not in itself, but only as a means for formulating and resolving problems in physics. Another characteristic that he developed quite young was his interest in physics experiments, which he attempted to reproduce with the primitive means at his disposal.

During his high-school years, Fermi devoted himself intensely to reading books on physics and mathematics that were in part borrowed and in part bought from second-hand book shops that he visited with his friend and study-companion Enrico Persico: As a result of these autodidactic studies, by the time he was seventeen, he had acquired

(*) Speech made at the presentation of the fuel fragment of uranium taken from the original Fermi reactor, donated by the President of the Republic, Giuseppe Saragat, to the Academy, and the inauguration of the medallion with the image of Enrico Fermi, by maestro Corrado Cagli. Original text courtesy of The Accademia Nazionale dei Lincei, Celebrazioni Lincee, no. 12, April 20, 1968, Rome.

a culture in physics and mathematics that would easily have sufficed for a university degree in those disciplines. He had a profound knowledge of all of classical physics, from mechanics to electrology, to optics and thermodynamics, and had completely mastered the necessary mathematical tools.

In October 1917, following a competitive examination, Fermi was awarded a scholarship to the Scuola Normale Superiore of Pisa, and he registered as a physics student in that University. It was then that I met him and first had the chance to realize the amazing degree of maturity in physics and mathematics he had already reached.

Even during his four years at the university one could say that Fermi remained basically self-taught, since he was already thoroughly familiar with most of the material in the required courses of physics and mathematics and could follow the lessons and take examinations with practically no effort, using the greater part of his remaining free time to assimilate the strange new theories that were just beginning to reach Italy and were more or less unknown to the physics professors in Italian universities. I recall that he used to say that all the university had given him was some basic knowledge of chemistry and mineralogy and a certain experience in the techniques of experimental physics, mainly of optical and X-ray spectroscopy.

To give a better idea of the work of utter renewal that Fermi was about to undertake in Italy, though still an unknown third- or fourth-year student appreciated only by a handful of professors and school-mates, I should explain the revolutionary new concepts in physics being developed at the beginning of the nineteen twenties. At the end of the preceding century, it was held that the theories we now call classic, which describe phenomena as a space-time development that obeys differential equations, were in principle able to explain all phenomena relating to non-living matter. Mechanics and electrodynamics are the best examples of this type of theory. Even the theory of relativity, formulated by Einstein in 1905, while thoroughly upsetting the commonly held concepts of space and time, was still a theory of the classical type. But at the same time —the first decade of the twentieth century— the study of energy exchanges between radiation and matter led Einstein himself and Max Planck to formulate the revolutionary hypothesis of the photonic nature of light. Then, in 1913, Niels Bohr, on the basis of an atomic model proposed by Rutherford, succeeded in explaining for the first time the simplest of the atomic excitation spectra, that of hydrogen, with a profoundly "non-classic", or "quantistic", as it was called, hypothesis on the possible movements of the electron in the atom, and the energy exchanges between electron and electromagnetic field corresponding to one photon energy.

A year after this discovery, which was destined to revolutionize physics as a whole, the first world war exploded, almost wiping out the exchange of ideas between the two opposing camps. Quantum theories went on developing brilliantly, and were verified by means of ingenious spectroscopic experiments, in Germany and in neutral countries linked to German culture such as Holland and Denmark, while little news of these discoveries filtered into the allied countries and just about nothing into Italy. At the end of the war, when various volumes of German physics reviews, bursting with theoretical and experimental work on quantum phenomena arrived all at once in Italy, very few physics professors in Italian universities thought them worthy of a glance. The little that they read was found repulsive, and they refused to look into it further. It should be admitted that the quantistic considerations of that period seemed purposely made to disgust a scientist accustomed to the logical and elegant constructions of classical theories. Actually, the new theories were not as elegant as the classical theories, where many specific results are derived with impeccable, mathematical logic from a few general principles; instead, they were more like cooking recipes with which, who knows how or why, it was possible to calculate, precisely, certain quantities that were typical of atoms.

Nevertheless, it seems almost unbelievable that a twenty-year-old student should be the only person in the Italian universities in those years to investigate quantum theories thoroughly, reading everything that had been published on the subject and even making his own original contributions to it. The theoretical physicist "type", in fact, did not exist in Italy. Physics professors were almost exclusively experimentalists, familiar with the current theories only in elementary form, while classical physics in a more elaborate guise, such as analytical mechanics or the theory of relativity, was cultivated chiefly by those having essentially mathematical interests, among them scientists like Volterra and Levi-Civita, who could not however be receptive to that hybrid mixture of theory and experimental facts that was the new quantum physics. Fermi was the first, and in that moment the only, Italian theoretical physicist of the century.

It can also be said that Fermi began his work as maestro in those years, since Luigi Puccianti, director of the Physics Institute, a really unpretentious person, often invited Fermi to hold conferences for teaching quantum theories to professors, assistants, and the other students. By the spring of 1922, after four years of university, Fermi already had to his credit important publications in the classical fields of theoretical physics such as electrodynamics and the theory of relativity, each of which would have been more than worthy to serve as the subject for a degree thesis. But, precisely because of the situation I have described, he was unable to use them for his thesis, since no one on the faculty was sufficiently well-versed in the field to be able to discuss them. Instead, he presented an experimental thesis regarding an ingenious method for producing images using X-rays.

From the first, in fact, Fermi was far from being exclusively a theoretical physicist, the kind who interpret only experiments done by others, as is the case for most theoretical physicists, even for the best-known. Bohr, Einstein, Heisenberg, Pauli, Dirac, for example, to mention just a few of the present-day "greats", have never carried out an experiment. Fermi was a complete physicist practically from the start, interested just as much in doing experiments as in theories, and he alternated the two forms of activity throughout his life. One could not say which was the greater achievement: the theory of the statistics of the ideal gas and that concerning the emission of beta-rays, or the experimental discovery of radioactivity produced by neutrons and the realization of the nuclear reactor. Only a few have reached great heights in both fields: Galileo, Newton, Lord Kelvin and Helmholtz are among the rare physicists that may be compared to him in this regard.

Orso Mario Corbino, who already knew Fermi and appreciated his unusual gifts, offered him on graduating a job as assistant and professor of mathematics at his institute in Rome. But Fermi was eager to know the great theoreticians of other countries, whose works he had studied to learn those quantum theories that were by now the unquestioned vanguard of physics that, in Italy, no one could have discussed with him on an equal basis. He therefore secured scholarships first to Göttingen and then to Leiden. At Göttingen an active school for theoreticians was flourishing under the guidance of Max Born, and there Fermi met some of his contemporaries, daring and ingenious spirits such as Dirac, Heisenberg, Jordan and Pauli, who shortly after, along with Born himself, became the creators of the new quantum mechanics using the matrix method. For obscure reasons, the exchange of ideas between Fermi and the other young theoreticians was not very productive then. Fermi's stay at Leiden was much more profitable for him: there he was appreciated for his real merits by that master of statistical mechanics, Ehrenfest, and by two of his young students, Uhlenbeck and Goudsmit, who remained good friends with Fermi for all his life.

On returning to Italy Fermi obtained, through the interest of Antonio Garbasso, an appointment to teach Mathematical Physics at the University of Florence, and there, in 1926, he formulated and published the statistical theory of a gas composed of particles that obey Pauli's exclusion principle, particles which we now call "fermions" in his honour. In Italy he was still known only to a limited number of physicists and mathematicians; only after the value of his discoveries was recognized abroad did his fame spread. In September 1927 there was an international physics conference at Como to celebrate the centennial of the death of Alessandro Volta. The entire "Almanach de Gotha" of world physics was present; one could count a dozen Nobel Prize winners, and the great architects of quantum physics —Bohr, Planck, Compton, Laue, Sommerfeld, Heisenberg and Pauli— also participated. Sommerfeld, the authoritative maestro from the Münich school presented a series of results in which he and his students demonstrated that all the strange and classically unexplainable phenomena of electrons in metals could be easily interpreted within the context of Fermi's new statistics. This was a genuine triumph for him, and many Italian professors were astonished that this young twenty-six year old, scarcely known in Italy, could be so famous in Germany.

At the time of this conference, however, Fermi already (since the end of 1926) held a chair of Theoretical Physics in Rome, the first in Italy, founded at Corbino's initiative. Unlike some other Italian physicists at that time, Corbino realized that the future of this science lay in the study of the atom, interpreted according to the quantum theories, and he proposed to create, in Rome, a center where these new disciplines could be developed and taught. Fermi was obviously the only person capable of carrying out such an enterprise. A small group of researchers formed around Fermi, of which this speaker had the honor to be the first, and young students quickly joined: Edoardo Amaldi, Emilio Segrè and others who subsequently contributed greatly to the progress of physics in Italy and abroad. I would also like to mention Ettore Majorana, the only one of the group who could compare to Fermi for exceptional insight in theoretical analysis, and who unfortunately died very young.

Without this providential initiative of Corbino's, it is almost certain that Fermi would have soon left Italy, since he was not lacking offers from foreign universities where he would have found an environment favorable to his studies. For this reason we should pay homage again to the memory of the distinguished maestro who made it possible for Fermi to carry out his work in Italy until just before the outbreak of the second world war.

The extraordinary atmosphere that grew up around Fermi, with the backing of Corbino, who saw to running the Institute so as to remove any administrative worries from the shoulders of the young people, has often been evoked by those who were fortunate enough to have been part of it. Fermi's unique personality, the minimal difference in age between teachers and pupils, their affinity in scientific interests and even in their diversions outside the university, created bonds of personal friendship among members of the Institute, a harmony of the kind that is rarely found among groups of researchers. There was no formality in the way in which Fermi taught us the latest physics theories, first among them the new quantum mechanics, developed by then into a logical scheme that for its elegance and perfection could rival the classical theories. Meetings were held that could be called seminars, but not according to any sort of schedule or other pre-established plan, on topics brought up at the moment by a question that one of us asked Fermi, or by some experimental result we had obtained which had to be interpreted, or by a problem that Fermi was studying and either had resolved or was trying to resolve. In any case, Fermi proceeded to explain the calculations he was writing on the blackboard with his constant, unhurried pace, neither accelerating in the easy steps nor slowing down noticeably before difficulties that would have brought to a halt anyone not having his incomparable technique and the intuition that allowed him to perceive the results even before having demonstrated them. Often we could not tell at the moment whether Fermi was expounding theories already well known to him or others, or whether we were witnessing a new step that he was taking at the confines between the known and the unknown. Many times we saw a new theory born, which Fermi developed while, you might say, he was thinking aloud.

Let me say that this unusual and highly effective teaching method during that happy period, recalled with nostalgia by all those who profited from it, has become almost unrealizable today. Aside from Fermi's genius and personality, one indispensable condition for creating an atmosphere like the school of Rome at Corbino's time was the small number of participants. The success of physics, and to a certain degree of all the sciences, in industrial and military applications, the resulting investment of considerable sums in every country, and the multiplication by a factor of ten or a hundred of the number of student researchers, have undoubtedly led to an increase in quantity in the crop of new results; but it is also undeniable that they have caused a deterioration in quality, not so much at the level of teaching and research as in the human relationships among those who devote themselves to it. An environment like that created by Fermi would be inconceivable in an Institute with fifty or a hundred students, also because of the huge amount of funding and relative administrative complications that modern research requires, and for its intensely competitive character.

The group of physicists in Rome was composed on the average of half-a-dozen professors and assistants and about a dozen students, but the restricted number was amply compensated by the enthusiasm of the participants, and above all by Fermi's incomparable guidance. The activity for the years 1927-31 was almost entirely carried out in the field of atomic and molecular spectroscopy, since this provides most of the information on the structure of these systems; also, we were well-acquainted with the techniques, and we had appropriate instruments. Fermi participated in the experiments and in the theoretical interpretation of the results. He was never a refined experimenter in building the best apparatus, but he was keenly intuitive regarding what the crucial experiments were for resolving a given problem, and he went straight to the target without bothering about unessential details. Similarly, in formulating theories, he utilized whatever means would take him most directly to the result, resorting to his mastery of analytical means if this was called for, otherwise turning to numerical calculations, heedless of mathematical elegance.

Among the theoretical and experimental contributions that Fermi and his collaborators made in those years were the important research projects on the Raman effect in molecules and crystals, on the absorption spectra of alkaline metals and on the hyperfine structures of spectral lines. In addition Fermi, together with Thomas, proposed a statistical theory of the electrons in the atom, and organized a project of numerical calculation of the eigenfunctions that practically all the students collaborated in, using adding-machines that were extremely primitive as compared to today's electronic computers. Nevertheless, a long series of numerical tables, published by the "Accademia dei Lincei" and highly useful for various applications to the properties of atoms, were compiled.

Around 1931 Fermi and others in the group began to feel that the future of spectroscopy, and of atomic physics in general, seemed rather limited, since the theory accounted for most of the phenomena observed and appeared not to leave space for new fundamental discoveries. Fermi foresaw that interest would shift from the atom's external parts to the interior, about a hundred thousand times smaller in diameter than the atom itself, that is called the nucleus.

Various properties of the nucleus were known at that time. It was known that most of the nuclei existing in nature are stable and others radioactive, the latter changing spontaneously into nuclei of different elements, altering the value of the electrical charge. The radioactive process takes place with the expulsion of an alfa-particle or helium nucleus; or with the expulsion of an electron, or beta-particle. Both phenomena are often accompanied by the emission of electromagnetic radiation in the form of gamma-rays. This demonstrated that the nucleus, like the atom, is a composite structure, and the initial problem was to investigate its components and the forces that hold them together. Since in those days the only basic particles known were the proton, or hydrogen nucleus, and the electron, it was believed that these were the materials forming the nuclei of all chemical elements. While the emission of alpha particles could be explained —more or less— by quantum mechanics, both the presence of electrons in the nucleus and certain details of their emission remained unaccountable.

We had already begun learning the techniques for studying radioactive phenomena, unknown to us and to most other Italian physicists before, when the work of Curie and Chadwick revealed the existence in the nuclei of a new particle which took the name of neutron; its electric charge was null and its mass about the same as that of the proton.

Majorana, independently of theoreticians in other countries, was among the first to suggest that the nucleus was composed solely of protons and neutrons, and he developed detailed theories on the subject which, unfortunately, despite Fermi's advice, he did not publish immediately, considering them premature. All difficulties regarding the presence of electrons in the nucleus were suddenly eliminated, but a new one arose: how, then, could the nucleus emit electrons? Pauli advanced, but timidly and without total confidence the hypothesis that the electron was created at the act of its emission, together with another invisible particle, with no charge, which Fermi named the neutrino.

In the fall of 1933 Fermi showed us an article that he had thought out and written in the early morning many days before, complete with all its mathematical developments, on a theory concerning beta-ray emission that was based on Pauli's neutrino hypothesis, from which specific results on the quantitative characteristics of the phenomenon were deduced. Few theories in modern physics have produced such prolific results and have resisted for over thirty years the rush of progress like Fermi's theory on beta-rays, which still dominates not only the ordinary beta process (the transformation of a neutron into a proton, creating an electron and a neutrino in the process) but also many transformations of recently discovered unstable particles.

This theory is definitely Fermi's most important and original theoretical creation; on its own it would have been sufficient to immortalize any physicist. But a few months later, in April 1934, he made an experimental discovery that was no less important: that of radioactivity produced by bombarding nuclei with neutrons. Irène Curie and Frédéric Joliot had created radioactive nuclei among the lighter elements by bombarding them with alfa-particles; Fermi found that neutrons were far more effective, since the absence of a charge allowed them to penetrate even the nuclei of heavy elements. In a feverish bout of work, in which along with Fermi three other physicists —Amaldi, Segrè and myself— and a chemist —D'Agostino— participated, more than forty new radioactive isotypes were created and studied. The following autumn Bruno Pontecorvo joined our group. We quickly discovered the surprising effect that certain substances such as water and paraffin produce an increase in the radioactivity induced when they are close to the neutron source and the bombarded element. Not a day later Fermi had already found the explanation for these paradoxical effects in the deceleration that neutrons undergo in colliding many times with the hydrogen nuclei contained in water or similar substances. A new wave of discoveries of radioactive isotopes followed, and among other things we investigated at length the many products derived from bombarding uranium. We had, without realizing it, produced the phenomenon of nuclear fission, as Hahn and Strassmann demonstrated only four years later.

I like to remember that as soon as Fermi became aware of his great discovery, and urgently needing funds for pursuing the consequences of it, he promptly telephoned Dr. Magrini, the then secretary of the "Consiglio Nazionale delle Ricerche" (National Research Council), asking him for a subsidy of 20000 liras, which arrived in a few days and was followed by another 60000 liras several months later. These sums, which seem laughable today even if we consider the greater value of the lira at that time, were a real fortune, exceeding our Institute's entire yearly endowment, and allowed us to continue working on Fermi's discovery. At times, in scientific research, a small contribution made immediately and without too many restrictions as to its use, can yield infinitely more than a large sum that arrives only after a long wait. These happy circumstances would be difficult, if not impossible, to re-create today.

Fermi continued working in the field of neutrons for another three years; his extensive publications in collaboration with Amaldi on neutron diffusion in matter and the selective absorption of neutrons by various elements are of 1935-1936. In this way he acquired experience on the behavior of neutrons that was unique among all physicists, and was to prove of great value in developing the nuclear reactor.

But in those years the storm that was gathering over Europe and would later extend to the rest of the world, began to create difficulties for the activity of the group in Rome, whose ranks had already thinned out as people left for other countries, looking for a climate of liberty and peace more favorable to carrying on scientific research. When Italy's alliance with the monstrous tyranny that was dominating Germany and the subsequent racial persecution made the situation even more unbearable, some of us —including Fermi— decided to emigrate. When Fermi travelled to Stockholm to receive the Nobel Prize, awarded him for the discovery of radioactivity induced by neutrons, and at the same time was offered a chair at Columbia University in New York, he took the opportunity to leave Italy definitively.

I would like to mention briefly Fermi's activity following his Italian period. At Columbia University he immediately undertook a program of studies aimed at investigating the possibility of a nuclear chain reaction based on the fission of uranium, discovered shortly before his arrival in America. This research was later transferred to a center created expressly for it at the University of Chicago, where, on December 2, 1942, Fermi succeeded in setting in operation the first reactor, basis for all the practical applications of nuclear energy. This accomplishment called for the combined contribution of all those gifts as theoretical physicist, experimentalist and engineer that only rarely in the history of science were to be found concentrated to such a high degree in one person.

Fermi later transferred, as a consultant to the various research groups, to the top-secret laboratory of Los Alamos, where the first nuclear weapons were created.

In Chicago again after the end of the war, Fermi initially devoted himself to a series of experiments, completed by theoretical observations, on the optics of neutron beams diffracted by crystalline lattices, in which phenomena similar to those well known for X-rays are produced. As usual, this body of work was a model of perfection, and so complete that very little remained to be done on the subject.

Having finished his work on neutrons, Fermi also developed a theory on the origin of cosmic radiation which, if it cannot be considered definitive, certainly contains some of the most interesting and fertile ideas put forth in this field.

At this point in his career Fermi, just as he had left spectroscopy in the 'thirties to dedicate himself to the more promising study of nuclear phenomena, turned his attention to a brand-new field that was becoming the vanguard in the march toward knowledge of the mysteries of inanimate nature: the physics of elementary particles, that is, of the effects that appear in very-high-energy collisions. At first these phenomena could be studied only in cosmic radiation, which in fact in the period following 1937 led to the discovery of a whole series of new particles not to be found on the Earth because of their short life-span: no longer than a millionth of a second, for the more stable ones. The positive and negative muons and the positive, negative and neutral pions were discovered successively, and later the families of "strange" particles as well, many of which have a larger mass than that of the protons.

With the development of high-energy accelerators, it became possible to create the new particles in the laboratory, with the great advantage of obtaining very high intensities in well-defined beams, instead of the few particles falling in casual place, time and direction, as it is the case with cosmic rays in nature. Among the very first big accelerators to function was the synchrocyclotron in Chicago, built expressly for Fermi in the new Institute for Nuclear Studies that now bears his name, and capable of producing pions in great quantity. Fermi threw himself with his usual enthusiasm into the experiments and the theory of these new phenomena, and in a short time had obtained fundamental results on the interaction between pions and nucleons (that is, protons and neutrons, which make up the nucleus).

By the beginning of the fifties, Fermi had reached a position of singular authority and prestige in physics throughout the world; he could claim some of the most essential contributions to the theories of the atom, the nucleus and elementary particles, and at the same time, to have opened the way, first with the discovery of the properties of neutrons, then by realizing the chain reaction, to the new atomic era. Universities, academies and other scientific organizations on two continents competed for his participation in congresses, symposia and special summer courses, since his presence, and above all the clarity and depth of his explanations, the enthusiasm that he was able to arouse among the physicists with his amazing ability to analyze and often resolve the most disparate theoretical or experimental problems that were brought up, were enough to make one of these manifestations a memorable event. He returned to Italy various times during this period, holding series of conferences on particle physics, his last being that of the School of the Italian Physical Society at Varenna in the summer of 1954. Not long after his return to America, he succumbed to the illness that had long been undermining his normally robust health, and he died on November 29, 1954, just fifty-three years old.

Having briefly summarized Fermi's scientific formation and his creative activity in both theoretical and experimental physics, I would like to highlight his qualities as a man, since they were equally important as those of the scientist for the essential role they played in the renewal that he brought about in Italian physics.

His fundamental traits struck even those who knew him superficially and were confirmed all the more by whoever had the chance to work with him. First of all, he possessed inflexible will power, an incredible capacity for physical and mental work and a degree of self-sufficiency —scientific and probably moral as well— that this speaker has never known in any other person. Not only did he have no need of teachers, studying mathematics and physics on his own, but later he even ceased to consult books or the original works of other authors. For him, it was enough to get an idea of an article's content from its title, and it was far easier and more convenient to reproduce the work himself, rather than follow someone else's treatise. But I would add: although he was surely pleased that his works were appreciated by his fellow-physicists, although he loved discussion, and his teaching above all gave him satisfaction and pleasure, Fermi would have had no need of these additional motivations for dedicating his whole life to physics.

Fermi shied away from honors of any sort (although they came to him without his searching them out) and even more so from appointments that were not strictly scientific and might steal part of the precious time devoted to research and teaching. Thus when the University of Chicago created the Institute of Nuclear Studies for him, Fermi, despite pressure from authorities and colleagues, firmly refused to assume its direction, which would have entailed administrative duties. He confined himself to his teaching position and to reserving a few rooms for research. If —rarely— he did accept assignments that were not involved with teaching or research, it was not for his own pleasure but rather from his sense of duty, which was remarkably strong: duty towards the family, his friends and pupils, and the institutions he was part of. His sense of duty was linked to an inflexible spirit of honesty and integrity; any form of injustice or moral compromise repelled him. Nothing disgusted him more than the immoral manoeuvres, unfortunately not always foreign to the sphere of the Italian University, of candidates and their protectors in university competitions for a post. Fermi used his influence to re-establish, at least in the field of physics, conditions under which personal merit would be the only valid qualification, regardless of schools.

Under Fermi's influence, a generation of young physicists came to maturity who were dedicated to science and teaching and extraneous to the intrigues of university politics, creating in their environment the conditions needed for a high scientific standard, which meant not only gathering together the most intelligent and productive young people, but also encouraging them with recognition in proportion to each one's merit. Viceversa, it is well known that favoritism is a highly effective means of selecting, in the university as in any organization, the second-rate elements.

Fermi's fame from Rome spread to the other Italian universities, where nuclei of young researchers and teachers eager to venture into the new fields of physics were beginning to form. Many of them came to Rome for more or less lengthy visits, or even remained for years. Among these young physicists who later contributed so much to the progress of physics in Italy or abroad were Gilberto Bernardini, Giuseppe Cocconi, Ugo Fano, Giulio Racah, Bruno Rossi, and Gian Carlo Wick. The Physics Institute of Rome had also become the object of pilgrimages for many of the brightest young American and European theoreticians.

With Fermi's fame, the interest in modern physics penetrated and spread within the sphere of the Italian university, while the influence of the traditions of old-fashioned experimental physics diminished further each day. Many of the most intelligent young people who registered for the university chose to study physics. They were not those looking for an "easy" degree or a comfortable career, but the few inspired by the sacred fire for a science that was in a phase of extraordinary discoveries and almost complete renewal.

In spite of Fermi's emigration and that of some of his students and followers, his tradition was not lost even during the second world war and in the hard years that came after, thanks to the courageous and tireless work of those who had remained. Overcoming difficulties of every sort, the small group of Italian physicists continued making contributions in the most modern fields, among them that of the new particles, and continued to grow, attracting new generations of brilliant students.

Although many young Italian physicists today who with their work uphold our country's name in the world may have only a vague recollection of Enrico Fermi, or did not know him at all, they do know that the tradition of integrity, of scrupulous scientific honesty, of high-level research, that now reigns in Italian physics, was introduced by him, and that it is thanks to Fermi if the gap between Italy and the more scientifically evolved countries, which in physics seemed unbridgeable at the beginning of the century, is smaller today than in many other fields of the natural sciences.

Enrico Fermi and Solid State Physics

Franco Bassani

1. – Introduction

The name of Enrico Fermi is one of the most frequently cited in Solid State Physics text books, along with those of W. Pauli, A. Einstein, W. Heisenberg, E. Schroedinger, F. Boch, R. Peierls, and L. Landau. He can in fact be considered one of the founding fathers of the modern theory of solids, although he rarely dealt with specific problems of Solid State Physics.

His contributions mainly involve the theoretical foundations, and it is based on these foundations that understanding of the most important physical phenomena of solid bodies has evolved. They are the results of his research in the period following his graduation, before his interests turned to studying the atomic nucleus at the beginning of the thirties, and later to investigating elementary particles. He reflected on these theoretical problems, however, throughout his life, as witnessed by his occasional contributions and popular articles, as well as by calculations and documents of the Los Alamos period, some of which appeared in 1955 after his death, and as his numerous students recall.

Fermi's main contributions, which the most recent advances in Solid State Physics are still based upon, are essentially the following: first, the formulation of the law of the statistical distribution of particles which obey the Pauli exclusion principle; next, the atomic model based on electronic density and statistical distribution (Thomas-Fermi Model); finally, the theory of pseudopotential, introduced in order to interpret the Rydberg states in gaseous systems and subsequently elaborated in a different form to calculate valence and excited electronic states in crystals. To these we should add the theory of the energy loss of fast electrons in condensed matter and, after high neutron flux was available in the nuclear reactor, the initial studies of neutron diffraction by crystals, which opened a new chapter in Solid State Physics.

The origin of all these contributions is amply documented in books by Emilio Segrè and Bruno Pontecorvo, who are his main biographers, and the historic-personal aspects of his life can be found in the book by Laura Fermi, "Atoms in the Family" which was published immediately after his death [1]. Fermi's scientific papers were re-published in the period 1961-65 in two volumes by the Accademia dei Lincei and by the University of Chicago Press, in chronological order and according to subject, complete with comments by co-workers and pupils still living at that time (referred to hereafter as FNM n) [2]. His work also includes a considerable number of books, review articles and conference lectures [3], from which we should start if we are to appreciate the continuity of his thought and his dedication as an educator and disseminator of science. In the following I will touch on certain points of particular importance for Solid State Physics, and try to demonstrate their current relevance.

2. – Fermi's statistics

Fermi's interest in statistical mechanics developed while he was still a student, and it never abandoned him. His graduating thesis for the Scuola Normale Superiore, presented on June 22, 1922, is entitled "Un teorema di calcolo delle probabilità ed alcune sue applicazioni" (A Theorem on Probability Calculus and Some of its Applications). The manuscript was discovered by Giovanni Polvani in the archives of the "Scuola" and was published post mortem (FNM 38b). It considers one of Laplace's theorems (which gives the probability that a number of unknown variables have a total value within two prefixed limits), and extends it to the probability that at least one of the variables be greater than a prefixed value. It is interesting to note that the young student does not confine himself to the mathematical aspect, but applies the theorem to two practical cases: the probability that in a game of chance, after a number n of throws, the player who disposes of unlimited financial resources bankrupts the player who can count only on a finite sum, and the probability that a comet, after a certain time, either moves away parabolically from a celestial body around which is orbiting or collides with it. The tendency to apply the general aspects of a theory to specific problems is evident here, and will remain a characteristic of all of Fermi's activity.

Other work followed, regarding problems in statistical mechanics. A first paper on the ergodicity of a normal mechanical system demonstrates that every normal system is quasi-ergodic, *i.e.* in an unlimited time it passes infinitely close, in phase space, to each point of predefined energy ("Nuovo Cimento", **25** (1923) 367 and "Phys. Zeits." **24** (1923) 261). A second paper concerns Richardson's statistical theory on the photoelectric effect, with the due corrections to Planck's distribution formula ("Nuovo Cimento", **27** (1923) 97).

There are however two papers that are very significant for statistical mechanics, one regarding the absolute constant of the entropy of the perfect gas (FNM 16) and another published in Nuovo Cimento in 1924 under the title "Comments on the Quantization of Systems that Contain Identical Elements" (FNM 19). In the latter he clearly shows that a new principle is required for understanding the behaviour of identical particles which obey the Bohr-Sommerfeld quantum-mechanical conditions, in particular the vanishing of entropy at absolute zero. Pauli formulated this new principle, the exclusion principle, the following year. Fermi was extremely cautious and concrete in his approach to physics, and perhaps for this reason he did not have the "flight of fancy" needed for formulating the exclusion principle, and he rather regretted this, as his biographers tell us. He understood its importance immediately, however, and on February 7, 1926 he submitted to the Accademia dei Lincei, through one of its members, A. Garbasso, the formulation of the law for the statistical distribution of a quantum gas of particles that obey Pauli's exclusion principle (FNM 30).

For a historical and technical analysis of the events linked to this discovery, one can consult the articles by F. Cordella and F. Sebastiani that recently appeared in "Il Nuovo Saggiatore" and "Il Giornale di Fisica" [4]. We need only note that in the 1926 paper and the following one, published in "Zeitschrift für Physik" (FNM31), the counting of states is performed in a harmonic potential for convenience only, but it is shown that *"the result is independent of the particular method adopted"*. It is clear therefore that Fermi understands the generality of the exclusion principle, and in applying it to the gas of free particles he extends it to translational motions, while in the original formulation it referred to the electrons' internal motion within the atoms. We should also note that in these papers Fermi already deduces the law of the linear dependence of the specific heat on the temperature in metals, and the value of the entropy constant of gasses, which confirms his constant need, mentioned above, to apply the theory to practical problems.

The most eminent physicists of the time recognized the importance of Fermi's statistics at once, and later Dirac and Pauli established its connection with the symmetry of the wave function for the exchange of particles, and with the spin. At the famous Congress of Como in 1927, whose focal point was the presentation of the principles of the new quantum mechanics, we find, among the comments on Bohr's talk regarding the probabilistic interpretation, a lengthy exposition by Fermi, who explains the existence of two types of particles, those obeying Bose-Einstein statistics (bosons) and those obeying the exclusion principle (fermions). Among the latter, he focuses on electrons, and after citing Pauli's results on the anomalous paramagnetism of metals and those of Sommerfeld on thermal and transport properties, he concludes his prophetic comments in this way:

"We can also try to construct a theory of metals capable of accounting for the forces which bind the entire metal. For this purpose, we need only consider the positive ions as placed at the vertices of the metallic crystalline lattice and then calculate the distribution of valence electrons when acted upon by electrostatic forces using a method similar to that applied by Debye and Hückel in their theory for strong electrolytes, the new statistics naturally replacing the classic one. The numerical calculations needed for this theory, though, are extremely long and not completed yet" [5].

These words trace a path of studies that will be at the base of Solid State Physics for the rest of the century. To tackle the highly complex problem of calculating electronic states in crystals however one still needed the Bloch theorem, which was formulated the following year. This states that the Schroedinger wave function for a periodic system is of the type

$$\psi_n(\vec{r},\vec{k}\,) = \exp\,[i\vec{k}\cdot\vec{r}\,]\,u_{n,\vec{k}}(\vec{r}\,)\,, \tag{1}$$

where $u_{n,\vec{k}}(\vec{r})$ is a periodic function with the lattice symmetry, and $\vec{k}$ is a wave-vector in reciprocal space. In principle, this makes it possible to calculate the electronic states for each value of the quantum number $\vec{k}$, in each crystal.

Wigner and Seitz, in 1933, dealt with this program for the first time in their famous article on the cohesive energy of metallic sodium [6], thus initiating the study of electronic levels in solids, the so-called "energy bands" $E_n(\vec{k}\,)$. It is well-known that the knowledge of such states and of their occupation according to Fermi's statistics makes it possible to understand the thermal, transport and optical properties of all crystals [7].

One could examine all the properties of solids and highlight the crucial role that the separation in energy between occupied states and "empty" states, the so-called Fermi level $E_{\rm F}$, plays in them. In reciprocal space this level, for metals, corresponds to a "Fermi surface"

$$E_n(\vec{k}\,) = E_{\rm F}\,, \tag{2}$$

with relative Fermi velocity, Fermi momentum, Fermi temperature, etc. For insulators and semiconductors the Fermi level is instead located within a forbidden interval of energy (energy "gap"). The shape of the Fermi surface determines the thermal and transport properties of metals while the position of the Fermi level inside the "gap" decides those of insulators and semiconductors, with all the applications that follow from them.

But Fermi statistics applied to solids generates specific consequences of great conceptual value [8]. One need only recall the explanation of superconductivity, rooted in the capacity of electrons at the Fermi surface to form pairs with zero total spin and momentum, because of their interaction with lattice vibrations. Such pairs are similar to "bosons", but their commutation rules, modified with respect to those for integer-spin particles, recall their fermionic origin. This is the reason for which, rather than condensing in the state of minimum energy (Bose-Einstein condensation) they create a new collective state at an energy lower than the minimum of the normal state (Bardeen, Cooper and Schrieffer or BCS superconducting state).

It is also worth remembering that the fundamental concept of electron "hole" derives from Fermi statistics. This concept was introduced by Peierls to explain the transport of current by positive charges, as detected in the anomalous Hall effect. The "hole" is similar to Dirac's "positron", but its nature is different, since it corresponds to the collective behaviour of electrons in a band $E_n(\vec{k}\,)$, when a state close to the maximum is unoccupied. It has positive elementary charge, but its effective mass and its dynamics are different from those of conduction electrons. We need not add that "holes" are essential in all of semiconductor and nanostructure electronics (transistors, microcircuits, etc.).

The explanation of optical properties is also linked to Fermi statistics, since they are due to electronic transitions [7]. In particular, the property of stimulated emission for electron-hole recombination (semiconductor "laser") results from the possibility of inverting the population of the states located at the junctions between different semiconductors by controlling the Fermi level.

3. – Thomas-Fermi model

After obtaining the statistical distribution function and using it in the first applications to specific heats and entropy, Fermi employed it to calculate the effective potential that acts on atomic electrons. The work was presented by O.M. Corbino to the Accademia dei Lincei in the session of December 4 1927, and is published in "Rendiconti Lincei" of the same year (FNM 43). Fermi was not aware of the work of L.H. Thomas, who had reached the same conclusions about a year earlier [9]. From a chronological point of view priority for the discovery should be given to Thomas, but the more detailed discussion of its limits, and subsequent applications by Fermi and his students fully justify the fact that this method is universally known as the "Thomas-Fermi model".

The fundamental idea consists of identifying the maximum kinetic energy of the electrons with that of a uniform gas, which can be expressed in terms of the electronic density. The total energy must be constant in each point, otherwise there would be a flow of electrons from one point to another, and therefore the potential energy, to within a constant value, depends on the position through its dependence on the electronic density at that point (it is proportional to $n^{2/3}$). Fermi also takes into consideration the temperature, while Thomas uses only the exclusion principle. Nevertheless, the difference is not very relevant because the electrons of the occupied atomic states, due to the limited space in which they are enclosed, have a high density, and form a completely degenerate gas; it follows that the correction to the effective potential is a small constant term proportional to T, which does not modify the equation at the limit $T = 0$.

The effective electric potential $eV(\vec{r})$ and the electronic density in each point $n(\vec{r})$ must be connected by Poisson's electrostatics equation and, by substituting the expression of the statistical model described above for $n(\vec{r})$, one obtains for the effective potential $eV(\vec{r})$ a non-homogeneous, second-order differential equation, with a term proportional to $V^{3/2}$. This equation allows us to calculate the potential of each atom having atomic number Z, with boundary conditions $rV(r) = -Ze^2$ for $r \to 0$, $V(r) = 0$ for $r \to \infty$, and with the further condition $\int n \mathrm{d}\vec{r} = Z$.

The method is very simple and the potential can be expressed in terms of a universal effective potential that can be adapted to every atom, with appropriate scale factors. The solution for the resulting general equation was obtained numerically by Fermi himself, and by Ettore Majorana using more refined mathematics, as Fermi's biographers mention in the description of their first encounter [1].

The effective potential emerging from the Thomas-Fermi model introduces an approximate, though simple and easily manageable, expression for screening the Coulomb potential of nuclear charges by the electrons as a whole, dependent only on their density. This expression has been universally used in solids throughout the second half of the century. The basic idea was subsequently extended to the contributions to the average potential resulting from the antisymmetry of the wavefunction (exchange terms), and to further contributions due to correlations, again expressed in terms of local density. Thanks to these advances average crystalline potentials could be evaluated, and from these one could calculate the electronic states in the various crystals, necessary for interpreting optical and transport properties [7] . For this reason, Gianfranco Chiarotti, in describing the indebtedness of Solid State Physics to Fermi, claims that his most direct contribution is the Thomas-Fermi model [10].

Although electronic level calculations using the Hartree-Fock self-consistent methods are far more accurate than those done with the Thomas-Fermi method, Chiarotti's assertion can certainly find consensus in view, among other things, of the subsequent evolutions of the density functional theory. The latter are based on Hohenberg and Kohn's famous theorem of 1964, in which the ground state of electrons in a crystal, and thus the effective potential which makes the calculation possible, is shown to depend solely on the electronic density and the bare external potential due to the spacial nuclear distribution. The electronic density is sufficient to determine the hamiltonian of the many-body system exactly, with inclusion of all the correlation effects, which are neglected in the Hartree-Fock independent electrons method.

The development of the concept of density functional, with its applications to atoms, molecules and crystals, earned Walter Kohn the Nobel Prize for Chemistry in 1998. Kohn himself, in presenting his theory at the "E. Fermi" School at Varenna in 1983, starts from the Thomas-Fermi model [11] and shows that it can be derived, by means of the variational principle, from a simple poisson-like form of the total energy as a function of the local electronic density $n(\vec{r})$. The concept at the base of the Thomas-Fermi model is therefore the same one that is later reformulated as an exact theorem. This theorem, used with approximate though more and more precise forms of the density funtional, among which expressions in terms of local density, has been essential to the exact calculation of the ground state energy in atoms and solids, including the correlation effects among electrons as well.

4. – The theory of pseudopotentials

Another chapter of Solid State Physics, which has greatly contributed to the knowledge of occupied valence states and excited states at higher energies, is the pseudopotential theory. This theory is based on the fact that for such states one can construct *ad hoc* a fictitious potential, simpler than the true average potential acting on the states. One way to construct it is to make use of the known eigenfunctions $\psi_i(\vec{r})$ and eigenvalues E_i of the innermost states and to force the valence and excited states to be orthogonal to them [7]. The function for such states can thus be written as

$$\psi_v(r) = |\varphi\rangle - \sum_i \langle\psi_i|\varphi\rangle\psi_i\,, \tag{3}$$

which, once replaced in the Schroedinger equation with crystal potential $U_c(\vec{r})$ and eigenvalues E yields the following equation for the "pseudofunction $\varphi(\vec{r})$":

$$-\frac{\hbar^2}{2m}\nabla^2\varphi + [U_c(r)\varphi + \sum_i (E - E_i)\langle\psi_i|\varphi\rangle\psi_i] = E\varphi\,, \tag{4}$$

where the pseudopotential is the sum of the two terms which act on the pseudofunction $\varphi(\vec{r})$, written within the square brackets.

One can notice that the condition of orthogonality to the inner states introduces a "non-local" repulsive term, which compensates almost completely the potential $U_c(\vec{r})$ near the nuclei and cancels its strong oscillations. The pseudopotential thus takes a much simpler form than the true potential and it makes it possible to calculate the valence and excited states when the inner states at lower energy are known. In the case of crystals, where the valence and excited states can be calculated from the pseudopotential Fourier coefficients corresponding to the reciprocal lattice vectors, it happens that only a few Fourier coefficients (those for the smallest vectors) are non-zero. These can be calculated directly with various approximations, in which the pseudopotentials are constructed starting from the atomic states of the crystal lattice; alternatively they can be used as a small number of available parameters.

The pseudopotential theory has introduced significant simplifications that allow us to calculate the electronic structure of metals, insulators and semiconductors, as well as excited states, optical transitions and optoelectronic properties.

It was interesting to discover, even *a posteriori*, that this theory can be retraced to Fermi, as M. L. Cohen clearly explained in an article of historical character [12] and in a talk on the applications of Fermi's atomic pseudopotential theory at the Varenna "E. Fermi" School [13].

Segrè recalls the origin of the theory in this way:

"Amaldi and I discovered absorption lines in atomic sodium gas that corresponded to enormous orbits. I called those excited atoms "swollen atoms", more scientifically but less expressively known as "Rydberg states" today. The extraneous gasses which we had introduced into our light absorption tubes so as to avoid distillation of the alkaline vapour, to our surprise, did not broaden the lines, but displaced them. We spoke to Fermi about this unexpected phenomenon; he thought it over for a bit and then said that it was probably due to the shielding produced by the dielectric constant of the added gas. It seemed obvious and I believe that we ourselves calculated the relative formula, which predicted a shift of the absorption lines toward the blue. For certain gasses, however, the line-shift went in the opposite direction! This came as a surprise and we had no choice but to ask Fermi again. This time the explanation did not come immediately. Only after a few days did Fermi find the reason for it, and he wrote an important paper which contains the idea of a pseudopotential for the first time, and a famous figure which will be found over and over in the most varied of Fermi's works" [14].

The experimental work Segrè refers to appeared in "Nuovo Cimento" in 1934, followed by Fermi's paper on the theory of pseudopotential [15]. This approach was discussed again by Chiarotti in his article published for the fiftieth anniversary of the nuclear reactor and there is no need to re-present it here [10]. It is enough to recall that the potential of the external atoms is described as the average of potential wells in a small region of space, where the wave function also varies little and can be replaced by an average value $\bar{\psi}(\vec{r})$. Fermi obtains, for the average potential due to external atoms, an expression that must be added to the potential of the atom where the electron resides, and is proportional to the density of the external atoms and to a characteristic length a, which can be either positive or negative depending on the type of atom. This accounts for the energy shift in the levels of the Rydberg states as the result of the pseudopotential applied to the average wave function $\bar{\psi}(\vec{r})$. The same form of pseudopotential was later used for the scattering of electrons by atoms and of neutrons by atomic nuclei, thus leading to identification of the characteristic length a as the length of the scattering cross-section.

The concept of Fermi's pseudopotential is similar to that represented by equation (4), although its form is much simpler and it is of different origin. It is worth noting, however, that also in crystals various simplified forms of the pseudopotential have been suggested and used. Among these we find the sum of atomic pseudopotentials in which the part close to the nuclei is given by a simple constant, whose value, positive or negative, is different for each type of atom. This particular form is fully analogous to that originally introduced by Fermi.

5. – Energy loss of electrons in condensed matter

After his arrival in the United States Fermi turned his interest to the problem of energy loss of charged particles as they travel through condensed gasses or condensed matter. The question at that time was whether muons (then called μ mesotrons) lose energy by spontaneous decay or by transfer to the material they cross. Rossi and others later showed experimentally that the first hypothesis was correct, but Fermi's work in 1940 (FNM 136) can be considered the origin of what later became a highly effective technique for studying the electronic properties of solids, *i.e.* the energy loss of fast electrons.

His phenomenological approach is still in use, and it consists in calculating the electric and magnetic fields produced by a charged particle having a given velocity $\vec{v}$ in a medium characterized by a complex dielectric function $\tilde{\varepsilon}(\omega)$. If the dielectric function is represented by means of Lorentz oscillators, the fields can be calculated analytically. These account for the energy loss through either the Poynting vector which moves away from the particle's path, or the work done on the current density $e\vec{v}\delta(\vec{r})$.

Fermi demonstrated that this results in most of the energy being given to the excitations of the medium, while a small part is released as Čerenkov radiation when $v > c/\sqrt{\varepsilon}$, in agreement with previous explanations of the Čerenkov effect by I. Frank and Ig. Tamm.

The theory of energy loss of fast electrons was completed by Paolo Budini and by L. D. Landau in the fifties, and has been widely used for studying elementary excitations in solids; in particular it made possible the detection of plasmons, excitons and interband electronic transitions.

In this area as well Fermi made a breakthrough that was later pursued by others, since his main interests were then pointing toward neutrons, nuclei and elementary particles.

6. – Neutron diffraction

Fermi's results in what was later called the Physics of Neutrons require a special analysis, because they represent a particularly significant portion of his contribution to physics. We need only note that the availability of intense neutron fluxes, such as those achievable in atomic piles starting with pile C.P.1 of December 1942 in Chicago, provided an investigative tool similar to that of X diffraction in crystals, and in many ways complementary to it [16]. In fact, the wave-length of properly thermalized neutrons is of the order of magnitude of crystalline interatomic distances.

The technique of neutron diffraction is now fundamental for determining the structure of materials. It is quite similar to X diffraction, except that in this case it is the nuclei rather than the atomic electrons that scatter the particle's waves. It is thus possible to determine the position of light atoms, especially the position of hydrogen atoms, which do not scatter other particles such as electrons, with important consequences for organic substances and the discovery of the hydrogen bond [16]. In addition, the magnetic interaction of neutrons with the orbital and spin magnetic moments of the electrons permits the determination of magnetic structures and symmetries in magnetic materials.

Fermi did not underestimate these potentialities, and he carried out crucial experiments using neutron beams of the C.P.3 reactor at Argonne National Laboratory immediately after the war, laying the basis for the theory and technology of neutron diffraction [17]. By using the type of pseudopotential described above for representing neutron scattering by nuclei, he found that the phase difference between diffused and incident neutrons has only two possible values: zero when the correction constant to the potential is negative and π when the correction constant is positive. This feature depends on the internal structure of the nucleus, and it was verified experimentally by Fermi and Marshall using a subtle but simple method [18]. In this method one considers various orders of Bragg reflections corresponding to parallel planes that are defined by different atoms, such as the planes (111) in sodium chloride formed of alternating planes of Na and Cl atoms. Different orders of reflection have different intensities depending on whether the phase differences for diffusion are the same in the two nuclei or have opposite signs. In other words, the structure factors of neutron diffraction depend on the phase changes of the diffused neutrons in a way that can be determined by properly comparing the intensities of the beams diffracted under different Laue conditions. Results of neutron diffraction in Be and BeO were quickly obtained and linked to the relative phase difference [19].

It is significant that Fermi pioneered this area of Solid State Physics as well, while his interests were drifting toward problems of quite a different nature, such as those of elementary particles. This further confirms the widely shared opinion that Fermi may have been the last exponent of the category of universal physicists, interested in every aspect of physics and capable of making contributions in the most varied fields.

As for Solid State Physics specifically, his interest was life-long, dating from his speech at the Como Congress mentioned above [5]. This continuity is documented by the book "Molecules and Crystals" of 1930 and informs even his papers from the Los Alamos period [20]. It is very likely that his collaboration and friendship with O. M. Corbino, one of the pioneers of Solid State Physics, may have influenced his interest in the field. In his paper on the debt that Solid State Physics owes to Fermi [10], G. Chiarotti concludes by recalling the words spoken by Corbino in 1929 at a meeting of the Society for Scientific Progress [21] in which he predicted a great future for the Theoretical and Experimental Physics of Condensed States, and anticipated its importance for applications. The seventy years elapsed have seen the realization of Corbino's dreams, both in fundamental physics and in technological applications, and have proven how decisive was the role that Fermi played in their evolution.

* * *

I would like to express my gratitude to GIORGIO SALVINI, GIANFRANCO CHIAROTTI and GIUSEPPE LA ROCCA for their valuable suggestions.

REFERENCES

[1] Enrico Fermi biographies: FERMI L., *Atoms in the Family* (University of Chicago Press, Chicago) 1954; published again in *The History of Modern Physics Series* Vol. **9** (AIP, Woodbury, NY) 1987.
SEGRÈ E., *Enrico Fermi Physicist* (University of Chicago Press) 1970.
PONTECORVO B., *Enrico Fermi* (Edizioni Studio Tesi, Pordenone) 1993. Translated from Russian. See also VERGARA CAFFARELLI R., *Fermi a Pisa* (E.T.S., Pisa) in press.

[2] Enrico Fermi scientific contributions: *Enrico Fermi, Note e Memorie (Collected Papers)* (Accademia Nazionale dei Lincei and The University of Chicago Press, 2 Volumes, Roma and Chicago) 1961-1965 (each paper is labelled by FNM followed by its presentation order number).

[3] Enrico Fermi books and manuals: *Introduzione alla Fisica Atomica* (Zanichelli, Bologna) 1928; *Fisica ad uso dei licei*, 2 Volumes (Zanichelli, Bologna) 1928; *Molecole e Cristalli* (Zanichelli, Bologna, 1934 and anastatic reproduction, 1988); *Meccanica Statistica*, entry in *Enciclopedia Italiana*, Vol. 32 (Istituto Treccani, Roma) 1936.
See also: *Meccanica Statistica*, selected papers, edited by ALTARELLI G. and CAPON G. (Edizione Teknos, Roma) 1988; *Nuclear Physics* (The University of Chicago Press, Chicago) 1949; *Conferenze di Fisica Atomica* (Accademia Nazionale dei Lincei, Roma) 1950; *Particelle Elementari* (Einaudi, Torino) 1958; *Termodinamica* (Boringhieri, Torino) 1958; *Notes on Quantum Mechanics* (University of Chicago Press, Chicago) 1961; *Notes on Thermodynamics and Statistics* (University of Chicago Press, Chicago) 1966.

[4] CORDELLA F. and SEBASTIANI F., Sul percorso di Fermi verso la statistica quantica, *Il Nuovo Saggiatore*, **16**, no. 1-2 (2000) 11; La statistica di Fermi, *Giornale di Fisica*, **41** (2000) 131.

[5] *Atti del Congresso Internazionale dei Fisici a Como* (Zanichelli, Bologna) 1928; see, *e.g.*, the paper by A. Sommerfeld, Vol. II, p. 449, and the extensive contribution by Fermi to the discussion following Bohr's communication, Vol. II, pp. 594-596.

[6] WIGNER E. and SEITZ F., *Phys. Rev.*, **43** (1933) 804; **46** (1934) 809; see also *Qualitative analysis of the cohesion in metals*, in *Solid State Physics*, Vol. **1** (Academic Press, New York) 1955, p. 97.

[7] See, *e.g.*, BASSANI F. and PASTORI-PARRAVICINI G., *Electronic States and Optical Transitions in Solids* (Pergamon Press, Oxford) 1975; BASSANI F. and GRASSANO U., *Fisica dello Stato Solido* (Bollati-Boringhieri, Torino) 2000.
[8] See, *e.g.*, SEITZ F., *Fermi Statistics*, in *Symposium Dedicated to E. Fermi on the Occasion of the 50th Anniversary of the First Reactor* (Accademia dei Lincei, Roma) 1993 p. 47.
[9] THOMAS L. H., *Proc. Cambridge Philos. Soc.*, **23** (1927) 542.
[10] CHIAROTTI G., *The Debt of Solid State Physics to Enrico Fermi*, in *Symposium dedicated to E. Fermi*, *op. cit.* ref. [8], p. 113.
[11] KOHN W., *Density functional theory: fundamentals and applications*, in *Punti focali nella teoria degli stati condensati*, *Proceedings of the International School of Physics "Enrico Fermi", Course LXXXIX*, edited by F. BASSANI, F. FUMI and M. P. TOSI (North Holland) 1985, p. 1.
[12] COHEN M. L., *Am. J. Phys.*, **52** (1984) 695.
[13] COHEN M. L., *Application of the Fermi atomic pseudopotential theory to the electronic structure of nonmetals*, in *Punti focali della teoria degli stati condensati*, *op. cit.* ref. [11], p. 16.
[14] SEGRÈ E., *A Mind Always in Motion* (The University of California Press) 1993.
[15] AMALDI E. and SEGRÈ E., *Nuovo Cimento*, **17** (1934) 145; FERMI E., *Nuovo Cimento*, **11** (1934) 157.
[16] BACON G. E., *Neutron Diffraction* (Pergamon Press, Oxford) 1975.
[17] BACON G. E., in *Fifty Years of Neutron Diffraction* (Hilger, Bristol) 1986.
[18] FERMI E. and MARSHALL L., *Phys. Rev.*, **71** (1947) 666.
[19] FERMI E., STURM W. J. and SACHS R. G., *Phys. Rev.*, **71** (1947) 589.
[20] FERMI E., PASTA J. and ULAM S., *Document LA*-1940, May 1955 (FNM 266).
[21] CORBINO O. M., *I Nuovi Compiti della Fisica Sperimentale*, *Atti della Società Italiana per il Progresso delle Scienze* XVIII Riunione Firenze, 1929.

About the Author

FRANCO BASSANI was born in Milano on October 29, 1929. He graduated in Physics at Pavia in 1952 as a student of the Ghislieri College. Researcher at the Universities of Milan, Illinois (USA), Pavia and at the Argonne National Laboratory, after 1963 he obtained full professorship and taught at the Universities of Messina, Pisa and Rome. He has been professor of Solid State Physics at the Scuola Normale of Pisa since 1980, and President of the Italian Physical Society since 1999. The author of three books and about 250 scientific publications, Franco Bassani is a Member of the Accademia dei Lincei, Doctor Honoris Causa of the Universities of Toulouse, Lausanne and Purdue (USA), and winner of the Somaini and Italgas prizes.

Fermi's statistics

Giorgio Parisi

The name of Fermi is often found in journals of contemporary physics, although in the derived form "fermion". Other times one enters directly into closely connected contexts, such as "Fermi's statistics" or "Fermi's momentum".

All bodies, especially elementary particles, are divided into fermions and bosons, depending on their quantum properties and on the statistics they obey (we shall see below what meaning physicists give to the word *statistics*). The term boson here is not taken from Bosone, duke of Borgogne, Charlemagne's relative, but from the Indian physicist Bose, who wrote the first fundamental paper on the statistical properties of light quanta (which are bosons). In brief, bosons obey Bose-Einstein statistics and fermions Fermi-Dirac one([1]).

Before retracing the history of the discovery of quantum statistics we must open a parenthesis to explain what physicists mysteriously mean by the word *statistics*. Let us consider a very simple case: two glasses and two small balls. We assume that the balls are macroscopic objects (and thus obey the laws of classical mechanics). Suppose we put the balls in random manner under the glasses; there are four possibilities:

- Two balls under glass A and none under glass B.
- Two balls under glass B and none under glass A.
- One of the two balls under glass A and the other under glass B.
- Same situation as above, but having exchanged balls.

([1]) Roughly speaking, if we confine ourselves to elementary particles, the particles that we consider as constituents of matter (electron, proton, neutron, neutrino, quark) are fermions, while those associated to the quantization of a field of force (photon, graviton, pion) are bosons. Composite particles are bosons, if (and only if) they are composed of an even number of fermions.

The corresponding probabilities are

$$P_{AA} = P_{BB} = 1/4; \quad P_{AB} = 1/2 \; , \tag{1}$$

where P_{AA} indicates the probability of having two balls under glass A; P_{BB} the probability of having two balls under glass B, and P_{AB} the probability of having one ball under glass A and one under glass B.

We can reach the same result by reasoning in a slightly different way. I put the first ball at random under one of the two glasses (the two possibilities are equiprobable), and then I add at random the second ball under one of the two glasses (here too the two possibilities are equiprobable): after easy calculations I will come up with the same result as before. This is all familiar to bridge players, who know that if East and West have only two *atout*, the *a priori* probability that the two *atout* are balanced (one East and one West) is fifty per cent.

The formula above can easily be generalized to the case of N balls in M glasses. For instance, if we have two glasses and we indicate with k_1 and k_2 the number of balls in the first and in the second glass respectively ($k_1 + k_2 = N$), we have that $P(k_1, k_2) = \frac{N!}{k_1! k_2!} (\frac{1}{2})^N$. For three glasses

$$P(k_1, k_2, k_3) = \frac{N!}{k_1! k_2! k_3!} \left(\frac{1}{3}\right)^N . \tag{2}$$

These classical results of the probability theory are referred to as the Boltzmann statistics since Boltzmann used them in his statistical deductions.

There could be some doubt regarding these results if the balls were really indistinguishable, but classically this never occurs: we can always follow (at least conceptually) the trajectory of the balls and identify the first ball, for instance, as the one that first was put in the glass.

In quantum theory things are different:

- Really indistinguishable objects do exist: all electrons are equal; there are none with a white spot, or slightly flattened at the poles or with a tiny scratch. Electrons do not have small signs of distinction like macroscopic objects(2).

- In quantum mechanics we cannot conceptually think of keeping the particles always under observation, following them continually so as to avoid the risk of getting them mixed up, unless we constantly perturb the system.

If the balls are indistinguishable, as in quantum theory, there are no impelling reasons for maintaining classical statistics (*i.e.* the one just described) and considering the instance in which the first ball is in glass A and the second in B as being different from the opposite situation. On the other hand there are not any obvious reasons for abandoning

(2) An undivisible object, that cannot be further split up, cannot have signs of distinction.

classical statistics and adopting quantum statistics either. Giving up the classical way of counting states implied a complete change of perspective, which was not at all easy to make. Actually, Bose took the first step unwittingly and Einstein, Fermi and Dirac quickly followed in his path.

Let us try to reformulate the same problem in quantum terms. There are two indistinguishable particles (or balls) which, by hypothesis (postulate) do not interact with each other. Each of these particles can occupy one of the two quantum states A or B with equal *a priori* probability (quantum states play the same role as the glasses(3)). We ask ourselves what are the probabilities of finding two particles in these two quantum states if the particles are randomly distributed.

If we open a manual of quantum mechanics we discover that, if the particles are bosons, for the previous case of two balls in two glasses

$$P_{AA} = P_{BB} = P_{AB} = 1/3 \,, \tag{3}$$

if we assume that the ball can stay in just one quantum state within each glass.

The previous formula can be easily generalized to the case of N balls in k glasses. For instance, if we have two glasses and let k_1 and k_2 indicate the number of balls in the first and second glass respectively, we have that $P(k_1, k_2) = (N+1)^{-1}$. For three balls

$$P(k_1, k_2, k_3) = C_3(N) \,, \tag{4}$$

where $C_3(N)$ is an appropriate function of N. Generally speaking, for Bose the probabilities do not depend on k but only on N, as proven by Dirac in the second half of 1926.

The previous result is easy to interpret. From a quantum viewpoint there are only three possibilities and not four: two balls in A, two balls in B and one in A with the other in B; indeed, since the balls are indistinguishable, it makes no sense to differentiate between which ball is in A and which in B. If the *atout* were indistinguishable, which they are not(4), the probability of a balanced hand would be only 1/3. An inflexibly classical thinker would interpret this phenomenon as an intrinsic attraction between the particles, which tend to remain in the same state longer than we would expect classically(5). Particles that follow this quantum statistics (Bose-Einstein's) are called bosons.

There are however other particles, those called fermions, which obey Pauli's exclusion principle: no more than one fermion at a time can occupy the same quantum state; multiple occupations of the same state are prohibited. The exclusion principle is at the basis of the almost-incompressibility of solid matter; in addition, it implies that not all

(3) An example of quantum state is an orbital of an electron around a nucleus.

(4) The *atout* are macroscopic objects and moreover a 4 of the *atout* seed is different from a 3 of the same seed.

(5) This attraction is at the basis of many important quantum phenomena, for example the laser effect, superconductivity and superfluidity.

the electrons rotating around a heavy atom occupy the most internal orbit, but begin to populate the external ones; chemistry originates from this effect.

If our particles are fermions, for two particles in two quantum states the only possibility is to place the two particles in different quantum states, and we have

$$P_{AB} = 1 \ . \tag{5}$$

Obviously things become more complicated if there are a larger number of particles and of quantum states. In this case formulas are slightly more complicated; nevertheless for fermions each quantum state can either be occupied by a single particle or remain free.

Having illustrated what quantum statistics is, let us try to set the fundamental papers in which this concept was introduced in their historical context.

At the beginning of the 'twenties the situation of quantum theory was confused. Planck's black-body radiation theory, according to which each normal mode of the electro-magnetic field can only have energy as multiple of $\hbar\omega$ was consolidated. Planck's theory had been generalized and one could calculate the energy levels of quantum systems, whose classical motion was periodic or quasi-periodic, by using the quantization formula of Bohr-Sommerfeld. The case of a generic classical motion eluded quantization. For a generic potential it was therefore impossible to find the energy levels of the system and, in a more general sense, to calculate the statistical properties. One of the few instances in which calculation was possible was that of thermal vibrations in a solid; one could thus deduce theoretically the specific heat of a solid, which is known to be constant at high temperatures, but drops toward zero at low temperatures, because of quantum effects. This result is fundamental and at the basis of the third principle of thermodynamics([6]).

The perfect gas, however, evaded every analysis; the specific heat was that of classical mechanics, *i.e.* independent of temperature. This result was extremely embarrassing as it disagreed with the third principle of thermodynamics which implies that the specific heat be zero at zero temperature. No progress could be made: the fundamental idea needed to remedy this state of affairs was missing and until 1924 the situation remained stalled. Einstein wrote about the matter in 1912: *"This research is just groping along without any precise bases. The more quantum theory succeeds, the less serious it appears. How the laymen would laugh if they were able to follow the strange course of these conceptual developments"*.

Everything changed with Bose's paper in 1924. According to the textbooks, Bose guessed that because the particles were indistinguishable the Boltzmann statistics was no longer valid, and should be replaced by a different system. In reality it is probable that Bose guessed almost nothing about statistics. Let us look at his article in detail.

Bose's proposal consisted in dividing the phase space (position and momentum) into cells whose volume was $\hbar^3$ ($\hbar$ is the Planck constant). Next he described the state of the

([6]) In first approximation the third law of thermodynamics states that entropy at zero temperature is finite. This law, in a slightly different version, was suggested as conjecture by Nernst around 1910. For strange reasons this conjecture (which can be demonstrated by using quantum mechanics) has gone down in history as Nernst's theory.

system by indicating how many cells contained a given number (k) of particles. With varying k this quantity was indicated as n_k. For instance, in the case of two particles in two glasses we can have:

(a) $n_1 = 2$ and all the other n's equal zero: in this instance the two particles are in different glasses: the number of glasses that contain one particle is 2.

(b) $n_0 = 1$ and $n_2 = 1$ and all the others equal zero: in this case the two particles are in the same glass: the number of glasses which do not contain any particle is 1 and the number of glasses that contain two particles is 1.

Bose proposed that the probability of having a given sequence of values of n be written as

$$\frac{C}{n_0!n_1!n_2!\ldots}, \tag{6}$$

where C is an appropriate quantity for making the total probability equal to 1.

If we recall that $0! = 1$, we see that for the first case (a) the probability is $C/2! = C/2$, while in the second case (b) the probability is $C/(1!)^2 = C$. Since the sum of the two probabilities must be equal to 1, we have that $C = 2/3$ and with a simple calculation we return to the previous results of the Bose-Einstein statistics (eq. (3)).

The revolutionary novelty of the article lies in the formula (6), which represents a radical break with classical statistics. Bose however does not waste a single word to justify it since he considers it "evident". Most likely, as Pais suggested [1], Bose wrote the formula in analogy to that of classical probability (see eq. (2)) for the number of particles under the glasses, without realizing that the laws of classical probability did not involve his formula (6) at all. To put it bluntly, Bose had been superficial, had written a formula which seemed obvious to him without thinking it over and unwittingly made a mistake (an error as fruitful as that of Colombus).

Bose himself later wrote *"I was not aware of having done something really innovative. [...] I was not expert enough in statistics to realize that I was doing something very different from what Boltzmann would have done, from Boltzmann's statistics."*

The suggestion was revolutionary and, for reasons that are hard to understand, it yielded the correct result for black-body radiation. The article was rejected by the "Philosophical Magazine", which probably behaved like the Salamanca wise men (it would be interesting to know their motivations). To get his paper published, Bose turned to his equivalent of Queen Isabel, *i.e.* Einstein, by writing him a letter in June 1924. Einstein at once grasped the importance of Bose's article, translated it personally from English to German in order to publish it in a German Journal and immediately wrote an article in which he drew further conclusions from Bose's proposal and applied his formula to the case of a monoatomic gas.

What was Fermi's role in this story? At the beginning essentially that of a spectator. The problems involved interest him, and he wrote various articles based on quantum statistics, for instance: "Sopra la teoria di Stern della costante assoluta dell'entropia in un gas perfetto monoatomico" (On Stern's Theory of the Absolute Constant of Entropy in a Perfect Monoatomic Gas) (1923), "Sulla probabilità degli stati quantici" (On the Probability of Quantum States) (1923), "Considerazioni sulla quantizzazione dei sistemi che contengono degli elementi identici" (Considerations on the Quantization of Systems Containing Identical Elements) (1924), "Sull'equilibrio termico di ionizzazione" (On the Thermal Equilibrium of Ionization) (1924), "Sopra la teoria dei corpi solidi" (On the Theory of Solid Bodies) (1925), interesting articles that contain intelligent observations but nothing extraordinarily new (see [2]).

Fermi's fundamental contribution came just after the introduction of Pauli's exclusion principle in 1925. Fermi instantly realized the profound consequences that Pauli's exclusion principle had for statistical mechanics and he quickly wrote two articles. The lengthy paper in German "Zur Quantelung Des Idealen Einatomigen Gas" is preceeded by a shorter article in Italian, with the same title ("Sulla quantizzazione del gas perfetto monoatomico" (On the Quantization of a Perfect Monoatomic Gas)) published in the Rendiconti dell'Accademia dei Lincei, presented to the Accademia in February 1926(7) [2].

The article in Italian is very short, 5 pages. Fermi considered a monoatomic gas in the presence of a harmonic potential(8), correctly calculated the quantum levels and by establishing that each level can be occupied by no more than one atom, he obtained the various thermodynamic properties of the perfect gas. At high temperatures he found the classical results again, while he discovered that at low temperatures the specific heat is no longer constant, but goes to zero proportionally to the temperature. The entropy at zero temperature is zero, while at high temperatures the entropy —to Fermi's great satisfaction— is equal to that assumed by Tetrode and Stern, who used a heuristic procedure. The paper in German contained more details, but the article in Italian contained all the new ideas.

The impressive rapidity with which Fermi grasped the consequences of Pauli's article is due, according to Pontecorvo [1], to the fact that, *"Fermi had been nursing the idea of this work for a long time: what was missing however was Pauli's principle. As soon as the principle was formulated, he sent his article to the printer. One could say in this regard that Fermi was rather chagrined at not having been able to formulate Pauli's principle himself, although —as can be seen from his work— he came very close to it."*

In any case Fermi's paper offers a conclusive version of his statistics within the sphere of early quantum mechanics. It is interesting to note that Fermi assumed that all atoms obey Pauli's exclusion principle, while we know now that this happens only to those composed of an odd number of fermions (those composed of an even number of fermions are bosons). Fermi, who could not know all this, erroneously applied his statistics to the

(7) For reasons of priority, Fermi often published his work quickly, in an abbreviated form in Italian, later writing an expanded version in German or English.

(8) Fermi could easily have put the gas in a cubic box, as he had in the past, but the adoption of a harmonic potential offers a technical advantage, since in the region where the potential is high the density is low and the behavior is similar to classical behavior. This subtle difference makes it possible to achieve the needed comparison with the classical limit in an extremely simple way.

helium atom which, being composed by six fermions, is a boson.

In the second half of 1926, the problem of quantum statistics within the old quantum theory (that of Planck, Bohr, Sommerfeld) was not very clear: on the one hand there was an outlandish proposal (Bose's) to evaluate the statistical weights(9), which, according to Einstein, reflected *"a well-defined hypothesis of influence among molecules, [...] an influence of a totally mysterious nature"*; on the other there was Fermi's lucid proposal, whose derivation from Pauli's exclusion principle was crystal-clear. Why the first should be used for photons and the second for electrons was beyond comprehension. In addition, the link between the two proposals was not at all evident.

The solution to the problem and the decisive formulation of quantum statistics as part of the new quantum wave theory arrived in August 1926 with a work by Dirac (in which Dirac, independently of Fermi, formulates the statistics for those particles which obey Pauli's principle).

In the new quantum mechanics a system composed of one particle was represented by a one-variable wave function $\psi(x)$, while a system composed of two particles was represented by a two-variable wave function $\psi(x, y)$. Dirac noted that if only symmetric wave functions ($\psi(x, y) = \psi(y, x)$) are considered admissible, the particles satisfy the Bose-Einstein statistics, while on the contrary if only antisymmetric wave functions ($\psi(x, y) = -\psi(y, x)$) are considered admissible one obtains the Pauli exclusion principle and Fermi statistics. Moreover, Bose's statistics, when reformulated in terms of occupation numbers for each quantum state, implied a very simple formula (as described by eq. (4)).

The theory of quantum statistics was formulated, and Fermi was rightly considered one of its principal authors. As an example of Fermi's fame, we can quote a letter from Einstein to Lorentz, in which Einstein declines the invitation to talk on quantum statistics at the Solvay congress in 1927 since he is not sufficiently qualified, and suggests that *"perhaps Mr. Fermi of Bologna*(10) *[...] or Langevin [...] could do a better job"*.

Immediately following the original papers, the applications got underway: the first was a statistical treatment of the innermost electrons of a heavy atom with many electrons, in articles by Thomas (end of 1926) and Fermi (end of 1927), which produced the Thomas-Fermi theory that made it possible to calculate quantistically the different properties of heavy atoms (for instance, the radius) as a function of the atomic number. Subsequently the theory of metals was formulated: here the electrons form an almost perfect, high density gas of fermions (at almost zero temperature) where the quantum effects are dominant.

As early as 1927 Pauli used Fermi's statistics to explain the paramagnetism of alkaline metals, and Sommerfeld began a systematic study of metals starting from the electrons' contribution to specific heat (proportional to the temperature).

Fermi did not take a direct interest in these applications of his theory, on which generations of physicists will work, but turned instead to what was in that moment

(9) Einstein wrote *"his deduction is elegant, but the substance remains obscure."*

(10) Fermi never worked in Bologna; he wrote the two articles on statistics while Associate Professor at Florence.

the new frontier of theoretical physics: quantum electrodynamics, with all its problems linked to the emission and absorption of photons, and he wrote a series of magistral works, widely admired for their extreme clarity. The experience gained by studying quantum electrodynamics will be of great use to him in writing his article of 1933, *Tentativo di una teoria dell'emissione dei raggi beta* (*Proposed Theory on the Emission of Beta Rays*), which is perhaps Fermi's most essential contribution to theoretical physics.

REFERENCES

[1] PAIS A., *Subtle is the Lord* (Oxford University Press) 1982; PONTECORVO B., *Enrico Fermi* (Edizioni Studio Tesi) 1993.
[2] The complete collection of Enrico Fermi's papers can be found in *Note e Memorie (Collected Papers)* (Accademia Nazionale dei Lincei, Roma, University of Chicago Press) 1961, 1965.

About the Author

GIORGIO PARISI was born in Rome on August 4, 1948. Full Professor since 1981, he presently teaches Calculus of Probability at the University of Rome "La Sapienza". A member of the Accademia dei Lincei, of the Accademia dei XL, of the French Academy of Science and of the National Academy of Sciences of the US. He was awarded the Boltzmann Medal in 1992 and the Dirac Medal in 1999. He has written more than 400 scientific papers.

Classical mechanics and the quantum revolution in Fermi's early works

GIOVANNI GALLAVOTTI

Fermi's papers written between 1921 and 1926 on Classical and Statistical Mechanics are analyzed in the context of the contemporary developments that led to the establishment of Quantum Mechanics. Fermi's position greatly influenced Italian Physics throughout the twentieth century.

1. – Introduction

Fermi's early papers, developed between 1921 and 1926 (the year of the fundamental "*Fermi statistics*") offer material for reflection because, among other things, they were carried out during a period when Physics was undergoing radical changes of which the young Fermi was well aware in spite of the isolation due to the almost non-existent involvement of Italian science.

His interests in and his mastery of General Relativity, at the time a new and novel theory, and of Electromagnetism are revealed by various papers on corrections to the masses and the equations of motion for charged bodies moving in electromagnetic or gravitational fields. At the same time he was also attracted by the planning and theoretical interpretations of experiments in Atomic Physics, seen as a way to obtain verification of the consequences of new basic laws which were being proposed in Europe. His attention to problems of Mechanics and Probability was also strong: already in his thesis he had solved a problem of "stop times" in random walks (to employ modern terminology), applying it to the theory of Jupiter's influence on comet orbits [1].

Here, however, I will focus on a series of papers regarding Mechanics, linked to questions generated by the emerging Quantum Mechanics.

2. – Adiabatic invariants and the quasi-ergodic hypothesis

It is surprising that Fermi, who on many occasions manifested his enthusiastic appreciation for Bohr's mechanics, did not in some way participate in the decisive moments when the transition took place from the empirical and often contradictory theory, based on Bohr-Sommerfeld rules, to the new mechanics of Heisenberg, Born, Jordan.

Fermi was not alone in not knowing (as he never quotes it) Einstein's paper [2, 3] on quantization: this paper was also essentially ignored by all his contemporaries [3], possibly because it was overshadowed by the more famous work [4] on Einstein's new derivation of Planck's law, or because at that time the papers by the Ehrenfests and by Burgers [5, 6] (also dated 1917 and considered a key reference) seemed to provide a sufficiently general background for the theoretical foundations of the new Mechanics.

Einstein's 1917 paper, contemporary to his other paper on Planck's law, contains a prescription for "intrinsic" quantization, *i.e.* a coordinates independent quantization rule which is a considerable extension of Bohr-Sommerfeld's quantization [7]: a rule which is still considered nowadays and applied (with different notations and when it is a reasonable approximation). In this same paper Einstein raises questions on a key point of Bohr-Sommerfeld's Mechanics *i.e.* on the quantization of the actions, casting doubts on the very existence of the appropriate action variables (the doubts are expressed in the form of a few clear final comments but refraining from strongly advocating the need of a reformulation of the Physics of atoms). He rightly claims that *generically* the actions that one wanted to quantize might actually not exist at all, at least not as well-defined Classical Mechanics objects.

Fermi, independently, soon concerned with this problem, showing that *even* in systems in which it was possible to define the action variables the quantization rules might be ambiguous or arbitrary; and he presented simple *concrete* examples in which the "principle of the adiabatics" for quantization rules [7], was not applicable [8].

He endeavoured to find and stress the great difficulties that would be met in attempting to provide solid bases for Bohr-Sommerfeld's theory, and was perhaps thus distracted from participating more actively in the widespread debate which, a few months after his lengthy stage in Göttingen, *i.e.* in the center of the development of the new ideas, generated the celebrated works on Matrix Mechanics.

As an autodidact he was probably inclined, at the time, to require strict logical and formal coherence in his reasoning and deductions, and thus to have an overly-developed critical sense (advantage and defect also common among less-known "*normalists*", *i.e.* graduates of the "Scuola Normale" of Pisa). As he said, in Septemper 1925, he found it excessive to renounce understanding what really happens: "*for my taste it seems that they are really starting to exaggerate in the tendency to give up understanding things*", pag. 24 in [9]. Perhaps this belief led him to dissociate himself from the trends and ideas which must have permeated the discussions that took place in Göttingen where shortly before, in 1923, he stayed for about six months; and possibly he considered them unscientific. Heisenberg's [10] paper appeared in July 1925, followed by the papers by

Born-Jordan [11], Dirac [12] and, in November 1925, by the "three men" paper (Born, Heisenberg and Jordan) [13].

During his stage in Göttingen, Fermi became interested in the question of the existence of the adiabatic invariants: quantities that were being considered (amid difficulties *cf.* [2, 14]) as essential for Bohr-Sommerfeld's quantization rules. In February 1923 he published an important paper [15] whose German version was followed by one in Italian, practically identical to the German original but divided into two articles [16, 17]: the first exclusively devoted to an Analytical Mechanics theorem (*cf.* below) and the second, entitled "*Proof that a mechanical system is, in general, quasi-ergodic*", to an application of the Mechanics theorem.

The paper provides a basic, and in a certain sense decisive, criticism of the idea of making adiabatic invariants the fundamental entities on which to base the quantization rules. He develops in great technical detail the difficulties already raised in the final comments of Einstein's (independent) paper cited above [2, 3]. *However*, the adiabatic invariants themselves are neither mentioned in this paper (in German), written in Göttingen in February 1923 [15], nor in its Italian version, also written in Göttingen in April 1923 [16], which deals with the above-mentioned mechanical theorem and reproduces almost *verbatim* the corresponding sections of the German version. The second shorter part of the paper [15] is *separately* reprinted in Italian [17]. The adiabatic invariants are explicitly mentioned and immediately criticized in detail in the subsequent papers [18, 19] which were also written in Göttingen but *only in Italian*.

The bilingual publication of [15] and the splitting of the Italian version into a "mathematical" part [16] and a brief "physical part" [17] indicate the importance that Fermi must have attributed to this part of his research. It is remarkable that, rather than presenting it as a destructive blow to the attempts at rationalizing Bohr-Sommerfeld's rules and the principle of the adiabatics, Fermi presents his results as if they were dedicated to the foundations of Statistical Mechanics; and more precisely to a solution to the basic problem of the existence of *quasi-ergodic* systems: *i.e.* systems whose constant energy surfaces are densely covered by the orbits of each of their points, with the possible exception of a zero area set, see below; this was a problem formulated by the Ehrenfests in order to save what they believed (erroneously, *cf.* §1.9 in [20]) to be Boltzmann's ergodic hypothesis (an hypothesis which was affected, in the Erhenfests' mathematical interpretation, by (obvious) contradictions, as pointed out by many authors).

3. – The two parts of Fermi's proof of the quasi-ergodic hypothesis

Fermi presents a "proof " that a Hamiltonian system with f degrees of freedom is generically quasi-ergodic: in the sense that, given two *arbitrary* surface elements σ and σ'' on the $(2f-1)$–dimensional constant energy surface, trajectories do exist that begin in σ and cross (in due time) σ''. The two surface elements must be "*endlich*" (in this case meaning "open") as can be deduced from the analysis and as Fermi was obliged to state explicitly in answering a critical comment [21].

This would be an extremely interesting result, even apart from the fact that, in general, it implies *non-existence* of Bohr-Sommerfeld's action integrals and *therefore* the impossibility of applying the principle of adiabatics for the quantization of generic systems. It is licit, I believe, to think that realization of the *latter impossibility* was the true motivation of his research; but it was probably advisable, particularly for a young scholarship-holder at Göttingen, to let the reader deduce the consequence by himself while hiding it behind an "innocent" although highly important result which served to confirm the foundations of Statistical Mechanics, a theory that was by then taken for granted. In fact, Ehrenfest was very impressed by these results, a circumstance which generated important contacts and collaborations between Fermi and his colleagues of Ehrenfest's school, Uhlenbeck among them. On the other hand, as already mentioned, Fermi published in the same period various articles on adiabatic invariants where the difficulties observed were explicitly mentioned and critically evaluated [18, 19].

The *first part* of the paper extends one of Poincaré's theorems [22] (which had been properly cited by Einstein [2] as implying a critique of the principle of the adiabatics): it is a purely mathematical analysis inspired by Poincaré's original papers [22]. Here Fermi studies a f degrees of freedom Hamiltonian system, with canonical coordinates $(p, q) \in R^{2f}$, and Hamiltonian $H(p, q, \mu)$ dependent on a parameter μ and which, for $\mu = 0$, is reduced to a system integrable by quadratures: the question is whether a *non-constant* analytic function $\Phi(p, q, \mu)$ can be defined on a constant energy surface $H(p, q, \mu) = E$ and vanish identically along the trajectories beginning on the surface if initially it is $\Phi = 0$.

Poincaré, instead, considered functions $G(p, q, \mu)$ that are analytic over the surface $H = E$ and in μ, and maintain the initial value (*whether it be zero or not*) along any trajectory of the system: in other words functions that are "integrals of motion" for all values of the parameter μ at least in a small vicinity of the origin. His result was that "generically" no such non-trivial integrals exist (non-trivial means not expressible as functions of H).

Fermi's theorem can be rephrased by saying that, if δ is Dirac's delta function, then functions taking the form $G(p, q, \mu) = \delta(\Phi(p, q, \mu))$ with Φ non-trivial cannot exist as integrals of motion as μ varies, at least under very general conditions on H. This Fermi's theorem is an important *complement* to that of Poincaré and, given the way it is proved, it can be considered as a generalization of Poincaré's (and Fermi himself considered it in this perspective).

The proof implies that the points (p, q) of the $(2f - 1)$–dimensional energy surface $H(p, q, \mu) = E$ are, for each μ, separated into two disconnected $(2f - 2)$–dimensional regions by the regular (*i.e.* analytic in (p, q) and μ) surface $\Phi(p, q, \mu) = 0$. One can see that an adiabatic invariant would really be (if existent) a function that embodies the properties of the (non-existent) Φ.

The *second part* of reference [15] is more interesting from the physical point of view: if σ, σ'' are two surface elements of the surface at constant energy then trajectories that visit both σ and σ'' must exist. It is well known that Fermi's argument for reaching this conclusion is based on the hypothesis that if σ' is the region scanned by the trajectories that start in σ then its frontier S_μ is a surface separating σ' from its complement, *with "surface" meaning a* $(2f-2)$*-dimensional surface which is also analytic in* (p, q, μ).

The regularity (in fact analyticity) hypothesis on S_μ, whose necessity Fermi realized perhaps too late and acknowledged in a footnote added in proof at the end of the paper, is extremely restrictive and, in view of what was learnt subsequently [23], so strong as to make the result of little interest, because it is this regularity that *should be demonstrated and not assumed.*

We know today that (if $f > 2$) there will be many invariant surfaces of lower dimension (precisely of dimension f) none of which, however, divides the constant energy surface into disconnected parts.

Modern perturbation theory of Hamiltonian systems [24, 23, 25] explains why, *within Fermi's assumptions*, the claims that most trajectories (*i.e.* all except a few covering a set of zero total area on the constant energy surface) are dense on the constant energy surface, or that the ergodic hypothesis holds in the sense in which Boltzmann formulated it [20] are definitely false. In a certain sense, in fact, one finds a situation of "trivial non-ergodicity" opposite to that assumed by Boltzmann.

In the systems considered by Fermi, the invariant surfaces of dimension f always occupy a very large region on the energy surface in phase space if μ is small, but its boundary is most likely "fractal" and therefore lacking the properties of regularity usually expected when speaking of a "surface". *In spite of all this, it is still possible that dense trajectories exist* (although occupying a small area of the constant energy surface) or, given two arbitrary and open regions of the constant energy surface, that there are trajectories connecting them (which is the sense that Fermi gives in his work to the property of "quasi-ergodicity" and whose validity he believed he had proved): the possibility, instead, is still widely investigated today, and belongs to the set of problems known as "diffusion in phase space" or "Arnold's diffusion" phenomena.

The above difficulty limits the interest of the second part, and thus of the conclusions, of Fermi's paper (this part extends over just one page in the paper [15] which, as mentioned, was separately reproduced in Italian [17]). The problem was almost immediately pointed out to Fermi, who addressed the critiques in a rather elusive and certainly unconvincing manner, *cf.* comment on objection 1 in [21].

4. – The FPU experiment and the confutation of the quasi-ergodic hypothesis

The problem must have remained present in Fermi's mind, since shortly before his last days he returned to it, redeeming his 1923 "error" with a paper of paramount importance: with J. Pasta and S. Ulam, he approached the question through an experiment (the "*FPU experiment*") that can be seen as an experimental version of the contemporary (but independent) theory of Kolmogorov. The latter indirectly made clear how essentially untenable it would be to claim that the second part of the paper [15] really proved the validity of the quasi-ergodic hypothesis.

The experiment, performed in 1954 (see ref. [26]), aimed to solve the equation of motion for a chain of anharmonic oscillators numerically, satisfying the hypotheses of

Fermi's mechanical theorem [16], so as to verify whether quasi-ergodic trajectories really existed according to the quasi-ergodic hypothesis. In this way one could have obtained at least an experimental justification for the quasi-ergodic theorem, that was claimed but not proven in his 1923 paper, and for its fundamental role in Statistical Mechanics.

The result, *cf.* [27] for a detailed analysis of the paper and of its influence, was that the system in question *did not behave* as it would have been expected if the ergodic hypothesis had been correct (both in the sense of the Ehrenfests' formulation and —more interesting and physically relevant— in the sense of Boltzmann's, *cf.* [20]).

This result is in perfect agreement with Kolmogorov's theory, which appeared independently and almost simultaneously: this is remarkable because it faced physicists with the fact that classical physics yielded wrong results at low temperatures or in the black-body theory, not only because in these domains Classical Mechanics ceased to be valid and needed to be replaced by Quantum Mechanics, but also because the principles of Statistical Mechanics (such as equipartition) clearly could no longer be considered valid in these systems, even from a Classical Mechanics approach *cf.* [28-30, 20]. Moreover, the results in [26] were significant because they represented the first realization of a numerical experiment in Statistical Mechanics, which opened the way for the vast amount of subsequent research on equilibrium and non-equilibrium Statistical Mechanics and on Fluid Mechanics based on numerical simulations, *cf.* for instance [31, 32].

5. – Theoretical physics in Italy and Fermi's initial uneasiness about Matrix Mechanics

Returning to Fermi's early works we stress that his paper on the generic validity of the quasi-ergodic hypothesis was presented as dealing with the foundations of Statistical Mechanics and as such it cannot be termed a success. If, however, we consider that very likely —as I have tried to show here— the work originated and was intended as a severe criticism of the basis of Bohr-Sommerfeld's theory on adiabatic invariants, then it achieved its purpose quite well.

Shortly afterward, Fermi made the discovery of "*Fermi's statistics*" and fully adopted Schrödinger's ondulatory mechanics, leaving behind his research on adiabatic invariants (after a last paper in which he tried interpreting the invariants within the framework of wave mechanics [33]). With this achievement Fermi returned to the front line of research on quantum mechanics after having risked being overtaken by events, having lost the chance offered him by Corbino's far-sighted award of the scholarship for Göttingen, to be a protagonist in the rise of matrix mechanics (the equivalent of ondulatory mechanics).

All this has left a deep trace in Italian Physics which, assuredly under Fermi's influence, still adheres to teaching quantum mechanics in its ondulatory form: virtually every generation of Italian physicists still continues to overlook, during its formative years, the study of Matrix Mechanics, losing everything of conceptual importance it has to offer, *cf.* [34].

While we can only be grateful to Fermi for having, with his example and prestige, kept many of us —at least while we were students— at a distance from the interminable discussions and analyses regarding the foundations of the new Mechanics, we must in a sense regret that the Matrix Mechanics of Heisenberg and his collaborators has actually never been taught in basic courses in the leading Italian universities. This proves how even physicists like Fermi — and even young— can fail to appreciate the significance of changes that are taking place under their own eyes. This is a phenomenon that continues to repeat itself in Science: although it may not necessarily be negative since it is very useful in limiting the influence of "fashions" that are in vogue for a few years and then end up forgotten, or almost so.

For the sake of clarity, we should point out that in other countries too, especially in its native land, Matrix Mechanics has always been and still is essentially ignored in basic courses, in favor of the ondulatory approach. Why this should be, poses an interesting historical problem.

Fermi's limited interest for the "Göttingen Physics" might have an analogue in the negative attitude that quite a few scientists have today towards Strings Theory or showed in the 1960's towards S–Matrix Theory. Progress in Physics is not always logical and coherent; it consists in a sequence of new ideas, few of which turn out to be fundamental and which even the most illustrious physicists may overlook as passing fashions.

* * *

I am grateful to CARLO BERNARDINI for his suggestion that I write an article on Fermi's early works for the volume celebrating the "Fermi centennial"; to SANDRO GRAFFI for having acquainted me with Einstein's work [2] and to Professor GIORGIO SALVINI for his comments on the manuscript and the explanations he kindly contributed.

REFERENCES

[1] FERMI E., *Un teorema di calcolo delle probabilità ed alcune sue applicazioni, Habilitation thesis at the Scuola Normale Superiore*, Pisa (1922), unpublished. Printed in [35], paper No. 38b.

[2] EINSTEIN A., *Zum Quantensatz von Sommerfeld und Epstein*, in *Verhandlungen der Deutsche physikalische Gesellschaft*, **19** (1917) 82. Reprinted in Italian in [3].

[3] GRAFFI S., *Le radici della quantizzazione*, Quaderni di Fisica Teorica, University of Pavia, 1993, ISBN 88-85159-09-05.

[4] EINSTEIN A., *On the quantum theory of radiation.* English version reproduced in [34].

[5] EHRENFEST P., *Adiabatic invariants and the theory of quanta, Philos. Mag.*, **33** (1917) 500. Reprinted in [34].

[6] BURGERS J. M., *Adiabatic invariants of mechanical systems, Philos. Mag.*, **33** (1917) 514.

[7] I summarize here the *main points* of the *principle of the adiabatics* and the definitions associated with it. Consider a Hamiltonian with f degrees of freedom of the form $H_t(p,q) = H_0(p,q) + \frac{t}{T}K(p,q)$, where H_0, K are functions of the canonical coordinates $(p,q) \in R^{2f}$ and T is the time scale over which variations in the Hamiltonian with time t are observed. The Hamiltonians H_0 and H_T are said to be *adiabatically connected in the limit* $T \to \infty$ or that H_0 is transformed in an infinitely slow manner (as $T \to \infty$) into $H_0 + K$. Let (p_t, q_t) denote the solution of the equations of motion $\dot{p} = -\partial_q H_t(p,q), \dot{q} = \partial_p H_t(p,q)$ with initial values (p_0, q_0). Let us also consider, for each *fixed* $t \in [0,T]$, a first integral $J(p,q,\ H_t)$

(assumed to be a continuous function of p, q, t) for the Hamiltonian H_t). $J(p, q,\ H_t)$ is said (Ehrenfest's definition [5]) to be an adiabatic invariant if, for each initial value (p_0, q_0), it is

$$J(p_0, q_0; H_0) = \lim_{T\to\infty} J(p_T, q_T; H_T)\,,$$

or, a slightly stronger requirement which is sometimes preferred, if

$$\lim_{T\to\infty} T \max_{0\le t\le T} |\frac{\mathrm{d}}{\mathrm{d}t} J(p_t, q_t; H_t)| = 0\,.$$

Suppose that for each given t the Hamiltonian H_t is integrable by quadratures and the canonical transformation which integrates it is regular in p, q, t; let $J_i(p, q; H_t) = \oint_{\gamma_i(t)} p\cdot \mathrm{d}q$ be the actions of the i-th cycle $\gamma_i(t)$ of the invariant f-dimensional torus on which the motions of the system evolve under Hamiltonian H_t at fixed t. Then it can be proved that J_i are adiabatic invariants (Burgers' theorem [6]); actually Burgers proved the theorem in a particular case, assuming the integrability by quadratures to be possible "by separation of variables", and Ehrenfest, and later Bohr, Kramers and "all the others" made use of it to formulate and to apply the principle of adiabatics, *cf.* below, until the principle was abandoned in 1925. It was Einstein [2,3] who provided the definition of adiabatic integral for the most general systems that could be integrated by quadrature. For convenience I recall that a system is integrable by separation of variables when it is possible to change the coordinates q into q' in such a way that, with the new variables, the momentum conjugated to q'_k can be expressed as a function of q'_k only and of f integrals of motion. A system integrable by quadrature, instead, is a system for which one can define a *canonical transformation* $(p, q) \to (p', q')$ (in which, in general, q' *does not depend solely on* q*'s but also on* p's) which makes the system integrable by separation of variables: this is an intrinsic geometric definition on which Einstein proposed (unheeded [3]) to lay the foundations of the principle of adiabatics. The *principle of adiabatics* (Ehrenfest's [5]) states that, if H_0 and H_T are two Hamiltonians adiabatically connected by a family of Hamiltonians H_t that are integrable by quadratures, then, if the quantization of H_0 selects the motions for which $J_i(p_0, q_0, H_0)$ has certain values (for instance $J_i(p_0, q_0, H_0) = n_i h$, with n_i integers), the quantization of H_T also requires the same rules for the $J_i(p_0, q_0, H_T)$ (*i.e.* in the example also $J_i(p_0, q_0, H_T) = n_i h$) and this allows us to formulate a general rule of quantization for the systems adiabatically connected to the harmonic oscillator. For instance, $H_t(p, q) = p^2/2m + m\omega^2 q^2/2 + \frac{t}{T}\,(-k/|q| - m\omega^2 q^2/2)$ *written in polar coordinates* yields the quantization rule for the hydrogen atom ($t = T$) by starting from that for the harmonic oscillator ($t = 0$). Fermi shows in an example [18] and in general [19] that the $J_i(p_T, q_T; H_T)$ *are not necessarily equal to the* $J_i(p_0, q_0; H_0)$ *if* H_0, H_T *are integrable by quadratures but* H_t *is not such so* $0 < t < T$. The main difficulty in applying the principle of the adiabatics to quantum theory lays in the non-integrability by quadratures of the Hamiltonians relevant for Atomic Physics, with the exception of the harmonic oscillator, the hydrogen atom, the free gas, the lattices of harmonic oscillators and a few other cases: for instance, the classic helium atom is not integrable by quadratures. The central problem of celestial mechanics *"post-Laplace"* [36] thus reappeared in atomic mechanics, and in a form with many more easily observable consequences.

[8] FERMI E., *Il principio delle adiabatiche ed i sistemi che non ammettono coordinate angolari, Nuovo Cimento*, **25** (1923) 171. Reprinted in [35], paper No. 12.

[9] DE MARIA M., *Fermi, un fisico da via Panisperna all'America, Le Scienze*, Collection *I grandi della scienza*, **8** (1999) 1.

[10] HEISENBERG W., *Quantum theoretical reinterpretation of kinematic and mechanical relations, Z. Phys.*, **33** (1925) 879. English version in [34].

[11] BORN M. and JORDAN P., *On quantum mechanics, Z. Phys.*, **34** (1925) 858. English version in [34].

[12] DIRAC P. A. M., *The fundamental equations of quantum mechanics, Proc. R. Soc. London, Ser. A*, **109** (1926) 642. Reprinted in [34].

[13] BORN M., HEISENBERG W. and JORDAN P., *On quantum mechanics, II, Z. Phys.*, **35** (1926) 557. English version in [34].

[14] FERMI E., *Sui principi della teoria dei quanti, Rendiconti del Seminario Matematico Università di Roma*, **8** (1925) 7. Reprinted in [35], paper No. 22.

[15] FERMI E., *Beweis dass ein mechanisches normalsysteme im algemeinen quasi ergodisch ist, Phys. Z.*, **24** (1923) 261. Reprinted in [35], paper No. 11a.

[16] FERMI E., *Generalizzazione del teorema di Poincaré sopra la non esistenza di integrali di un sistema di equazioni canoniche normali, Nuovo Cimento*, **26** (1923) 101. Reprinted in [35], paper No. 15.

[17] FERMI E., *Dimostrazione che in generale un sistema meccanico è quasi ergodico, Nuovo Cimento*, **25** (1923) 267.

[18] FERMI E., *Il principio delle adiabatiche ed i sistemi che non ammettono coordinate angolari, Nuovo Cimento*, **25** (1923) 171. Reprinted in [35], paper No. 12.

[19] FERMI E., *Alcuni teoremi di meccanica analitica importanti per la teoria dei quanti, Nuovo Cimento*, **25** (1923) 271. Reprinted in [35], paper No. 13.

[20] GALLAVOTTI G., *Statistical Mechanics* (Springer Verlag, Berlin) 1999.

[21] FERMI E., *Über die existenz quasi-ergodisher systeme, Phys. Z.*, **25** (1924) 166. Reprinted in [35], at the end of paper No. 11a.

[22] POINCARÉ, H., *Les méthodes nouvelles de la Mécanique Céleste*, Vol. III (Gauthier-Villard, Paris) 1899.

[23] KOLMOGOROV N., *Preservation of conditionally periodic movements with small change in the Hamilton function*, in *Stochastic Behavior in Classical and Quantum Systems*, edited by G. CASATI and J. FORD, *Lect. Notes Phys.*, Vol. **93** (Springer-Verlag) 1979.

[24] GALLAVOTTI G., *Teoria delle perturbazioni*, entry in *Enciclopedia della Fisica* (edizioni dell'Enciclopedia Italiana, Rome) 1994.

[25] GALLAVOTTI G., *The elements of mechanics* (Springer-Verlag, Heidelberg, 1983 and Boringhieri, Torino 1986).

[26] FERMI E., PASTA J. and ULAM S., *Studies of nonlinear problems*, Los Alamos report LA-1940 (1955) Vol. II, pp. 978-988. Reprinted in [35].

[27] FALCIONI M. and VULPIANI A., *Enrico Fermis's contribution to non-linear systems: the influence of an unpublished article*, this volume, p. 271.

[28] GALGANI L. and SCOTTI A., *Planck-like distributions in classical nonlinear mechanics, Phys. Rev. Lett.*, **28** (1972) 1173.

[29] BENETTIN G., GALGANI L. and GIORGILLI A., *Boltzmann's ultraviolet cut-off and Nekhoroshev's theorem on Arnold diffusion, Nature*, **311** (1984) 444.

[30] BENETTIN G., GALGANI L. and GIORGILLI A., *The Dynamical Foundations of Classical Statistical Mechanics and the Boltzmann-Jeans Conjecture*, edited by S. KUKSIN V. F. LAZUTKIN and J. PÖSCHEL (Birkhauser) 1993.

[31] EVANS D. J. and MORRISS G. P., *Statistical Mechanics of Nonequilibrium Fluids* (Academic Press, New York) 1990.

[32] BOHR T., JENSEN M. H., PALADIN G. and VULPIANI A., *Dynamical Systems Appproach to Turbulence* (Cambridge University Press) 1998.

[33] FERMI E. and PERSICO E., *Il principio delle adiabatiche e la nozione di forza viva nella nuova meccanica ondulatoria, Rendiconti Lincei*, **4** (1926) 452. Reprinted in [35], paper No. 37.

[34] VAN DER WAERDEN B. L., *Sources of quantum mechanics* (Dover) 1968 (this is a collection of the main papers on Matrix Mechanics with an important critical introduction).
[35] FERMI E., *Note e Memorie (Collected papers)* (Accademia dei Lincei and University of Chicago Press) vol. I, 1961 and vol. II, 1965.
[36] GALLAVOTTI G., *Quasi periodic motions from Hypparchus to Kolmogorov, Rendiconti Accademia dei Lincei, Matematica e applicazioni*, **12** (2001) 125.

About the Author

Born on December 29, 1941, GIOVANNI GALLAVOTTI graduated in Physics at the University of Rome in 1963. He was appointed Full Professor of Mechanics in 1971 and "Lefschetz professor" at the Department of Mathematics of Princeton University in 1982, was a member of the Princeton Institute for Advanced Study during the II semester of the Academic year 1984/85 and has been a corresponding member of the Accademia dei Lincei since July 1994. He was awarded the "President of the Republic National Prize" on June 18, 1997, and participated in the ICM 98 plenary conference in Berlin in August 1998. Gallavotti is the author of 196 publications in English and 4 monographs in Italian, of which three were published also in English and one further monograph in English (see `http://ipparco.roma1.infn.it`).

On the adiabatic invariants(*)

TULLIO LEVI-CIVITA - Rome

Of the canonical systems

$$\frac{dp_i}{dt} = -\frac{\partial H}{\partial q_i}, \quad \frac{dq_i}{dt} = \frac{\partial H}{\partial p_i} \qquad (i = 1, 2, \ldots, n),$$

with characteristic function H, independent of t, which contains slowly varying parameters a, two particularly conspicuous types of adiabatic invariants are known:

1° (Gibbs-Hertz's theorems). The volume V enclosed in phase space by a generic isoenergetic manifold

$$H = E \qquad (E \text{ constant}),$$

which applies to quasi-ergodic systems; systems that do not allow other uniform integrals apart from $H = E$ (see, *e.g.*, nos. 3-5 of this paper).

2° (Burgers's Theorem). Sommerfeld's n loop integrals

$$J_i = \oint p_i \, dq_i \qquad (i = 1, 2, \ldots, n),$$

which are adiabatic invariants for (Stäckel's) systems that are integrable by means of separation of the variables and that admit in total n integrals, (quadratic in the p's).

Here we have two extreme cases, which correspond respectively to the minimum (*i.e.* one) and, in a certain sense, to the maximum (*i.e.* n) of uniform integrals in the assumed conditions.

(*) Translated from the Italian "Sugli invarianti adiabatici", in "Resoconto del Congresso Nazionale dei Fisici", Como, 11-20 September 1927, on the occasion of the commemoration of the first centenary of Alessandro Volta's death, vol. II, pp. 475-513 (Zanichelli Editore, Bologna).

As far as I am aware no such precise result has been obtained for intermediate cases, *i.e.* for canonical systems which possess, apart from $H = E$, a certain number, let us say m, of uniform (independent) integrals

$$F_r = c_r \qquad (r = 1, 2, \ldots, m).$$

Indeed Fermi has pointed out (no. 7) that, at least in general terms, the definition of the adiabatic variation of the constants E and c_r which would seem to be the most spontaneous, based only on the principles of statistical mechanics in accordance with the quasi-ergodic case, is not acceptable.

I plan to show (nos. 8-15) how, in the hypothesis that the m integrals F_r are in involution between themselves, the classical methods of analytical mechanics, and in particular the consideration, which dates back to Morera, of certain systems associated to total differentials, suggest a different criterion (which is also inspired by a strict analogy with the quasi ergodic case) to impose the adiabatic relationship (between the variations of E and c_r and those of the parameters a) in such a way that the conditions of integrability are automatically satisfied. From this one deduces (no. 14) the *existence of* $m + 1$ *adiabatic invariants, which can be constructed by means of quadrature.* Burgers' theorem is contained in this proposition (no. 16) as a very special case. Moreover this new demonstration of the adiabatic invariance of the integrals J_i also embraces, without exceptions, those cases of partial or total commensurability (of certain periods) for which, as is well known, direct demonstrations required minutely detailed complementary discussions and various analytical supports. Finally I have fleetingly touched upon further applications and possible extensions (no. 17).

1. *Recent atomic theories and their formulations.* – According to the laws of classic mechanics, the motions that a (holonomous) system with n degrees of freedom can assume under the action of fixed forces depend continuously on $2n$ constants (initial conditions) liable to assume (in a certain field) all the possible values (of that field).

Niels Bohr (1) based his theory of the atom on the premise of ordinary mechanics (indeed, for the hydrogen atom, on the problem of two bodies), but he nevertheless introduced into it, as an audacious combination, an extraneous postulate which derives instead from Planck's quantum concepts, and he therefore brings the discontinuum into play.

Briefly, everything comes down to the introduction of privileged orbits, which correspond to simple values in arithmetical progression, more precisely of the type $nh/2\pi$ (n integer number, h Planck's constant) of some appropriate combinations J_0, $J_1, \ldots$, J_m, (only one in the simpler case originally considered by Bohr) of the integration constants.

The theory, developed by Bohr himself and by other eminent scientists with a fervour worthy of its remarkable consequences, received wondrous spectral confirmation and also, mainly due to Sommerfeld (2), an immediate systematic approach, kept up to date (up

---

(1) See, *e.g.*, *Les spectres et la théorie de l'atome.* Paris, Hermann, 1923.

(2) *Atombau und Spektrallinien.* Braunschweig, Vieweg, 1922; 4ª ed., 1924.

to last year) by later editions of Sommerfeld's book, and by new works, such as Born's and Hund's [3], Andrade's [4], Juvet's [5], and also by papers rich in original ideas, not only from the authors already quoted, especially Jeans [6], Jordan, Heisenberg, Kramers, Slater [7], etc.

Many physicists, and not only traditionalists, disliked the combination mentioned above of Newtonian mechanics with a selective principle of quantum discontinuity, hence, on the one hand, the simultaneous efforts of Heisenberg, Born, Jordan, the latter alone, and Dirac [8] to eliminate from atomic theory any element not amenable to direct experiment and to construct a mechanics of periodic phenomena on a clearly discontinuous basis (matrix calculus); on the other hand the brainwave of returning to the model of the vibrations of continuous media through De Broglie's [9] and Schrödinger's [10] wave mechanics, according to which (in a no less perfect agreement with the experimental results) the explanation of the discontinuous behaviour of the spectral lines derives from eigenvalues and eigenfunctions of differential equations which define the state of the medium; and finally the more general concepts of Hilbert, Von Neumann and Nordheim, which embrace both viewpoints [11].

If the efforts of theoretical physicists are now directed towards this new path, so attractive and promising, it would not be right to abandon the middle way, in other words: the hybrid approach, and in that sense Sommerfeld's approach, that had been reached by the theory of the atom by associating a single quantum principle to ordinary mechanics, "displeasing God and his enemies", but which is undoubtedly attractive, corresponding to elementary and concrete forms of physical intuition, and above all being suitable to lead to quantum relationships in the simplest way with the usual procedures of analytical mechanics.

2. *Adiabatic invariants according to Ehrenfest* [12] *and their speculative importance in Sommerfeld's systematization.* – Fundamental, from this eclectic point of view, is the study (for the dynamic systems connected to the various types of atoms) of those

---

(3) *Vorlesungen ueber Atomdynamik.* Berlin, Springer, Bd. I, 1925.
(4) *The structure of the atom.* London, Bell, 1923; 3ª ed., 1927.
(5) *Mécanique analytique et théorie des quanta.* Paris, Blanchard, 1926.
(6) *Atomicity and quanta.* Cambridge University Press, 1926.
(7) In several articles, particularly in "Zeitschrift für Physik", 1924-1927.
(8) See, above all for the German authors, the years 1926 and 1927 of the already quoted "Zeitschrift für Physik", and, for Dirac's papers, Vol. 112, 1926, of the "Proc. of the R. S. of London".
(9) *Ondes et mouvements.* Paris, Gauthier-Villars, 1926.
(10) *Abhandlungen zur Wellenmechanik.* Leipzig, Barth, 1927.
(11) See a paper by these three authors *Ueber die Grundlagnen der Quantentheorie* in "Math. Ann.", B. 98, pp. 1-30; and also VON NEUMANN, *Mathematische Begründung der Quantenmechanik*, "Göttinger Nachr.", 1927, pp. 1-57.
(12) *Adiabatic invariants and the theory of quanta*, "Phil. Mag.", vol. XXXIII, 1917, pp. 500-513.

combinations of integration constants,

$$J_0,\ J_1, \ldots,\ J_m$$

to which the values $nh/2\pi$ (integers apart from the universal factor $h/2\pi$) must be attributed.

Ehrenfest called them *adiabatic invariants*, and we will follow this designation; while Smekal, who contested the use of the adjective *adiabatic*, proposed calling them more generally *invariant parameters*. Regardless of the name, the physical view linked to it, due precisely to Ehrenfest and generally known as the principle of adiabatics, is essential. It comes down to this. Let us suppose that in the mechanical model of a given atomic system, masses, conditions or forces susceptible of varying with continuity come into play: this translates mathematically into the hypothesis that the dynamic equations, or, if you prefer, the characteristic function H of the corresponding canonical system depend on a certain number (it is not necessary to specify it) of parameters $a_1,\ a_2, \ldots$, that we will denote altogether by a.

If, as these parameters a vary continuously, the qualitative feature of the mechanical system does not change, so that it constitutes at every stage the model of an atomic system (for which, for example, the gradual modification of the values of the parameters can be interpreted physically as due to alterations in temperature, environment, electrical state, etc.) it is obvious that the characteristic combinations

$$J_0,\ J_1, \ldots,\ J_m$$

must, on the one hand, also vary with continuity, and on the other must preserve integer values (apart from that constant factor). This is only possible if such combinations remain constant.

This is Ehrenfest's principle, which proposes, even to the surviving devotees of pure mechanics, the abstract study, extremely interesting in itself and for its applications, of adiabatic invariants.

3. *The case of canonical systems. Volume in phase space.* – Let us limit ourselves, in order to clarify our ideas, to the consideration of a canonical system of degree $2n$

$$\frac{dp_i}{dt} = -\frac{\partial H}{\partial q_i}, \quad \frac{dq_i}{dt} = \frac{\partial H}{\partial p_i} \qquad (i = 1, 2, \ldots, n), \tag{1}$$

whose characteristic function H depends on the p's, on the q's and on the parameters a, but not explicitly on t.

Let us suppose that (in the field in which the parameters a will vary) the isoenergetic manifolds

$$H = E \tag{2}$$

are *closed surfaces* (more precisely closed manifolds with $2n-1$ dimensions) in the phase space Φ_{2n} (representing the $2n$ conjugated variables p's and q's). We will indicate by V and we will call *volume* (even if it is a field with more than three, or with only two dimensions) the Euclidean extension of the phase space Φ_{2n}, enclosed by a generic σ. Well, if the canonical system (1) is *quasi-ergodic*, *i.e.* if, in the absence of other uniform integrals beyond (2), all or "almost all" ([13]) the trajectories along which E has an assigned value fill *practically* (*i.e.* in the well known sense) the corresponding σ, the volume V within it is an *adiabatic invariant.*

This fine property is virtually implicit in some considerations amply dealt with by Gibbs in his famous book on statistical mechanics ([14]), but it is not explicitly enunciated there. The credit for bringing this to light, linking it specifically to the adiabatic processes (in a very precise sense that we will recall in a moment) belongs to Paul Hertz ([15]).

4. *P. Hertz's demonstration of the adiabatic invariance of V.* – Given the interest of the result, also with regard to the aim we have in view, it is worth the trouble to outline rapidly its deduction.

First let us recall how one defines, in the case of quasi-ergodic systems, the average value $\overline{F}$ of any (continuous) function of the position, $F(p|q)$, on a given isoenergetic surface σ, assumed to be entirely contained in a regularity field of the function $H(p|q)$.

At every *ordinary* point (*i.e.* not multiple) M of the said surface, at least one of the $2n$ partial derivatives of H is different from zero. Let us indicate by z that (or one of those) p or q for which (at the ordinary point in question) $\frac{\partial H}{\partial z} \neq 0$. Let us then indicate by x the remaining $2n-1$ p's, q's and by dX the product of their $2n-1$ differentials. Thanks to the relationship

$$H = E \tag{2}$$

it is possible to substitute, as $2n$ independent variables, the same $2n-1$ variables x and E for the original p's e q's, or, if you wish, for the x's and z. The x's can be looked upon, in the neighbourhood of M, as coordinates of the points of the hypersurface σ. And it is immediately seen from the transformation of multiple integrals, that the element of (Euclidean) volume dV of phase space can be put in the form

$$dV = dp_1 \dots dp_n \, dq_1 \dots dq_n = dz \, dX = \frac{dE dX}{\left|\frac{\partial H}{\partial z}\right|} \,. \tag{3}$$

([13]) "Almost all" is defined as follows: imagine a generic trajectory identified by the initial values p_i^0, q_i^0. Let us attribute, on a generic manifold (2), to a set σ_1 the points (p_i^0, q_i^0), from which a trajectory exits that is dense in all σ_1, and to the complementary set σ_2 the points from which instead a periodic trajectory, or one that excludes some portion of σ, exits. The (hypersurface) measure of σ_2 must be zero.
([14]) *Statistical mechanics.* Yale University Press, 1902.
([15]) See, WEBER-GANS - *Repertorium der Physik.* Leipzig, Teubner, 1916, Bd I, N. 270, pp. 535.

According to the principles of statistical mechanics, attributing a uniform density to the phase space, every elementary field dX surrounding a generic point M of an isoenergetic surface σ makes an elementary contribution to the average value of a function proportional to

$$\frac{F(M)dX}{\left|\frac{\partial H}{\partial z}\right|}.$$

If σ does not have multiple points, it will be sufficient to divide it into a finite number of pieces choosing, from amongst them, an appropriate z (amongst the $2n$ p's and q's) so that an integral of the type

$$N = \int_\sigma \frac{F(M)dX}{\left|\frac{\partial H}{\partial z}\right|} \tag{4}$$

is defined and is univocally determined (independently of how one proceeds with the division into pieces).

Given in particular

$$D = \int_\sigma \frac{dX}{\left|\frac{\partial H}{\partial z}\right|}, \tag{5}$$

we assume as the average value $\overline{F}$ of the function F

$$\overline{F} = \frac{N}{D}. \tag{6}$$

If there are multiple points, a rather more profound discussion is required, but the consideration of integrals of the type D and N is still justified and therefore the notion of the average value of a (finite and continuous) function $F(M)$ is still valid.

Taking all the above as given, let us return to the hypothesis that H depends, not only on the p's and q's, but also on certain parameters a, and let us make them vary so slowly —this is the justification of the adjective adiabatic— that in the meantime the point M of σ, representing the act of motion, moving along a generic trajectory (amongst those dense in σ), has essentially invaded the whole surface $H = E$.

If we attribute arbitrary increments da to the parameters a, in a specific point M of σ (defined by the coordinates p, q), the value of the corresponding increment $H(p|q)$ is obviously

$$d_a H,$$

with p, q remaining unchanged.

However, under the hypothesis that (even an elementary) increment of the a's occurs while the representative point $M(p, q)$ of the dynamic system essentially invades the whole σ, it is only natural to think that H undergoes, not *the local increment* $d_a H$, belonging to the initial phase or to another instantaneous phase, but the average value relative to a time interval long enough for the whole σ to be able to contribute to it. Hence, it is assumed that the average value $\overline{d_a H}$ formed according to (4), (5), (6) be subordinated, as the induced variation of the function H, to an elementary adiabatic variation of the parameters a.

This average increment of H, which depends only on a and da (not on p, q) should thus be regarded as a definition of the change $d_a E$ undergone by the total energy E of the system (belonging to a generic solution) as a consequence of the adiabatic variation, in the sense now declared, of the parameters a, on which the mechanism of the system depends by means of the characteristic function $H(p|q|a)$. We are therefore led to put

$$d_a E = \overline{d_a H} = \int_\sigma d_a H \frac{dX}{\left|\frac{\partial H}{\partial z}\right|} : \int_\sigma \frac{dX}{\left|\frac{\partial H}{\partial z}\right|}, \tag{7}$$

which does not lead to comments in the case of a single parameter a, but, in the case of several parameters, can be considered justified only on condition that the second member is an exact differential with respect to the arguments a ([16]).

Now it is easy to recognize (Paul Hertz's theorem) that this really occurs, indeed that the function $E(a)$, defined by the total differential equation (7) coincides with that defined in finite terms as

$$V(a|E) = \text{const.}, \tag{8}$$

V designing, as already agreed, the volume of phase space enclosed by a generic surface $H = E$.

To establish this it is sufficient to evaluate, on the basis of its geometric meaning as a volume, the change undergone by the function (8), when arbitrary increments da, dE are attributed to a and to E.

Let us consider x and E as independent variables in phase space, focusing our attention on a generic point M of σ and on a surrounding element of it dX. First however let us make another observation on the equation

$$H = E, \tag{2}$$

treating the a, p, q, as independent variables and E as their function. In such an approach, attributing increments da to a (leaving p, q unchanged) means passing (in the neighbourhood of the generic point M to which the values of the p's and of the q's refer) from the isoenergetic surface $H = E$ to the analogous one $H = E - d_a H$. It follows that, as a consequence of the increment given to a, in the proximity of M, E is incremented

---

([16]) Indeed, if this were not the case, the energy E could not be regarded as a uniform function of a; even though these vary adiabatically from an initial value a^0 to a final value a^1, E would undergo a change ΔE that in addition depends on the path along which the a's pass (in the space that represents them) from point a^0 to point a^1.

by $-d_a H$, while the x's remain unchanged, and the z varies in the way dictated by (2). Therefore, at the surface element dX, the volume V undergoes the increment (3), where the following value is substituted for dE

$$\frac{-d_a H\, dX}{\left|\frac{\partial H}{\partial z}\right|}. \tag{9}$$

Adding all these attributes we have

$$d_a V = -\int_\sigma \frac{-d_a H\, dX}{\left|\frac{\partial H}{\partial z}\right|}. \tag{10}$$

As for

$$d_E V = \frac{\partial V}{\partial E}\, dE, \tag{11}$$

this is simply the volume (of phase space) contained between $H = E$ and $H = E + dE$, which, evaluated as above, we find expressed as

$$d_E V = dE \int_\sigma \frac{dX}{\left|\frac{\partial H}{\partial z}\right|}. \tag{11}$$

If we consider E as a function of the a defined by (8), the total differential $dV = d_a V + d_E V$ cancels out; from (10), (11), writing $d_a E$ for greater clarity, we therefore have in place of the generic dE,

$$-\int_\sigma \frac{d_a H\, dX}{\left|\frac{\partial H}{\partial z}\right|} + d_a E \int_\sigma \frac{dX}{\left|\frac{\partial H}{\partial z}\right|} = 0, \tag{7'}$$

which in fact coincides with (7) and shows that $d_a E$ is the exact differential of the function E of a defined by eq. (8). Viceversa, if we define $d_a E$ on the basis of (7), which translates, in the set circumstances, the adiabatic variation of the parameters a, on the basis of (7′), in accordance with (10) and (11), it follows that $dV = 0$, hence the fundamental result that the volume of phase space, enclosed by an isoenergetic surface $H = E$, is an adiabatic invariant, holds.

5. *Case of a single degree of freedom. Elementary examples.* – We will linger for a moment on the special case of dynamic systems with a single degree of freedom, even though these things have been said and repeated in several forms (by the quoted authors and by others). Since q is the only Lagrangian coordinate, $\dot q = \frac{dq}{dt}$; $T = \frac{1}{2} A \dot q^2$ the kinetic energy, with A a positive function of q; $U(q)$ the force function, we will have the momentum

$$p = \frac{\partial T}{\partial \dot q} = A \dot q, \tag{12}$$

and consequently the canonical expression of the energy

$$H = \frac{1}{2A} p^2 - U(q). \tag{13}$$

It is well known ([17]) that, when q is initially contained between two simple roots q', q'' of the equation

$$U = E,$$

the motion is periodic. The trajectories, in the Cartesian phase p, q, are the closed curves

$$H = E.$$

The system is manifestly quasi-ergodic because the trajectories of a given energy E, in this case, actually coincide with the isoenergetic manifolds $H = E$, so that they fill them fully (without gaps during a single period).

The invariant of Gibbs-Hertz is the volume, in this case the area, V enclosed by $H = E$.

Considering p as a function (with two values) of q defined by the quadratic equation $H = E$, the expression for V may be put in the form

$$V = \oint p\, dq, \tag{14}$$

where the symbol $\circ$ indicates that the integral must be extended to the closed curve $H = E$, so that, since, as shown by expression (13) of H, the curve is symmetrical with respect to the axis of q, it is also possible to write

$$V = 2 \int_{q'}^{q''} p\, dq,$$

where p denotes the positive root of the quadratic equation $H = E$. If in (14) we introduce the time t as an independent variable and the period of motion is indicated by τ, we clearly have $V = \int_0^\tau p\dot{q}\, dt$, and, since $p\dot{q}$ is identified with the double $2T$ of the kinetic energy, we also have

$$V = \int_0^\tau 2T dt. \tag{14'}$$

In the second member we can recognize the Maupertuisian *action* ([18]) relative to a period of the motion: it is therefore, just like V, an adiabatic invariant.

([17]) Cf., *e.g.*, LEVI-CIVITA and AMALDI - *Lezioni di meccanica razionale.* Bologna, Zanichelli, vol. $(\mathrm{II})_1$, chapter I, § 6.
([18]) *Ibidem*, vol. $(\mathrm{II})_2$, chapter II, no. 13.

Introducing the average value $\overline{T}$ of the kinetic energy with respect to a period, and in addition the frequency $\nu = \frac{1}{\tau}$, again we have

$$V = 2\overline{T}\tau = \frac{2\overline{T}}{\nu}. \tag{14''}$$

In the case of an oscillator (material point subject to an elastic restoring force) we can set in eq. (13) $A = 1$ and $U = -\frac{1}{2}\,\omega^2 q^2$, assuming for example the mass of the mobile to be unitary, designating by q its abscissa and indicating by ω^2 the constant coefficient of the restoring force. All the motions thus defined are harmonic with the frequency constant ω, thus resulting in

$$q = r\cos(\omega t + \vartheta_0),$$

where r (> 0) and ϑ_0 represent the integration constants.

In the phase plane p, q the isoenergetic curves are the ellipses

$$\frac{1}{2}\,p^2 + \frac{1}{2}\,\omega^2 q^2 = E$$

and we therefore have for the enclosed area

$$V = \frac{2\pi E}{\omega}.$$

As can be seen, it is not the total energy E of the oscillator that is invariant with respect to the adiabatic influences, but instead the ratio between energy and frequency.

This could in any case also be deduced from (14″), bearing in mind that, in the case of an oscillator, the average values of the kinetic energy and of the potential energy are equal to each other, and each has $\frac{1}{2}\,E$; with on the other hand $\nu = \frac{\omega}{2\pi}$.

For the simple pendulum, if as usual we assume the deviation ϑ from the vertical to be the Lagrangian coordinate q, we have, from the definition of T and from the energy integral,

$$T = \frac{1}{2}\,l^2\dot{\vartheta}^2 = E + gl\,\cos\vartheta,$$

where g is the acceleration due to gravity, l the length of the pendulum, and we have assumed the mass of the pendulum itself to be unitary. Equation (14′) gives

$$V = l^2\int_0^\tau \dot{\vartheta}^2\,dt.$$

Let us substitute ϑ for t as the integration variable, and indicate by $-\vartheta_0$ and ϑ_0 the extremes of a simple oscillation (*i.e.* corresponding to half a period), which are defined on the basis of E, g, l by the equation

$$E + gl\cos\vartheta_0 = 0. \tag{15}$$

We can write

$$V = 2l^2 \int_{-\vartheta_0}^{\vartheta_0} \dot{\vartheta} d\vartheta = 2\sqrt{2}l \int_{-\vartheta_0}^{\vartheta_0} \sqrt{E + gl\cos\vartheta}\, d\vartheta. \tag{16}$$

This gives ([19])

$$\frac{\partial V}{\partial E} = 2\int_{-\vartheta_0}^{\vartheta_0} \frac{d\vartheta}{\dot{\vartheta}} = \tau. \tag{17}$$

In particular for $E = -gl$ we have $\vartheta_0 = 0$, $V = 0$.
On the other hand, if as usual we put

$$\sin\frac{\vartheta_0}{2} = k,$$

the definition (15) of ϑ_0 gives

$$E + gl = 2glk^2,$$

and therefore, also bearing (17) in mind,

$$\frac{\partial V}{\partial k} = \frac{\partial V}{\partial E}\frac{\partial E}{\partial k} = \frac{\partial V}{\partial E} 4glk = 4glk\tau.$$

Recalling the well known expansion of τ ([20])

$$\tau = 2\pi\sqrt{\frac{l}{g}} \sum_0^\infty {}_n c_n^2 k^{2n} \qquad \left(c_0 = 1,\, c_n = \frac{1.3.\ldots.(2n-1)}{2.4.\ldots.2n}\right)$$

and bearing in mind that V cancels out for $E = -gl$, *i.e.* for $k = 0$, we have, integrating the preceding expression of $\frac{\partial E}{\partial k}$ from 0 to k,

$$V = 8\pi\sqrt{l^3 g} \sum_0^\infty {}_n \frac{1}{2n+2}\, c_n^2 k^{2n+2} = \tag{17'}$$

$$= 4\pi\sqrt{l^3 g}k^2 \sum_0^\infty {}_n \frac{1}{n+1}\, c_n^2 k^{2n}.$$

([19]) It would not be inappropriate to point out that it is really necessary to derive with respect to E also the limits $\pm\vartheta_0$ (which depend on it), but the relative contribution is nil because the function under the sign goes to zero for $\vartheta = \pm\vartheta_0$.
([20]) See, *e.g.*, op. cit. ([18]), chapter I, no. 38.

6. *Systems that are integrable by separation of variables. Burgers' theorem.* – Returning now to the general case and considering how remarkable and productive the adiabatic invariant derived from the energy integral is, it is natural to ask whether analogous deductions might be possible when other integrals of the dynamic system under consideration become known.

In this context one should remember above all the brilliant application of Ehrenfest's principle, immediately carried out by Burgers ([21]) to Stäckel's integrable systems with the method of variable separation (and hence by quadrature). For such material systems Sommerfeld had introduced with brilliant success the postulate

$$J_i = \oint p_i dq_i = \text{integer multiple of } h/2\pi \qquad (i = 1, 2, \ldots, n).$$

It is to Burgers that the credit goes for having provided a rational justification for it, stating that, in the above case, each of the integrals J is an adiabatic invariant. Burgers gave two proofs of different types, one which exploited differential properties, the other the integral properties of Stäckel's ([22]): both procedures were ingenious and profound, while the second also seemed susceptible to some qualitative extension. However it did not connect to the rigorous approach of analytical mechanics, which would later appear to be the most suitable to provide meaningful generalisation.

7. *Canonical systems to be designated as non-primitive of order m, which allow other uniform integrals m beyond H = E. Fermi's negative result.* – Let us return to the consideration of a generic canonical system

$$\frac{dp_i}{dt} = -\frac{\partial H}{\partial q_i}, \quad \frac{dq_i}{dt} = \frac{\partial H}{\partial q_i} \qquad (i = 1, 2 \ldots, n) \tag{1}$$

with a characteristic function $H(p|q|a)$ independent of time and (as in no. 3) dependent instead adiabatically on an arbitrary number of parameters a.

Let us suppose that the system allows, as well as the energy integral

$$H = E, \tag{2}$$

other m uniform integrals, also independent of t,

$$F_1 = c_1, \; F_2 = c_2, \ldots, \quad F_m = c_m. \tag{18}$$

([21]) *Adiabatic invariants of mechanical systems*, "Phil. Mag.", volume XXXIII, 1917, pp. 514-520.

([22]) See in the latter case BURGERS'S dissertation (presented at the University of Leiden; Haarlem, 1918; in Dutch); alternatively BORN, loc. cit.([3]), pp. 98-148.

We shall call such canonical systems, for brevity, *non-primitive of order m*. The non-primitivity of order zero thus corresponds to the hypothesis that there exists only one uniform integral, the energy.

For formal convenience it will help us to introduce the designations

$$F_0 = H, \quad c_0 = E, \tag{19}$$

so that the known integrals (2) and (18) of our canonical system can all be included in the formula

$$F_r = c_r, \qquad (r = 0, 1, 2, \ldots, m). \tag{20}$$

Let us further suppose that eqs. (20) define in phase space Φ_{2n} a closed manifold (*i.e.* without a frontier) with $2n - (m+1)$ dimensions, which we will designate by τ; that other uniform integrals do not exist; and finally that nearly all the integral curves, each of which occurs on a fixed τ, are, in the usual meaning, dense on the corresponding τ; in other words verify a rather less restrictive condition (d'), which will be specified in the following no. 13.

Letting ourselves be guided by the same criteria that led us in no. 4 to the definition (7) of $d_a E_l$, we will now be led to introduce the adiabatic variations of the constants c_r under the form of average values ([23])

$$\overline{d_a\, F_r}$$

of the dF_r, with respect to the manifold τ (in the hypothesis that the integral curves essentially fill it).

([23]) The meaning to be attributed to such average values on τ is an obvious generalisation of that specified at no. 4 for the isoenergetic surface. And precisely, said M an (ordinary) point of the manifold τ, $d\tau$ a surrounding element, $m+1$ arguments (between the $2n$ conjugate p, q) in relation to which the (20) can be resolved will be indicated by $z_0, z_1, \ldots, z_m$ (overall by z). Therefore, the functional determinant

$$\Delta = \begin{pmatrix} F_0\, F_1\, \ldots\, F_m \\ z_0\, z_1\, \ldots\, z_m \end{pmatrix}$$

will be different from zero. Let us call x the other $2n-(m+1)$ arguments p, q; dX the product of the relative differentials. Considering x and c as $2n$ independent variables in the place of p, q,, we will have

$$dV = \frac{dX\, dc_0\, dc_1\, \ldots\, dc_m}{|\Delta|},$$

and the average value of a generic dF_r is defined as the relationship

$$\int_\tau \frac{d_a\, F_r\, dX}{|\Delta|} : \int_\tau \frac{dX}{|\Delta|}.$$

But, as Fermi pointed out (24), the variations, thus defined, are not generally exact differentials with respect to the parameters a, and they become exact differentials only under very particular circumstances; thus it seems we can exclude the need to engage in a general study of adiabatic invariants along this path.

8. *Elementary integrals* p_r = const. *Consequent reduction of the canonical system. Adiabatic invariant provided by the reduced volume.* – Let us concentrate for a moment our attention on the typical case in which some of the Lagrangian coordinates, let us say *e.g.*

$$q_1, q_2 \dots, q_m,$$

are *cyclic*, or, as is also said, *ignorable*, in the sense that they do not appear in the expression of the energy H. Thus, since

$$\frac{\partial H}{\partial q_r} = 0 \qquad (r = 1, 2, \dots, m),$$

the first m canonical equations (1) give the m integrals

$$p_r = \text{const.} = c_r \qquad (r = 1, 2, \dots, m). \tag{21}$$

Bearing them in mind, the remaining eqs. (1) divide into two groups:

a) The canonical system, in the $2(n-m)$ conjugate arguments

$$\begin{pmatrix} p_{m+1} \cdots p_n \\ q_{m+1} \cdots q_n \end{pmatrix},$$

$$\frac{dp_i}{dt} = -\frac{\partial \mathcal{H}}{\partial q_i}, \quad \frac{dq_i}{dt} = \frac{\partial \mathcal{H}}{\partial p_i} \qquad (i = m+1, \dots, n), \tag{22}$$

in which, to show that that the p_r should be substituted by their constant values c_r, I have written, in place of H,

$$\mathcal{H} = (H)_{p_r = c_r}. \tag{23}$$

b) The remaining m equations

$$\frac{dq_r}{dt} = \frac{\partial H}{\partial p_r} = \frac{\partial \mathcal{H}}{\partial c_r} \qquad (r = 1, 2, \dots, m), \tag{24}$$

(24) *Alcuni teoremi di meccanica analitica importanti per la teoria dei quanti*, "Nuovo Cimento", VII, vol. 25, 1923, pp. 271-285.

whose task is only to provide (by quadrature) the temporal expression of the ignorable coordinates q_r, once the reduced canonical system (22) has been integrated. In this the c's can be treated in the same way as the parameters a (which already appeared in the original H).

As long as only the isoenergetic manifolds of the reduced system (22)

$$\mathcal{H} = E$$

are closed surfaces with $2(n-m)-1$ dimensions of the space Ψ of the p_i, q_i $(i > m)$, the volume W delimited by it is an adiabatic invariant for arbitrary slow variations of all the parameters a, c ed E.

9. *Generalisation of the preceding result suggested by the theory of canonical transformations.* – If the known integrals (18), although without having the special form $p_r =$ const., are in involution between themselves, in other words if the $\frac{m(m-1)}{2}$ Poisson brackets

$$(F_r, F_s) \qquad\qquad (r, s = 1, 2, \ldots, m),$$

vanish, it is always possible ([25]) to reduce oneself to the elementary case aforementioned by means of a canonical transformation, *i.e.* by introducing, in place of the original arguments p_i, q_i, $2n$ independent combinations P_i, Q_i, of which the first m P's coincide with the $F_r(p|q)$'s, and the condition of canonicity

$$\sum_1^n {}_i P_i dQ_i = \sum_1^n {}_i p_i\, dq_i + \text{exact differential}$$

is also satisfied.

This ensures that, in the new variables P, Q, the differential equations (1) preserve the canonical form with the same characteristic function H (expressed for P, Q, instead of for p, q). And the system allows, by construction, the m elementary integrals

$$P_r = c_r.$$

Once the qualitative condition regarding the closure of the manifolds has been satisfied

$$\mathcal{H} = (H)_{P_r = c_r} = E \tag{25}$$

in the space of the $2(n-m)$ arguments P_i, $Q_i (i > m)$ —and that leads to observations which we will examine in a moment— *the volume W of the field delimited by $\mathcal{H} = E$ in this*

([25]) LIE-ENGEL, *Theorie der Transformationsgruppen*, B. II. Leipzig, Trubner, 1980, chapter X, pp. 207-209.

space constitutes an adiabatic invariant for arbitrary slow variations of the parameters a, and also of the integration constants E and c.

The existence of this invariant for every non-primitive canonical of order m is therefore —qualitative specifications apart— proved.

10. *Critical considerations. Analytical and constructive need to re-express the result in the original variables. Indication of the road to follow.* – Although theoretically possible, the introduction of new canonical variables (P_i, Q_i), of which the first m P coincide with the F_r, leads to various observations:

1°) Above all it depends on analytical operations of high order; generally higher than the integration of the canonical system, of which the integrals in involution $F_r = c_r$ are known.

2°) While the original variables p, q are, by hypothesis, in bi-univocal correspondence with the phases (acts of motion) of the mechanical system, it cannot be claimed *a priori* that this follows from the P, Q, since (with the exception for the P_r chosen to be equal to F_r) they will generally be *non* uniform functions of the p, q.

Therefore, while moving from the representative space of (p, q) to that of (P, Q) *locally* the topological characters are preserved, it is not certain that this applies in the whole field of (p, q) that must be considxsered. *E.g.* it is possible that the property of certain curves that they *practically* fill the manifolds does not have an invariant character when non-uniform transformations, etc. are considered.

3°) It is true on the other hand that, because of the specific nature of the canonical transformation between the (p, q) and the (P, Q), the (reduced) isoenergetic manifolds τ of equation

$$\mathcal{H} = (H)_{p_r = c_r} = E, \tag{25}$$

with respect to the auxiliary variables $P_{m+1}, \ldots, P_n$, $Q_{m+1}, \ldots, Q_n$, can be defined directly with respect to the original variables, by means of the $m+1$ (uniform) equations

$$F_r = c_r, \quad H = E\,,$$

or, which is the same thing, by means of eq. (20) of no. 7.

However, in the phase space Φ_{2n} of the (p, q), such $m+1$ equations determine a manifold of $2n-(m+1)$ dimensions, and it is not possible to see geometrically what is the invariant W, which, in the space of the $2(n-m)$ arguments $P_{m+1}, \ldots, P_n$, $Q_{m+1}, \ldots, Q_n$, is the volume enclosed by a surface (25).

To appreciate these difficulties more precisely, let us imagine (this does not constitute a substantial restriction) that the independent integrals m (18)

$$F_r = c_r, \qquad (r = 1, 2, \ldots, m)$$

can be solved with respect to $p_1, p_2, \ldots, p_m$, setting these m arguments, with respect to the remaining ones (and indeed of the c and of the a parameters) in the form of

$$(18') \qquad p_r = f_r \qquad (r = 1, 2, \ldots, m).$$

When we set aside the $p_1, p_2, \ldots, p_m$, *i.e.* we consider an auxiliary space Φ' with $2n - m$ dimensions representing the arguments

$$p_{m+1}, \ldots, p_n; \quad q_1, q_2, \ldots, q_n,$$

the eqs. (20) of the τ manifold, in the phase space Φ_{2n}, are reduced, to the single equation

$$(25') \qquad (H)_{p_r = f_r} = E,$$

which represents a surface (manifold with $2n - (m+1)$ dimensions) in Φ'. But it will still be necessary to free ourselves of a further m dimensions, in other words, formally speaking, of m arguments, which is done automatically by means of the canonical transformation, independently of the m ignorable coordinates $Q_1, Q_2, \ldots, Q_m$, conjugated to $P_r = F_r$.

Only by overcoming somehow such a difficulty, can we lower by m units the dimensions of a manifold (25′), in such a way that it presents itself as a surface in a space Ψ with $2(n-m)$ dimensions. And only then, bearing in mind the qualitative specifications and taking into account the metric to be attributed to the reduced space Ψ, would it become legitimate to speak of volume W enclosed by the surface (25′), and within it we would have the adiabatic invariant suggested by the canonical transformation.

Finally, keeping to the indicated path, the very demonstration of the existence of the adiabatic invariant W is not fully satisfactory because transformations come into play that in general are not uniform throughout the field to be investigated, and such transformations can alter some of the topological characters that are important to take into account. On the other hand, even we did not consider such a drawback, believing, as in realty it is, that the transformation at the very least makes the existence theorem rather plausible, the constructive problem remains, which, as we have just discovered, seems to require the previous determination of the ignorable combinations

$$Q_1, Q_2, \ldots, Q_m$$

and thus differential operations which may be of high order, while it is desirable to arrive at the explicit expression of W in the simplest possible way, which implies, as we shall see, only one quadrature.

We must therefore try to characterise directly (in other words without auxiliary transformations) the adiabatic invariant W.

In such an investigation a procedure we owe to Morera ([26]), who already derived from it a quick demonstration of Lie's theorem on the reduction of canonical systems, will be fundamental. For the aim we have in mind we will somewhat modify the procedure, making the introduction of that completely integrable system of total differentials even more spontaneous, which allows, as Morera showed, an agile discussion of the questions of reducibility.

11. *New feature of Lie's theorem on the reduction of canonical systems. Associated system* (A_0) *of total differentials.* – Let us put ourselves in the conditions already repeatedly stated, and refer to a non-primitive system of order m (no. 7), *i.e.* to a canonical system (1), of which, if the characteristic function H is independent of t, we know besides the energy integral

$$H = E, \tag{2}$$

m integrals

$$F_r = c_r, \qquad (r = 1, 2, \ldots, m), \tag{18}$$

also independent of t and in involution between themselves.

As is well known ([27]) this latter circumstance continues to exist even if we assume eqs. (18) in their solved form

$$p_r = f_r, \qquad (r = 1, 2, \ldots, m), \tag{18'}$$

and they translate formally into the identities

$$\frac{\partial f_r}{\partial q_s} - \frac{\partial f_s}{\partial q_r} + \{f_r, f_s\} = 0 \qquad (r, s = 1, 2, \ldots, m), \tag{26}$$

where the symbol { } denotes a Poisson bracket, limited to the arguments

$$\begin{pmatrix} p_{m+1} & \cdots & p_n \\ q_{m+1} & \cdots & q_n \end{pmatrix}.$$

The fact that eqs. (18) are as many integrals of the canonical system (1), and therefore eqs. (18′) are as many invariant relations, implies that the brackets $(H, p_r - f_r)$ cancel each other out, taking eqs. (18′) themselves into account.

([26]) *Intorno ai sistemi di equazioni a derivate parziali del I° ordine in involuzione.* "Rend. del R. Ist. Lombardo", vol. XXXVI, 1903, pp. 775-790.
([27]) Cf., *e.g.*, loc. cit. ([17]), vol. $(\text{II})_2$, chapter X, no. 29.

Introducing the reduced characteristic function

$$\mathcal{H}(p_{m+1},\ldots,\,p_n|q) = (H)_{p_r=f_r},$$

and bearing in mind that the partial derivative of H with respect to a generic argument x (whether this is one of the p_i's with index $> m$, a generic q, an a or a c) is

$$\frac{\partial \mathcal{H}}{\partial x} = \frac{\partial H}{\partial x} + \sum_1^m {}_r \frac{\partial H}{\partial p_r}\frac{\partial f_r}{\partial x}, \tag{27}$$

we arrive, by obvious transformations ([28]), at the recognition that the following relationships exist (identically, with respect to all the arguments that appear there)

$$\frac{\partial \mathcal{H}}{\partial q_r} + \{\mathcal{H}, f_r\} = 0 \qquad (r = 1,\, 2, \ldots,\, m). \tag{28}$$

Given all that, let us return to the differential equations

$$\frac{dp_i}{dt} = -\frac{\partial H}{\partial q_i},\quad \frac{dq_i}{dt} = -\frac{\partial H}{\partial p_i}, \qquad (i = 1,\, 2, \ldots,\, n). \tag{1}$$

The first m's of the first group can certainly be ignored, considering them substituted by the m integrals, or indeed even by the (18′) that express in finite terms $p_1,\, p_2, \ldots,\, p_m$, in terms of the other unknowns $p_{m+1}, \ldots,\, p_n;\ q_1, \ldots,\, q_n$ (of the constants c and of the parameters a).

The remaining $2n - m$ equations (1) can be split into two groups, of $2(n-m)$ and of m equations respectively, writing:

$$\begin{cases} dp_i = -\frac{\partial H}{\partial q_i}\,dt, \\ dq_i = \frac{\partial H}{\partial p_i}\,dt \end{cases} \qquad (i = m+1, \ldots,\, n); \tag{29}$$

$$dq_r = \frac{\partial H}{\partial p_r}\,dt \qquad (r = 1,\, 2, \ldots,\, m). \tag{30}$$

If in the second members of eqs. (29) we introduce, instead of the derivatives of H, their expressions (27) by means of the corresponding derivatives of the reduced function $\mathcal{H}$, and account for eqs. (30), we can set eqs. (29) themselves in the equivalent form

$$\begin{cases} dp_i = -\frac{\partial \mathcal{H}}{\partial q_i}\,dt + \sum_1^m {}_r \frac{\partial f_r}{\partial p_i}\,dq_r, \\ dq_i = \frac{\partial \mathcal{H}}{\partial p_i}\,dt - \sum_1^m {}_r \frac{\partial f_r}{\partial p_i}\,dq_r, \end{cases} \qquad (i = m+1, \ldots,\, n). \tag{A_0}$$

([28]) *Ibidem*, no. 30.

This system of equations, associated to eq. (30), form, as already in eqs. (29), (30), an ordinary differential system of rank $2n - m$. But there is the remarkably favourable circumstance that *by themselves eqs.* (A_0) *form a completely integrable system of total differentials in the unknown functions* p_i, q_i $(i > m)$ *of the variables* $t, q_1, q_2, \ldots, q_m$, *considered as independent.* This will be called *associated system* of the original non-primitive system of order m, with respect to its m integrals in involution (18), or, which is the same thing, to eqs. (18′).

The conditions of complete integrability of the system (A_0) are indeed expressed, as could be verified in an obvious way ([29]), from eqs. (26) and (28) which formally translate our hypotheses. This allows us to consider in isolation the system (A_0) in the $2(n - m)$ unknown functions p_i, q_i $(i > m)$, which, being completely integrable, can be re-expressed as a differential system of rank $2(n - m)$.

Once the system (A_0) has been integrated, and hence the functions p_i, $q_i(i > m)$ of $t, q_1, q_2, \ldots, q_m$ have been obtained, ([30]), it is enough to think of the $q_1, q_2, \ldots, q_m$, no longer as independent variables, but as functions of t which satisfy eqs. (30), having thus essentially assigned the general integral of the original canonical system. In order to characterise such functions $q_r(t)(r = 1, 2, \ldots, m)$ in such a way as to verify also eqs. (30), everything is obviously reduced to substituting in the second members of eqs. (30) themselves, instead of the $p_i, q_i, (i > m)$, their expressions as a function of the $t, q_1, q_2, \ldots, q_m$ which result from the integration of the system (A_0), hence the determination of the $q_r(t)$'s comes to depend on an ordinary differential system of rank m, and therefore seems to require an operation of this order. In reality one could still observe that, using Jacobi's method for the integration of the (A_0), the last operation of rank m can be avoided, substituting it with a simple quadrature. But this is of no interest for our current aim, while it is important to note that every integral of the system of total differentials (A_0) (in which the q_r are looked upon as independent variables together with t) is also an integral *a fortiori* for the ordinary differential system consisting altogether of eqs. (A_0) and of eqs. (30), and therefore also for the original canonical system (in which the q_r should be looked on as convenient functions of t). Not only that, but the very fact that it is certainly possible to arrive at the canonical system from (A_0) also holds for any integral invariants of the system (A_0).

For now let us give an example of the first observation, reserving until the following no. the illustration of the second one which is connected, as we will show immediately after, in the most direct and clear way, to the theory of adiabatic invariants.

The integral of the system of total differentials (A_0) that we now wish to point out is

---

([29]) Anyone seeking guidance on the general theory of systems with total differentials can consult our *Lezioni di calcolo differenziale assoluto*, collected by Prof. E. Persico, Roma, Stock, 1925, chapter II.

([30]) Such functions p_i, q_i will also depend on $2(n - m)$ integration constants, and precisely on the initial values (arbitrary, at least within a certain field) p_i^0, q_i^0 that we wish to attribute to the p_i, $q_i(i > m)$, corresponding to also arbitrary initial values $t_0, q_1^0, \ldots, q_m^0$ of the independent variables $t, q_1, \ldots, q_m$.

none other than that of the energy [reduced by means of the other known integrals (18) or (18′)]

$$\mathcal{H} = E. \tag{25'}$$

To recognize that it really is an integral of (A_0) it is necessary, and it is sufficient, to verify that

$$d\mathcal{H} = \sum_{m+1}^{n}{}_i \left(\frac{\partial \mathcal{H}}{\partial p_i} dp_i + \frac{\partial \mathcal{H}}{\partial q_i} dq_i \right) + \sum_{1}^{m}{}_r \frac{\partial \mathcal{H}}{\partial q_r} dq_r,$$

goes to zero for any determination of the differentials dt, dq_r of the independent variables when the expressions (A_0) of the dp_i, dq_i are substituted. That this is so follows with certainty from eqs. (28). In this particular example returning to the ordinary differential system does not provide anything new as indeed it gives again the (reduced) integral of the energy.

12. *The space Ψ with $2(n-m)$ dimensions of the phases p_i, $q_i (i > m)$. The corresponding Euclidean volume as an integral invariant of the system (A_0). Application to the volume W_0 enclosed in Ψ by an $\mathcal{H} = E$.* – Let us consider a generic solution

$$\begin{cases} p_i = p_i(t|q_1, q_2, \ldots, q_m) \\ q_i = q_i(t|q_1, q_2, \ldots, q_m) \end{cases} \qquad (i = m+1, m+2, \ldots, n) \tag{31}$$

of the system of total differentials (A_0), which are completely integrable, and let us represent the determinations $t, q_1, \ldots, q_m$ of the independent variables separately as points P of a space Σ with $m+1$ dimensions, the determinations p_i, q_i of the functions as points M of another space Ψ with $2(n-m)$ dimensions (reduced phase space). The geometrical fact that M is a univocally determined function of P, (as, from the analytical aspect, we have had occasion to point out in the preceding no.), as soon as the position M_0 of M which corresponds to a particular point P_0 has been assigned, is equivalent to eqs. (31). We can write, imagining P_0 as given,

$$M = M(P|M_0). \tag{31'}$$

With these geometrical pictures it becomes easy to characterise in words a typical integral invariant (with $2(n-m)$ dimensions) belonging to any system (A_0). This is what is involved.

Let us fix arbitrarily (within the values in which the system (A_0) behaves regularly) a finite portion C_0 of Ψ, let M_0 be one of its generic points. Let us consider the solutions (31′) defined by the single points M_0 of C_0 (considered as initial, *i.e.* assumed for a prefixed P_0). Let us imagine that we vary P starting from P_0, always remaining within the regularity domain of (A_0).

To each such P a field C of Ψ, the place of the positions of the M points which, initially, occupied C_0, remains subordinated.

I say that the volume W_0 belonging to C, in so much as we attribute a Euclidean metric to the space Ψ, is independent of P. That is to say we have in

$$W_0 = \int_C dp_{m+1} \dots dp_n \, dq_{m+1} \dots dq_n \tag{32}$$

an integral invariant of the system of total differentials (A_0).

To justify this I will begin, to avoid any ambiguity, by substituting ∂ for d in the differentials that are under the sign of the multiple integral, reserving the symbol d for the differentials which refer to the system (A_0), both of the independent variables $t, q_1, q_2, \dots, q_m$, and of the functions p_i, q_i $(i > m)$.

Whatever the initial field C_0, the second member of eq. (32) becomes, once the integration has been carried out, a well determined function of P, *i.e.* of $t, q_1, q_2, \dots, q_m$. We must establish that such a function reduces to a constant, *i.e.* that its differential dW is zero. Using the expression (32) of W_0, which, setting for brevity

$$\partial C = \partial p_{m+1} \dots \partial p_n \, \partial q_{m+1} \dots \partial q_n, \tag{33}$$

can be written as

$$W_0 = \int_C \partial C, \tag{32'}$$

we obtain, by the usual algorithm,

$$dW_0 = \int_C \partial C \sum_{m+1}^{n} {}_i \left(\frac{d\partial p_i}{\partial p_i} + \frac{d\partial q_i}{\partial q_i} \right),$$

where everything behaves formally as if the relations

$$\frac{d\partial p_i}{\partial p_i} = \frac{\partial dp_i}{\partial p_i}, \quad \frac{d\partial q_i}{\partial q_i} = \frac{\partial dq_i}{\partial q_i},$$

had the meaning of the corresponding partial derivatives

$$\frac{\partial(dp_i)}{\partial p_i}, \quad \frac{\partial(dq_i)}{\partial q_i}.$$

The dp_i, dq_i must be thought to have been substituted by their expressions (A_0). Assuming the canonical form (with respect to each of the independent variables) of the differentials dp_i, dq_i, provided by (A_0), every binomial

$$\frac{\partial dp_i}{\partial p_i} + \frac{\partial dq_i}{\partial q_i}$$

cancels out, and with it W_0, *q.e.d.*

Let us now add the qualitative hypothesis that, for a generic determination of the $q_1, q_2, \ldots, q_m$ (I omit t, which as a datum never enters explicitly), the (reduced) isoenergetic surfaces

$$\mathcal{H} = E \tag{25'}$$

be closed in the space Ψ of the $p_{m+1}, \ldots, p_n; q_{m+1} \ldots, q_n$, while still being able in general to vary with $q_1, q_2, \ldots, q_m$.

In conformity, let us indicate by σ one clearly determined surface among these closed surfaces, which must be thought as dependent on $q_1, q_2, \ldots, q_m$, and also, as usual, on E, on c and on a. Since $\mathcal{H} = E$ is an integral of the system (A_0), every point M that initially belongs to σ, in other words for any determination of the $q_1, q_2, \ldots, q_m$, belongs to it for any other determination. Therefore the field C enclosed by σ (which will also vary in general with the $q_1, q_2, \ldots, q_m$) always remains the correspondent, in the sense specified at the beginning of this no., of its initial determination. But the volume of such a field is an integral invariant, therefore we have the important corollary that the (Euclidean) *volume W_0 enclosed, in the space Ψ of p_i, q_i $(i > m)$, by a generic σ of equation*

$$\mathcal{H} = E$$

is (unlike the σ itself) *independent of $q_1, q_2, \ldots, q_m$, and therefore a function only of the integration constants E and c, and also of the parameters a* (which may occur in the characteristic function H of the assigned canonical system).

13. *Wider hypothesis concerning the density of the integral curves. Fundamental properties of W_0 being an adiabatic invariant.* – The essential interest of the preceding considerations lies in the fact that, just as for the $2n$-dimensional volume V for the almost ergodic systems, so for the non-primitive canonical systems (1) of order m, the $2(n-m)$-dimensional volume W_0 is an *adiabatic invariant*.

This is verified immediately, as long as we suppose (see no. 7) that *nearly* all the integral curves of the canonical system are dense:

d) on the manifold τ with $2n - (m+1)$ dimensions defined by the $H = E$, $F_r = c_r$ $(r = 1, 2, \ldots, m)$, or, more symmetrically, by

$$F_r = c_r \qquad (r = 0, 1, 2, \ldots, m); \tag{20}$$

or even only (this is the less restrictive condition mentioned in no. 7)

d') on any of the manifolds σ with $2(n-m)-1$ dimensions $H = E$, of Ψ, which are obtained by attributing, in H, arbitrary determinations (constant, or even functions of t) to the $q_1, q_2, \ldots, q_m$.

Naturally, if *d*) is verified, so is in particular *d'*), but not viceversa.

Anyway, from the fundamental property (preceding no.) of the integral invariant W_0 of being independent of the $q_1, q_2, \ldots, q_m$, we are able, based on *d'*) to repeat identically, for the $2(n-m)$-dimensional volume W_0 in Ψ the reasoning developed in no. 4 about the $2n$-dimensional volume V in Φ_{2n}.

And so the announced adiabatic invariance of W_0 is proven.

14. *The various associated systems* (A_α) $(\alpha = 0, 1, \ldots, m)$. *Common properties. Corollaries.* – For the non-primitive canonical system (1) of order m we find the m conditions

$$(H, F_r) = 0 \qquad (r = 0, 1, 2, \ldots, m)$$

in that we suppose that the $F_r = c_r$ are as many integrals, and moreover we find the $\frac{m(m-1)}{2}$ equations

$$(F_r, F_s) = 0 \qquad (r, s = 1, 2, \ldots, m)$$

due to the hypothesis that the F themselves are in involution. With the definitions (19) of no. 7 ($F_0 = H$, $c_0 = E$) the two groups combine in the single scheme

$$(F_r, F_s) = 0 \qquad (r, s = 0, 1, \ldots, m), \tag{33}$$

in which all the F's behave in the same way.

We have arrived at this by expressing the necessary and sufficient conditions for a canonical system with characteristic function F_0 (independent of t) to admit the m integrals in involution (also independent of t) corresponding to the remaining F. But, assuming complete symmetry, we can attribute the role of F_0 to another generic F, let us say for example

$$F_\alpha,$$

and state that every non-primitive system of order m can be linked to other m systems, which have $F_1; F_2; \ldots; F_m$ respectively as characteristic function and, each time, the remaining F as integrals in involution. For each of these ordinary differential systems we have (under qualitative specifications of regularity, solvability, etc.) an associated system (A_α) of total differentials, constructed according to the criterion of no. 11. Accounting for the different construction algorithm, the various (A_α) in general turn out to be different from one another. The difference however is not so profound as might be thought *a priori*: *all these associated systems* (A_α) *admit —as we will show in a moment— the same integrals* $2(n-m)-1$ *independent of* t. In other words the $2(n-m)$ expressions (31) of the $p_i\, q_i$ $(i > m)$ extracted from them by integration, if they do not actually coincide, give rise, when t is eliminated, to the same $2(n-m)-1$ consequences.

To establish such a property we will consider together the aforementioned $2(n-m)-1$ integrals independent of t of a generic system (A_α), the (A_0) of no. 11 to clarify our ideas, and the m equations

$$p_r = f_r \tag{18'}$$

which define, we can say, the $p_1, p_2, \dots, p_m$, or which is the same, the equivalent equations

$$(18) \qquad F_r = c_r \qquad (r = 0, 1, \dots, m),$$

showing that this set of $2n - (m+1)$ integrals coincides with the one defined by the Jacobian system

$$(34) \qquad (F_r, F) = 0 \qquad (r = 0, 1, \dots, m).$$

This system, symmetrical with respect to all the $m+1$ F, possesses precisely $2n-(m+1)$ integrals F independent of each other.

To this aim let us start from the following general observation: let $F_1, F_2, \dots, F_m$; $G_1, G_2, \dots, G_\mu$ be two groups of functions independent of the $2n$ conjugated variables p_i, q_i $(i = 1, 2, \dots, n)$, all in involution between themselves, which is formally expressed by

$$(35) \qquad (F_r, F_s) = 0 \qquad (r, s = 1, 2, \dots, m);$$

$$(36) \qquad (F_r, G_j) = 0 \qquad (r = 1, 2 \dots, m;\ j = 1, 2, \dots, \mu);$$

$$(37) \qquad (G_j, G_l) = 0 \qquad (j, l = 1, 2, \dots, \mu).$$

Let us suppose that (c_r denoting constants) the m equations

$$F_r = c_r$$

are solvable with respect to the same number of p's —the first m's— under the form

$$(18') \qquad p_r = f_r \qquad (r = 1, 2, \dots, m).$$

By means of a well known lemma, already mentioned in no. 11, see ref. ([27]), eqs. (35) imply

$$(35') \qquad (p_r - f_r, p_s - f_s) = 0 \qquad (r, s = 1, 2, \dots, m),$$

and (36)

$$(36') \qquad (p_r - f_r, G_j) = 0 \quad (r = 1, 2, \dots, m;\ j = 1, 2, \dots, \mu),$$

the first ones being identically satisfied (after the expression in brackets has been calculated), and the second ones with the understanding that any surviving p_r $(r = 1, 2, \dots, m)$ be substituted by f_r. That given, let us call $\mathcal{G}$ what a generic G becomes after being reduced by means of (18) [or (18')], *i.e.* let us set

$$\mathcal{G}_j = (G_j)_{p_r = f_r} \qquad (j = 1, 2, \dots, \mu).$$

If x denotes any one of the arguments $p_{m+1}, \ldots, p_n; q_1 q_2, \ldots, q_n$, we have [as for $\mathcal{H}$ in no. 11; see eq. (27)]

$$\frac{\partial \mathcal{G}_j}{\partial x} = \frac{\partial G_j}{\partial x} + \sum_1^m {}_s \frac{\partial G_j}{\partial p_s} \frac{\partial f_s}{\partial x}$$

which can be written, as long as x is different from one of the p_s,

$$\frac{\partial G_j}{\partial x} = \frac{\partial \mathcal{G}_j}{\partial x} + \sum_1^m {}_s \frac{\partial G_j}{\partial p_s} \frac{\partial (p_s - f_s)}{\partial x}. \tag{38}$$

But this also applies to x coincident with one of the $p_1, p_2, \ldots, p_m$, *e.g.*, p_r, since in that case it is reduced to the identity

$$\frac{\partial G_j}{\partial p_r} = \frac{\partial G_j}{\partial p_r}.$$

With the expressions (38) of the derivatives of a G_j (with respect to any one of the $2n$ arguments p_i, q_i), eqs. (36′), bearing in mind (35′), become

$$(p_r - f_r, \mathcal{G}) = 0 \tag{36''}$$

with the usual understanding with respect to the surviving $p_1, p_2, \ldots, p_m$. But, expanding the bracket, we obtain the analogous equation of (28) for $\mathcal{H}$, *i.e.*

$$\frac{\partial \mathcal{G}_j}{\partial q_r} + \{\mathcal{G}_j, f_r\} = 0 \quad (r = 1, 2, \ldots, m; j = 1, 2, \ldots \mu), \tag{36'''}$$

and here we are still dealing with identities, since none of the $p_1, p_2, \ldots, p_m$ appears any longer.

Using again (38) and bearing in mind (36″) and (35′), eqs. (37) assume the reduced form

$$(\mathcal{G}_j, \mathcal{G}_l) = 0 \qquad (j, l = 1, 2, \ldots, \mu), \tag{37'}$$

or, if preferred, since the $p_1, p_2, \ldots, p_m$ are absent,

$$\{\mathcal{G}_j, \mathcal{G}_l\} = 0 \qquad (j, l = 1, 2, \ldots, \mu). \tag{37''}$$

On the basis of these formal equivalences it is now rather easy to ascertain that, for every solution F of (34), the relationship F = const., by means of a reduction using (18′), in other words

$$\mathcal{F} = (F)_{p_r = f_r} = \text{const.},$$

is actually an integral (independent of t) of the system (A_0) linked to F_0 according to the construction of no. 11.

For the other (A_α)'s the same proof will naturally hold, except for a cyclic substitution on the indices $0, 1, \ldots, m$ of the F's, with all its consequences.

Here is the demonstration.

Equations (34), associated with (20), altogether express that the $m+2$ functions

$$F_1, F_2 \ldots, F_m; \quad F_0 = H, F$$

are in involution between themselves.

Treating the last two in the same way as done with G just above, we can certainly state that, for $\mathcal{F}$, the equations corresponding to (36‴) and (37″) will equally apply, *i.e.*

$$\frac{\partial \mathcal{F}}{\partial q_r} + \{\mathcal{F}, f_r\} = 0 \qquad (r = 1, 2, \ldots, m),$$
$$\{\mathcal{H}, \mathcal{F}\} = 0.$$

These are exactly the conditions for $\mathcal{F} =$ const. to be the integral of the system (A_0) of no. 11,

q.e.d.

It is therefore proved that the various auxiliary systems of total differentials (A_α) $(\alpha = 0, 1, \ldots, m)$, associated with the equations (that reduce to eqs. (18) for $\alpha = 0$),

$$F_0 = c_0, \ldots, F_{\alpha-1} = c_{\alpha-1}, F_{\alpha+1} = c_{\alpha+1}, \ldots, F_m = c_m \tag{39}$$

admit the same integrals, independent of t, $F =$ const.; and precisely all and only all those defined by the Jacobian system (34). Now in any case, this includes the equation

$$(F_0, F) = (H, F) = 0.$$

Thus such integrals all belong also to the original canonical system.

Besides, even without formulae, we can reach the same conclusion combining the fundamental property of the (A_α) systems, just established, with an observation of no. 11 concerning the system (A_0). Indeed on the one hand the various systems of total differentials (A_α) $(\alpha = 0, 1, \ldots, m)$ admit all the same integrals independent of t; on the other hand, as found in no. 11, every integral of (A_0) belongs in particular to the original canonical system. As a consequence, the same can be said for any integral independent of t of a generic (A_α), as long as it too belongs to (A_0).

This way of reasoning is susceptible to an important extension because, from the fact that the various systems (A_α) admit the same integrals independent of t, it follows that they share every other property equally independent of t; in particular, every integral (single and multiple) which is invariant for one of them is invariant for all the others. Here too, since at no. 12 it was noted that every integral invariant for (A_0) is also invariant *a fortiori* for the original canonical system, we conclude that *every invariant integral, in particular if adiabatic, of any one of the systems of total differentials* (A_α) *is also invariant for the assigned canonical system.*

15. *Existence for every non-primitive system of order* m *of* $m+1$ *adiabatic invariants.* – In no. 12 [formula (32)] an adiabatic invariant W_0 of the system (A_0) has been defined.

For every other (A_α) we can (if the due qualitative circumstances are satisfied) construct analogously an adiabatic invariant W_α. Because of the special way in which the function (F_α) intervenes in such a construction, we have different results each time, at least in general; hence the theorem:

A canonical non-primitive system of order m, allows in general $m+1$ adiabatic invariants, each of which presents itself, as specified in no. 12 for W_0, as a volume with $2(n-m)$ dimensions of a certain field, characterized (in the space Ψ of the $p_{m+1}, \ldots, p_n; q_{m+1} \ldots, q_n$, treating the $q_1, q_2, \ldots, q_m$ as parameters) by the $m+1$ known integrals

$$F_r = c_r \qquad (r = 0, 1, \ldots, m).$$

16. *Special cases. Burgers' theorem as an immediate corollary of the preceding result.* – In Liouville's classic case in which the order of non-primitivity is $m = n-1$, a total of n integrals independent of t and in involution are known; and the integration of the canonical system can be traced back to quadrature ([31]). *Such a system allows in general* (we mean under the conditions of regularity, independence, etc., in their turn specified, and to be ascertained case by case) n *adiabatic invariants.*

Naturally this is also valid for the special Stäckel type in which the integration can be carried out by separating the variables. Indeed this type is one of the non-primitives of order $n-1$, as it has, as well as the integral of the kinetic energy, other $n-1$ quadratic integrals in the p's, all in involution amongst themselves ([32]).

At least for this case let us develop the explicit calculation of the n's adiabatic invariants, according to the general theory above.

For convenience of notation as will appear in a moment, we will attribute to the $2n$ conjugate variables p_i, q_i the indices $0, 1, \ldots, n-1$, instead of $1, 2, \ldots, n$: in other words we will denote by p_0, q_0 the conjugate pair denoted so far by p_n, q_n. With this understanding, we can assume, as characteristic of the Stäckel type, the following expressions of the F_r's:

$$F_r = \sum_0^{n-1}{}_h \phi_{rh} \left(\frac{1}{2} p_h^2 - U_h \right) \qquad (r = 0, 1, \ldots, n-1),$$

([31]) Loc. cit. ([17]) vol. (II), chapter X, nos. 44, 45.
([32]) *Ibidem*, no. 64.

where every U_h is a function only of the variable q_h, and the ϕ^{rh} (with two indices that can both vary between the same limits, thanks to the convention we have just adopted) are reciprocal elements ([33]) derived from n functions

$$\phi_{rh}(q_h) \qquad (r, h = 0, 1, \ldots, n-1)$$

each of which depends only on the argument indicated.

Let us proceed to the construction of W_0 according to the rule of nos. 11 and 12. We must imagine that we have solved the equations

$$F_r = c_r \qquad (r = 1, 2, \ldots, n-1) \tag{18}$$

with respect to $p_1, p_2, \ldots, p_{n-1}$, and bring the values that are obtained in this way into F_0, which, thus reduced, is denoted by $\mathcal{F}_0$ and comes to correspond to the H of no. 11.

Similarly, the equation $\mathcal{F}_0 = c_0$ that here corresponds to (25′) appears as the result of the elimination of $p_1, p_2, \ldots, p_{n-1}$ amongst all the n equations

$$F_r = c_r \qquad (r = 0, 1, \ldots, n-1), \tag{20$_s$}$$

where the F_r have Stäckel's explicit form (40).

Therefore the equation mentioned above (which is of greatest interest to us, since the adiabatic invariant W_0 must be deduced from it) is necessarily equivalent to the result of the elimination of $p_1, p_2, \ldots, p_{n-1}$ amongst the n equations $(20)_s$; or again to the expression of p_0 (in terms of the q and of the c) extracted from the resolution of eqs. $(20)_s$ themselves. Using the reciprocal elements ϕ_{r0} of the coefficients of $\frac{1}{2}p_0^2$ in the various equations $(20)_s$ we immediately have the resolvent

$$\frac{1}{2}p_0^2 = U_0 + \sum_0^{n-1}{}_s c_r \phi_{r0} \tag{41}$$

which is equivalent to $\mathcal{F}_0 = c_0$ and would be reduced in fact to such a form dividing by ϕ_{00} and isolating c_0.

While in general the $\mathcal{H}$ of no. 11 might also depend on the $q_1, q_2, \ldots, q_m$, here there is the special circumstance that in $\mathcal{F}_0$ only q_0 appears. The space Ψ is now the plane p_0, q_0, and the invariant W_0 is thus reduced to the area of this plane enclosed by a curve (41).

Here we find again Sommerfeld's loop integral

$$J_0 = \oint p_0 \, dq_0 \tag{42}$$

as an adiabatic invariant. A cyclic substitution of the indices is obviously sufficient to find the $n-1$ other J_α $(\alpha = 1, 2, \ldots, n-1)$.

([33]) *I.e.* algebraic complements divided by the value of the determinant.

As qualitative conditions everything clearly reduces to:

1°) the regularity of the functions (40) and the solvability of $(20)_s$ in the field of the values under consideration; which, starting from $\phi_{rh}(q_h)$, assumed to be regular, is ensured by the non-vanishing of their determinant $||\phi_{rh}||$;

2°) the n curves analogous to (41) being closed in their respective plane p_i, q_i.

As you can see, the difficulties and the complementary discussions that seemed necessary in certain cases of commensurability, according to the ordinary demonstrations, do not occur at all on the basis of our general theory.

17. *Indication of further research.* - It not possible for us to linger now on other examples relating to dynamic problems of smaller non-primitivity ($m < n-1$), which in a certain sense appear even more interesting, in that they are not integrable by quadrature; and we must also restrict ourselves to the simple statement that some of the previous results can be extended to differential systems of any form.

About the Author

TULLIO LEVI-CIVITA (1873-1941) was a mathematical physicist and author, together with Gregorio Ricci Curbastro, of the "Calcolo differenziale assoluto", an essential mathematical instrument in the development of General Relativity. Levi-Civita tackled, and solved, many important problems of analytical mechanics including the problem of adiabatic invariance. A complete edition of his collected works has been published by the Unione Matematica Italiana.

Fermi's coordinates and the Principle of Equivalence

BRUNO BERTOTTI

A few months before his "Laurea" in Physics at the Scuola Normale Superiore of Pisa in July 1922, Fermi, then 21 years old, submitted to the "Rendiconti of the Accademia dei Lincei" a note entitled "Sopra i fenomeni che avvengono in vicinanza di una linea oraria" (On the phenomena that occur in the neighbourhood of a time line); it was subsequently published in three parts at pages 21, 51 and 103 of volume 31 (1922)(1). This note acquired, and still retains, an extraordinary importance for the physics of gravitation, not only with respect to its foundations, in particular the Principle of Equivalence, but also in many practical applications. It reveals great technical ability in mathematical physics, in particular in the geometric representation of curved hypersurfaces (*i.e.* the generalization of ordinary curved surfaces to a number of dimensions greater than two; to be precise, the Riemannian manifolds). Fermi does not treat the physical and epistemological meaning of his work at length, although in the first paragraph he does demonstrate a full comprehension of its essence. In 1922 Einstein's theory of General Relativity, within whose frame of reference Fermi's memoir was written, was just becoming known, the subject of controversy and doubt, and experimental tests were lacking. Of course Fermi had thoroughly studied the works by T. Levi-Civita on Riemannian manifolds, which provided the mathematical basis of the theory; and certainly he knew (and quoted) the treatise by Hermann Weyl "Raum, Zeit, Materie" (Space, Time and Matter) published by Springer in 1921, an introduction to the theory of relativity of extraordinary physical

(1) A previous note is linked to this, "Sull'elettrodinamica di un campo gravitazionale uniforme e sul peso delle masse elettromagnetiche" (On the electrodynamics of a uniform gravitational field and on the weight of electromagnetic masses), which appeared in vol. **22** (1921) of "Nuovo Cimento", at pages 176-188; in a subsequent note, "Sul peso dei corpi elastici" (On the weight of elastic bodies), Rend. Acc. Lincei, **14** (1923) 114-124, he studies the influence of elasticity on weight.

and mathematical clarity. But just as certainly his remarkable creativity enabled him to venture alone into unknown regions and make relevant and original contributions.

1. – The Principle of Equivalence

In order to properly understand Fermi's work, in its cultural as well as scientific context, we must briefly introduce the new concept of gravity that A. Einstein presented in 1916 with the theory of General Relativity.

The question: "Why does a body have weight?" is usually answered: "Because it is attracted by the earth". Nowadays we know that this reply is false and misleading, and requires, as do all radical changes in our thinking, a reflective and critical pause. Suppose we have a spring balance (a dynamometer) in a laboratory inside a rocket that moves in a straight line in empty space, very far from the earth, and suppose that its engines produce an acceleration g directed toward the nose; what is the result of a dynamometric measurement carried out on a body with mass m? As happens for a train that is departing and is therefore subject to an acceleration, the bodies inside it are pushed in the direction opposite to the motion. In the rocket the balance measures a *weight* of the body with value mg , identical to that measured on the surface of the earth when the acceleration of gravity is g; both quantities are proportional to the mass. But then, if the observer in the rocket's laboratory knows nothing about the dynamical condition of the rocket, or where the earth is, this measurement is perfectly equivalent to that on the surface of the earth, when the acceleration of gravity is also g. We must conclude that *acceleration creates weight.* Again, let's imagine a small laboratory in free fall toward the centre of the earth within a vertical well; a dynamometer inside it will measure a weight exactly equal to zero, precisely what happens within a rocket at rest in empty space. The acceleration of the laboratory compensates exactly the acceleration of gravity. We must therefore conclude that the weight of a body is determined by the state of motion of the laboratory in which it is measured (fig. 1). Among all possible states of motion of a laboratory, there is one class —like straight, uniform motions in interplanetary space, far from the earth, for instance, and free fall within a well— for which the weight is zero. The weight measured inside a laboratory is a consequence of its *unnatural* state of motion, whether this be due to the rocket's engine, or to the solid platform which prevents an object on the earth's surface from falling below it. The states of motion of a laboratory in free fall determine the privileged reference systems —the inertial systems— for which the principle of inertia holds: a body inside not acted upon by forces continues indefinitely in its rectilinear and uniform motion. With respect to an accelerated frame of reference the motion appears to be subject to a force (correctly called "apparent"), which, however, is indistinguishable from that normally associated to a gravitating body in the vicinity. The Principle of Equivalence, which A. Einstein made the foundation of the theory of General Relativity, states that apparent and gravitational forces at one point are indistinguishable.

We cannot fail to mention here the extraordinary influence that the phenomenon of weight has exerted on biology and culture. Obviously biological evolution, especially the transition to an erect position from the apes to man is fundamentally conditioned by

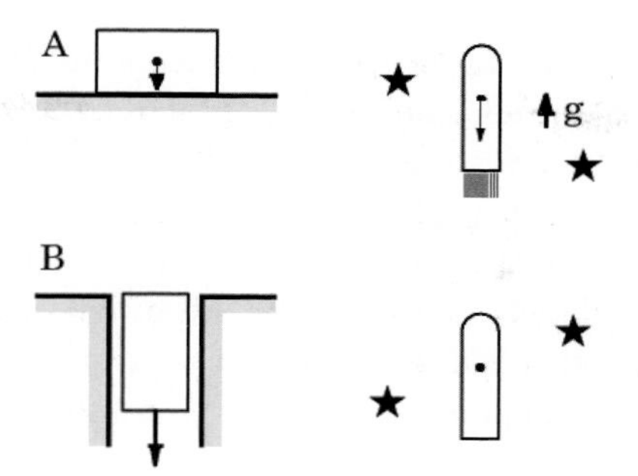

Fig. 1. – The Principle of Equivalence. It is impossible to distinguish the situations on the right from those on the left: (A) the weight in a laboratory on the surface of the earth is indistinguishable from that inside an accelerated rocket; (B) the same lack of weight occurs in a laboratory in free fall in a well, and inside a rocket in empty space, not rotating and with no running engines.

weight, and would have been totally different, for instance, on a small asteroid or inside a large spaceship with its engines off; with much smaller or even zero weights, culture and evolution would have been completely different. To raise a 70 kilograms object to the height of a meter, 700 joules of energy are needed; since the human body produces about one kW of power, it can fall and get up again, say, once a second. The tension linked to the erect position and the associated psychological mechanisms thus depend in an essential way on the value of gravity acceleration, around 981 g/cm^2 on earth. The weight determines a privileged direction; up and down are deeply, unconsciously and indelibly inscribed in the human mind. In the unconscious, the fear of death is often fear of falling: in the Iliad a common attribute of death is *aipús*, "steep". Religions place the good gods in the heavens, and relegate the evil ones to the underworld; and grace descends from above for our salvation. In the second and third millennium before the Christian era, neolithic civilizations of western Europe erected thousands of gigantic megaliths, perhaps to reassert their supremacy and victory over the slavery of weight. Architectural beauty is based on the contrast between the height of solid structures and the weight they are able to bear with often-surprising stability. Large, beautiful structures such as towers, bridges and columns, created by modern engineering, are the direct result of the need and desire to keep gravity under control. In philosophy, there is a complete inversion between Aristotelian cosmology, in which the *natural* state of motion of bodies is their fall toward the centre, to which their very essence aspires, and the Principle of Equivalence, for which *weight is an illusion, an artefact* caused by the wrong and unnatural reference system. Aristotelian cosmology is based on this error: it requires that the earth be at the centre — the lower world— and that it be gloriously crowned externally by its opposite, the divine empyrean. The contrast couldn't be greater, and it reminds us of the precariousness and impermanence of so many of our philosophical and moral concepts.

On the basis of the Principle of Equivalence, and without resorting to any other experimental observation, Einstein built the extraordinary physical and mathematical structure of General Relativity. From the practical point of view, this appears to be a theory of gravitation conceptually and wholly different from Newton's classic theory —two bodies attract each other with a force that is inversely proportional to the square of their distance. But under ordinary circumstances they have practically the same consequences, and it is difficult to distinguish between them experimentally. Extensive and costly programs have allowed us to study the differences between the observable consequences of the two theories and to conclude that the Newtonian model is incorrect; General Relativity at present has no significant and well-defined competitors.

2. – The geometrical nature of gravitation

The Principle of Equivalence states that motion in free fall cannot be distinguished *locally* from a direct and uniform inertial motion. Pay attention to the word "locally", which is crucial in what follows (see sect. **5**); obviously, the two types of motion are different if observed over wide regions of space and time. In an inertial system and in the absence of gravitation, the motion of bodies is controlled by the geometrical concept of a straight line, and therefore by Euclidean geometry; neither dynamical concepts nor mass are necessary. For consistency's sake, even in the presence of gravitation one must adopt a similar description of motion; this can be accomplished by generalizing the geometry and the concept of straight line.

In the last century geometricians were already aware that the well-known laws of Euclidean geometry (for instance, the sum of the internal angles of a triangle equals $180°$) allow for an obvious and powerful generalization, as suggested, in two dimensions, by the geometry of a curved surface (such as the sphere). Just as in Euclidean space a straight line segment represents the shortest path between two points, the same definition applies in general; on a sphere, for instance, the "line" segments are arcs of a great circle, intersections between the sphere and the planes passing through the origin. The routes of intercontinental flights follow such arcs, thus minimizing the covered distance. The "straight lines" so defined in a generic curved space are called *geodetic lines*. In this way we obtain a different geometry from the Euclidean one: for example, on a sphere the sum of the internal angles of a triangle constructed with the arcs of great circles is always *larger* than $180°$. In non-Euclidean geometries the structure of space depends on new and complex variables; in order to determine the geometry of a two-dimensional curved surface, for instance, we must associate the *radius of curvature* (which is constant on a sphere) to each point. Moreover, the correct framework for describing the motion is not space alone, but the combination of space and time, a geometrical entity in four dimensions: *spacetime*.

Eintein's great discovery was that trajectories of bodies (light enough to disregard their gravity) in the presence of other gravitating bodies can be described as *geodetic lines* in a properly chosen *curved spacetime*; in this sense, one can say that the gravitational motion is still the natural and geometric one, corresponding to straight and uniform motion in a non-curved spacetime. The variety of gravitational motions is attributed to the curvature, which characterizes the non-Euclidean nature of spacetime. The nature of gravitation is geometrical, not dynamical.

The geometrical structure of a surface or of a curved spacetime is completely determined when the distance (usually indicated by the symbol $\mathrm{d}s$) between any two points (or events) close to each other is given; in technical language, the *metric*. Other properties and geometrical quantities derive from it: parallelism, volumes, angles, etc. When two events are located on the trajectory of a material point, their distance is proportional to the elapsed time. General Relativity makes an explicit construction possible for the metric and the geometry of the spacetime in the presence of gravitation, particularly in the Solar System.

If the spacetime is not curved, there is a class of reference systems that move with respect to each other in straight and uniform motion —the inertial systems— for which the principle of inertia holds and all "weight" is eliminated. In the presence of gravitation, however, this is possible only locally; it is therefore interesting to construct local inertial systems in which the weight is, so to speak, eliminated as much as possible.

On a generic curved surface it is impossible to construct a global system of Cartesian coordinates; for instance, the meridians of a sphere intersect the equator perpendicularly, but all join at the poles. How can one build, in the vicinity of a point P_0 of a curved surface, a system of coordinates which best approximates a Cartesian system (in which the lines parallel to the axes intersect each other perpendicularly)? The answer is simple: the points P of the surface are projected onto the plane π tangent in P_0 and the Cartesian coordinates of the projection, originating in P_0, are taken as coordinates of P. If the neighbourhood is sufficiently small, since the geometrical properties of the tangent plane coincide locally with that of the surface, the geometry obtained in this way is Euclidean, and the distance between two nearby points $P(x, y)$ and $P'(x + \mathrm{d}x, y + \mathrm{d}y)$ is given by Pythagoras' theorem:

$$\mathrm{d}s^2 = \mathrm{d}x^2 + \mathrm{d}y^2. \tag{1}$$

However, if we want a meaningful result, we must know how, in these coordinates, the expression for the distance is affected by the curvature of the surface. Let us consider a point P on a sphere of radius R at a distance r from the pole P_0 much smaller than R, and another point P' very close to P. Let (x, y) and $(x + \mathrm{d}x, y + \mathrm{d}y)$ be the Cartesian coordinates of their projections on the plane π tangent at the pole; it is shown that the distance $\mathrm{d}s$ between P and P' is given by

$$\mathrm{d}s^2 = \left(1 + \frac{x^2}{R^2}\right)\mathrm{d}x^2 + \left(1 + \frac{x^2}{R^2}\right)\mathrm{d}y^2 + \frac{2xy}{R^2}\mathrm{d}x\,\mathrm{d}y \ (^2). \tag{2}$$

---

(2) For the skilful reader who wishes to obtain this formula: $\mathrm{d}s^2$ is larger than the square of the Euclidean distance $\mathrm{d}x^2 + \mathrm{d}y^2$ because the arc (or the segment, which is the same in this approximation) PP' in general has a component $\mathrm{d}z = R\mathrm{d}(\cos\theta) = -R\sin\theta\mathrm{d}\theta = -R\theta\mathrm{d}\theta$ orthogonal to the plane π. When $r/R = \theta \ll 1$ its square is

$$\frac{r^2}{R^2}(\mathrm{d}r)^2 = \frac{1}{R^2}(x\mathrm{d}x + y\mathrm{d}y)^2,$$

which leads to (2).

One can see that in this expression the coefficients differ from those for the Cartesian distance (1) by terms of order $(r/R)^2$. This result, in order of magnitude, holds for an arbitrary number of dimensions: the distance between P and P' differs from the distance (Euclidean or pseudo-Euclidean) on π by terms of order Kr^2; here K indicates the order of magnitude of quantities of the type $1/R^2$, which characterize the curvature in P_0. The coordinates constructed in this way are called Riemann's or *normal coordinates*.

For spacetime the geometry is more complex, since Pythagoras' theorem (1) has a different expression even when the curvature is zero: the particular nature of time appears through the use of minus signs in the expression of the "pseudo-Euclidean metric" (taking the velocity of light as equal to one)

$$\mathrm{d}s^2 = \mathrm{d}t^2 - \mathrm{d}x^2 - \mathrm{d}y^2 - \mathrm{d}z^2. \tag{3}$$

The squares of "lengths" can even be negative and the lengths themselves imaginary! In this case, too, the projection on a hyperplane (with four dimensions!) tangent at a point P_0 allows the metric of a curved spacetime to be approximated with the preceding pseudo-Euclidean expression and the corrections due to curvature K of order $(rK)^2$ to be estimated.

3. – Fermi's coordinates. Geometry

The technical description below is not strictly necessary for understanding what follows.

In order to define Fermi's coordinates within a laboratory in arbitrary motion one must first call to mind the concept of *world line*. In spacetime an event is defined by four numbers (t, x, y, z), of which t indicates the instant of time and (x, y, z) the location where the event takes place. To describe the motion of a material point, the trajectory $(x(t), y(t), z(t))$ is assigned as a function of time; the set of events $(t, x(t), y(t), z(t))$ is the world line. For instance, if a point is at rest at its origin, the world line is $(t, 0, 0, 0)$. If spacetime is flat and Euclidean geometry holds, the coordinates are conveniently defined with a system of Cartesian axes. For an event P_0 —the origin— one builds a set of four unitary and orthogonal vectors $(\mathbf{e}_t, \mathbf{e}_x, \mathbf{e}_y, \mathbf{e}_z)$ —the base; the coordinates of an event P are the projections of the vector P_0P over the base vectors:

$$t = \mathbf{e}_t \cdot P_0P, \quad x = \mathbf{e}_x \cdot P_0P, \quad y = \mathbf{e}_y \cdot P_0P, \quad z = \mathbf{e}_z \cdot P_0P.$$

(In this section a boldface letter indicates a spacetime vector.)

Let L_0 be a point inside the laboratory (for instance its centre of mass) and t the time indicated by a clock on it. At every event of its world line we build an orthogonal base of reference in the following manner. Let $(\mathbf{e}_t^0, \mathbf{e}_x^0, \mathbf{e}_y^0, \mathbf{e}_z^0)$ be a base for the event at the origin of time $t = 0$, chosen so that $\mathbf{e}_t^0$ be tangent to the direction $\mathbf{u}_0$ of the world line and, therefore, the space vectors $(\mathbf{e}_x^0, \mathbf{e}_y^0, \mathbf{e}_z^0)$ lie in the plane π_0 orthogonal to $\mathbf{u}_0$. To obtain the base for a nearby event at the time $\mathrm{d}t$, we first parallely transfer the base at the origin $t = 0$; for simplicity's sake in fig. 2 it is indicated with the same symbols. If

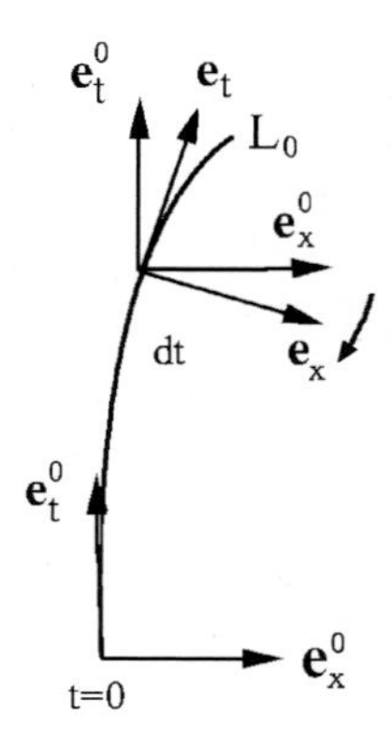

Fig. 2. – Fermi's base associated to the world line L_0 of a point in accelerated motion. A base $(\mathbf{e}_t^0, \mathbf{e}_x^0, \mathbf{e}_y^0, \mathbf{e}_z^0)$ at the origin of time, with $\mathbf{e}_t^0$ tangent to L_0, is tranferred to a nearby event and rotated in the plane $(\mathbf{e}_t^0, \mathbf{u})$ in such a way that the time vector is parallel to the tangent vector $\mathbf{u}$. The figure illustrates the case in which the vector $\mathbf{e}_x^0$ lies in this plane. The construction is repeated for all instants of time. (For the sophisticated and expert reader it should be added that this representation holds only when the metric is defined positive and the geometry is Euclidean. For spacetime the metric can be positive or negative; pseudo-Euclidean geometry holds and one must use a different representation.)

the point L_0 is not accelerated, the vector $\mathbf{e}_t^0$ is still tangent to the world line L_0. To formalize the Principle of Equivalence, however, we need to consider the general case in which the laboratory, and hence L_0 as well, is accelerated; then $\mathbf{e}_t^0$ is no longer parallel to the vector $\mathbf{u}$ tangent to L_0 at $\mathrm{d}t$. At this event, a plane is then defined by the two vectors $\mathbf{u}, \mathbf{e}_t^0$. In this plane we perform the infinitesimal rotation of the whole base in $\mathrm{d}t$ which brings $\mathbf{e}_t^0$ into the new position tangent to $\mathbf{u}$. Figure 2 shows the above-mentioned plane containing $\mathbf{e}_x^0$; in this case the rotation leaves unchanged the other two space vectors of the base $\mathbf{e}_y^0, \mathbf{e}_z^0$. In general, however, the whole base at $\mathrm{d}t$ is affected by the rotation and carried into a new position $(\mathbf{e}_t, \mathbf{e}_x, \mathbf{e}_y, \mathbf{e}_z)$ maintaining, of course, its orthogonality and its unitary character. By repeating this construction for each successive instant we obtain Fermi's base for each event P_t of L_0. This base is composed of a vector $\mathbf{e}_t$ tangent to it and a triad $(\mathbf{e}_x, \mathbf{e}_y, \mathbf{e}_z)$ in the three-dimensional orthogonal space π_t.

In order to construct Fermi's coordinates, let us suppose first of all that no gravitating body is present nearby and that spacetime is flat. Given an event P, let π_t be the space orthogonal to L_0, which contains it; the time t thus selected is the time coordinate of P. The space coordinates of P are the projections of PP_t over the reference triad in π_t (fig. 3):

$$x = \mathbf{e}_x \cdot PP_t, \quad y = \mathbf{e}_y \cdot PP_t, \quad z = \mathbf{e}_z \cdot PP_t.$$

If the event P lies on a world line L, its coordinates $(x(t), y(t), z(t))$ define the motion of L with respect to L_0.

As Fermi shows, extending this construction to the case in which the spacetime is curved is very simple and is based on immersion in a flat space with a higher number of

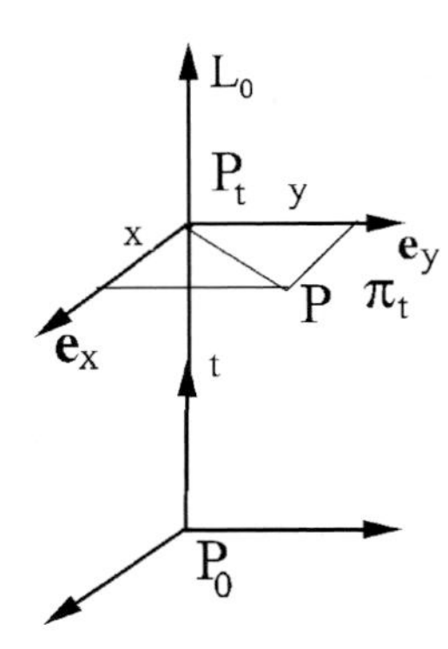

Fig. 3. – Fermi's coordinates relative to a world line L_0. The plane π_t for an event P cuts L_0 orthogonally at an event P_t, at a distance t from the origin P_0; t is the time coordinate of P. The space coordinates (x, y, z) are the projections of the segment PP_0 over the three spatial axes in P_t.

dimensions. As we shall see further on, however, this construction is valid only when the distance r is not too large, that is, within a tubular neighbourhood with L_0 as the axis.

4. – Fermi's coordinates. Physics

Before investigating more thoroughly the physical meaning of Fermi's coordinates, three observations should be made. First, let us consider the geometrical character of the construction. From the beginning, those studying relativity had two different techniques available: algebra and differential calculus based on specific coordinates and the explicit use of vectorial and tensorial components, often with complicated indexes; and the geometrical method, which uses only concepts of segment, angle, volume, etc. Without doubt, the second, introduced by H. Minkowski in 1908, if workable, is more powerful and appropriate for an essentially geometrical theory like General Relativity; the first, however, is simpler and closer to the style of reasoning of physicists, who are accustomed to using physical quantities measured in well-defined frames of reference. This duality is still present, and it must be said that the first method, in general conceptually easier and often the only one available, has many more followers. It has, however, led to numerous errors and false paths; the most conspicuous example is that of the Soviet physicist V. A. Fock, who in his book "The Theory of Space, Time and Gravitation" (published in English in 1959) reached the absurdity of attributing a physical meaning to particular coordinates defined only mathematically. Interestingly enough, Fermi, whose scientific background was essentially physical, promptly and decisively chose the geometrical route.

Second, Fermi's treatment holds for arbitrary values of velocity, even close to the velocity of light, and implicitly contains the relativistic effects of time dilation and length contraction. Essentially it defines, for each event of L_0, a "local" Lorentz transformation which varies from instant to instant; and all this in a geometrical fashion, without using muddled matrices.

Last of all, more thoughtful readers might wonder what role the laboratory's orientation around its centre of mass plays in this construction. Parallel transport, in fact, is defined in a purely geometrical way and no arbitrariness is allowed for; but in practice, how can Fermi's coordinates be obtained within a rotating laboratory? For instance, coordinates that refer to a laboratory located on the earth's (rotating) surface are certainly not appropriate. Fermi does not mention this problem and takes for granted that the geometry itself, through the parallelism in spacetime, defines the absence of rotation. An absolute element is thus surreptitiously introduced, which seems to be *a priori* independent of measurements and material objects, which must of necessity be used in order to verify its realization; a procedure contrary to E. Mach's epistemological principles, for instance. For brevity's sake, we can say that, in practice, absolute rotation is defined with respect to distant matter, in particular radio galaxies. This definition corresponds, within the limits of present (and very high) accuracy, to the local definition of geometrical parallelism.

Once the local coordinates (t, x, y, z) had been explicitly calculated, Fermi had no difficulty in obtaining the expression for the generalized "Pythagoras' theorem " in spacetime (3):

$$\mathrm{d}s^2 = (1 + \mathbf{a} \cdot \mathbf{r})^2 \mathrm{d}t^2 - \mathrm{d}x^2 - \mathrm{d}y^2 - \mathrm{d}z^2. \tag{4}$$

Here $\mathbf{a}$ is the vector acceleration of L_0 and $\mathbf{r}$ the vector with space coordinates (x, y, z). Note that by acceleration we mean a precise geometrical quantity which measures the deviation of the world line from a geodetic. This formula contains the dynamics of a non-inertial frame of reference (on the earth's surface or in an accelerated rocket): a free body, moving along a line which is not a geodetic, has the acceleration $-\mathbf{a}$.

Finally, one must define the error made by taking the metric in the previous form (4), in which the influence of the curvature is neglected. The result for a sphere (eq. (2)) is not very different from that for the tubular environment of a geodetic: the distance $\mathrm{d}s$ is still given by a quadratic form in the differentials of the coordinates $\mathrm{d}t, \mathrm{d}x, \mathrm{d}y, \mathrm{d}z$ in which the coefficients differ from those in (4) by terms that are time-dependent and of order Kr^2; here, however, r is the distance from the geodetic in the orthogonal space. For example, in a single spatial dimension x we have three corrections:

$$\mathrm{d}s^2 = [1 + K_{tt}(t)x^2]\mathrm{d}t^2 - [1 + K_{xx}(t)x^2]\mathrm{d}x^2 + 2K_{tx}(t)x^2 \mathrm{d}t\, \mathrm{d}x. \tag{5}$$

The functions $K(t)$ have dimensions L^{-2} and are determined by curvature of spacetime. The error made in neglecting the curvature in a laboratory on the earth is very small. The curvature of spacetime in the Solar System is determined essentially by the sun and at a distance D from it is of order

$$K = \frac{GM_\odot}{c^2 D^3} = 10^{-8} \left(\frac{1\ \mathrm{AU}}{D} \right) \frac{1}{D^2}, \tag{6}$$

where the Astronomical Unit (AU) is the earth-sun distance, 150 million km. For a laboratory at 1 AU with dimensions r, the error Kr^2 is of the order $10^{-8}\ (r/D)^2$; if r is the earth's radius the error is $\approx 2\times10^{-18}$! This is the order of magnitude of the influence that solar gravitation has on local phenomena.

It is interesting to note that the important question of the error committed by adopting Fermi's metric (4) was ignored until the '70s; the reader will find the explicit expressions for such corrections in [1] and [2]. The merit for the construction of Fermi's coordinates in a curved space, when the central line L_0 is not a geodetic, goes to Walker [3]. In [4] the geodetic coordinates have been extended to the case in which, in addition to the curvature produced by distant bodies, a contribution from gravitating bodies nearby, such as the earth, is present.

5. – The nature of gravity

We are now able to clarify the paradox of sect. **2**: if the weight is an illusion, what is the gravitational force, which also has such a fundamental role in the structure of celestial bodies and in planetary systems?

In the equations of motion of a free body the adimensional coefficients of the metric take on the role of gravitational potentials per unit mass (adimensional when the velocity of light is equal to one). For instance, when the velocity of the body is much smaller than that of light, the coefficient of $\mathrm{d}t^2$ is equal to $1-2U$, where U is the Newtonian potential (per unit mass). In Fermi's case, in fact, (eq. (4)) this coefficient is

$$(1+\mathbf{a}\cdot\mathbf{r})^2 = 1+2\mathbf{a}\cdot\mathbf{r}+\dots$$

and correctly corresponds to the potential $-\mathbf{a}\cdot\mathbf{r}$. The metrical corrections (5) therefore correspond to a *quadratic* potential in the coordinates (x,y,z).

In one-dimensional space (5), the relative motion, when the velocity is not too high, is determined by the potential $-K_{tt}x^2/2$, which corresponds to the acceleration $K_{tt}x$. In general, if the relative motion is described by Fermi's coordinates, it has a (relative) acceleration *linear* in the space coordinates, and of magnitude $\approx Kr$. This corresponds, in classical mechanics, to the relative motion of two bodies at $\mathbf{r}_1$ and $\mathbf{r}_1+\mathbf{r}$, in a potential per unit mass U, when their distance r is much smaller than the characteristic scale $D=U/|\nabla U|$ over which U varies; in this case their relative acceleration is

$$-\nabla U(\mathbf{r}_1+\mathbf{r})+\nabla U(\mathbf{r}_1) = -\mathbf{r}\cdot\nabla\nabla U(\mathbf{r}_1)+\dots .$$

The three-dimensional matrix $\nabla\nabla U$ is calculated at the position $\mathbf{r}_1$ of the first body and thus, in general, is a function of time. Accelerations of this form are those that govern tides: the effect of moon and sun on a particle of the ocean located at position $\mathbf{r}$ relative to the centre of the earth is described by the expression given above. In General Relativity the relative acceleration of two points in geodetic motion has the form

$$\mathsf{K}\cdot\mathbf{r}, \tag{7}$$

where the three-dimensional matrix K is determined by the curvature of the spacetime. In this case, too, we can speak of *tidal acceleration*. This law of motion takes the name

Fig. 4. – Only by observing the relative motion of two bodies can one detect, and measure, the gravity due to a body nearby: for instance, in a laboratory on the earth's surface the trajectories of two bodies initially at rest converge toward the centre of the earth, while within an accelerated rocket they are parallel.

of *equation of geodetic deviation*, inasmuch as it describes how two neighbouring geodetic lines move away from, or closer to, each other because of the curvature (fig. 4).

Therefore, although the acceleration of a single body cannot be observed, the relative motion of two bodies is influenced by gravity through tidal effects; only these effects make it possible to decide, within a closed laboratory, whether the measured acceleration is due to the nearness of the earth or the propulsion of a rocket. Suppose we remove the obstacles to the motion for two bodies at rest within the laboratory; if the latter is pushed by a rocket, their trajectories are parallel, but if it is at rest on the earth's surface the trajectories will converge toward its centre: a very small but crucial difference (fig. 4).

Thus, if we want to describe gravitational phenomena in a region that is small relative to the characteristic scale, Fermi's generalized coordinates are essential. Since they are defined in a geometric and invariant form, there is no risk that, in using them, one may reach conclusions regarding unobservable phenomena, such as weight; the mathematical formalism is focussed on the essential effect, the relative acceleration. This conceptual and computational tool is important nowadays, for example, for the design and operation of gravitational wave detectors: instruments whose sizes are, in general, much smaller than the wavelength of interest. They are basically used to determine the time-dependent curvature of spacetime through the geodetic deviation experienced by two neighbouring material points.

REFERENCES

[1] MISNER C. W., THORNE K. S. and WHEELER J. A., *Gravitation* (Freeman) 1973.
[2] MANASSE F. K. and MISNER C. W., *Fermi normal coordinates and some basic concepts in differential geometry*, *J. Math. Phys.*, **4** (1963) 735-745.
[3] WALKER A. G., *Relative coordinates*, *Proc. R. Soc. Edinburgh*, **52** (1932) 345-353.
[4] ASHBY N. and BERTOTTI B., *Relativistic Perturbations of an Earth Satellite*, *Phys. Rev. Lett.*, **52** (1984) 485-488.

About the Author

BRUNO BERTOTTI was Professor of Astrophysics at the University of Pavia; his main interests are currently focused on space physics; his research has also dealt extensively with gravitational physics, both theoretical and experimental. Early in his career, in the 50's, he was a student of E. Schrödinger at Dublin.

Fermi and quantum electrodynamics

Marcello Cini

1 – The fifth Solvay conference in 1927 [1] marks the final acceptance of non-relativistic quantum mechanics by the community of physicists through the formulation of Born, Heisenberg and Jordan [2] and with the Bohr interpretation as the definite and correct theory of the phenomena which take place at the level of the constitutive elements of reality. In this way Wolfang Pauli [3] remembered this turning point six years later:

"*With Heisenberg's indetermination principle* [...] *the initial phase of the development of the theory came to an end. The theory leads to the solution of the problem which has been sought for a long time and supplies a correct and complete description of the phenomena concerned. The solution is obtained by abandoning the causal and classical space-time description of nature.*"

It was the statement, to use Thomas Kuhn's [4] terminology of a new "paradigm" which was the beginning of a long period of "normal science", founded upon the application of formalism and of its interpretative rules to the problems of the structure of matter (atomic and molecular physics, solid state physics and nuclear physics). These rules can be summarized very schematically in the "condition of quantization" for the dynamic variables of a system, which transforms them from numerical variables into mathematical entities, which do not satisfy the commutative property of the product. The fundamental connection between classic conjugate variables, known as Poisson brackets, became in this way the fundamental connection among the correspondent conjugate quantum variables, "commutation relations".

Enrico Fermi had not taken part in the constitutive phase of the new theory. In the winter of 1923, he went to Max Born at Göttingen on a scholarship from the Ministero della Pubblica Istruzione (Ministry of Public Instruction), but his period there was not very profitable. *"It is not easy to understand why* —writes Emilio Segrè [5] in the introductory biographic note of the collection of "Note e Memorie" (Collected Papers) published by the Accademia dei Lincei and the University of Chicago Press— *but it is possible that his love for concrete problems and his diffidence for too general questions at that time vague, and worse still, mixed with philosophy, had discouraged him from speculation which in the end was to lead to Quantum Mechanics"*. It is also possible, adds Segrè [6] that at Göttingen, *"the physicists of his age who were therefore very young, like Heisenberg, Pauli, Jordan etc., all exceptionally clever, and who should have formed his group of peers, had not recognized Fermi's merits and had involuntarily put him aside."*

His following stay at Leida was more profitable. He went to work with P. Ehrenfest who rapidly recognized his merits and encouraged the shy Italian. In fact Heisenberg's work on the representation of quantum operators by matrices 1924-25 did not seem very clear to Fermi, who, only later through Schrödinger's wave mechanics, fully understood quantum mechanics. Even Fermi's first fundamental contribution to the knowledge of the atomic world —the distribution of the particles that obey Pauli's principle and which carries his name [7]— was born (February 1926) outside the new mechanics and certainly before he had mastered it. It is probable that Heisenberg's approach —founded on the substitution of the classical numerical variables with matrices which do not satisfy the commutative property of the product, seemed too abstract for his profound sense of physics. Be that as it may, *"Schrödinger's memoirs* — writes Segrè— *were the first to be really understood and caused great enthusiasm. Fermi soon explained them, first to his students and friends, then to Corbino and later he spoke about them at the Mathematical Seminary."*

But others will dwell on this phase of Fermi's activities more completely.

2 – The situation was very different as far as the solution of problems involving the electromagnetic field and its interaction with material charged particles was concerned. The extension of the quantistic paradigm to the relativistic systems, with infinite degrees of freedom, in fact presented difficulties which were much greater than those which had to be dealt with to apply the rules of quantum mechanics to resolve the "brain-teasers" of "normal science". This required dealing with unresolved conceptual problems in order to formulate an adequate theoretical outline —quantum electrodynamics— to interpret a new area of the physical reality. This undertaking lasted for twenty years —from 1927 to 1947— along a route that was anything but straight. Proposals that were in part reciprocally incompatible had to be confronted while new proposals were advanced.

Fermi contributed significantly to this long process of construction of the theory in '29 intervening with his usual lucidity and clear thinking in the debate —which had opened in '27 with the fundamental work of Dirac [8] and Jordan [9]— with the aim on the one hand of stressing the kind of difficulty that presented the development of a complete theory and on the other of identifying the concrete problems which, in spite of these difficulties could have been tackled and resolved. Nevertheless the importance of Fermi's contribution goes farther than these mere clarifications. In fact it was by transferring his original formulation of quantum electrodynamics to the problem of beta decay that Fermi managed to supply in '33, using Pauli's neutrino idea, the theory of this phenomenon that bears his name.

It is worth briefly mentioning the context in which it is placed before talking about this contribution in more detail.

The fundamental problem to be faced was that of explaining the corpuscular properties of the electromagnetic field (the existence of photons) and of their interaction with the electrically charged material particles introducing the principles of quantum mechanics in its classical description supplied by Maxwell's equations. From the beginning, this problem was faced from two radically different points of view: that of Jordan and that of Dirac. Jordan [8] starts off with the wave-like properties of the electromagnetic field, to demonstrate that by treating the amplitudes of the stationary waves of radiation in a cavity as dynamic quantum variables, the field acquires exactly the corpuscular properties that Einstein had found in 1905 using the formalism of statistical thermodynamics. In other words Jordan shows that the the corpuscular properties of electromagnetic waves derive from the conditions of quantization imposed on the wave amplitudes that is from the non-commutativity of the dynamic variables of the field. For Jordan within the limits of classical physics there are no particles, but only the classical physical entity endowed with spatial extension and continuity: the field with its wave-like properties. This view characterizes all the following developments in Jordan's research, in complete contrast with that which was to guide Dirac in his successive formulation of the theory.

For the latter, in fact, quantum mechanics had its roots in the Hamiltonian formalism of classical mechanics. As this formalism is deeply rooted in the theory of the movement of material bodies (rather than in the theory of electromagnetism) to construct a correct quantum theory of the electromagnetic field it is necessary to start from the *corpuscular* nature of radiation. According to this formulation Dirac begins his work in '27 [7] starting with the quantum description of a statistical ensemble of non-interacting particles obeying Schrödinger's equation for a single particle in the presence of a common perturbation. Nevertheless in this description the number of particles endowed with fixed energy is only an average value in the statistic ensemble which can be any real number. If it is thought, instead, that these numbers represent effective possible values, namely that they can only be integer numbers, it is essential that they become non-commuting variables and do not commute. It is therefore the number of particles of fixed energy which become quantum variables. With this they automatically acquire —and this is the essential point of Dirac's procedure— the property of the identical particles (Bosons) which satisfy the Bose-Einstein statistics. And since photons are bosons, this procedure gives us a method for dealing correctly with the process of emission and absorption of photons by the material charged particles without having introduced any specific reference to the classic electromagnetic field. It is only to obtain the explicit expression of the probabilities of these two processes that Dirac, in the final paragraph of his work, introduces the expression of the classical Hamiltonian of an atomic system in interaction with the electromagnetic field, and derives from this the value of the constant which links the field amplitude with the corresponding number of photons. In this way the author demonstrates, *a posteriori* that "the wave viewpoint is consistent with that of the light quanta, and supplies the value of the interaction coefficient in the quantum theory of radiation."

These two profoundly different points of view lead Jordan and Dirac to develop the theory in directions which appear to diverge. Jordan, in fact, in his successive works, demonstrates that even to explain the existence of particles that obey the statistics of Fermi-Dirac (fermions) it is possible to start with a *wave description* which assumes corpuscular properties when the amplitudes have to satisfy quantization conditions which make them non-commuting variables. For Jordan [10] therefore *"the natural formulation of quantum theory of electrons will be constructed in a way that light and matter will be simultaneously introduced as interacting waves in a three dimensional space"*. Dirac, on the contrary, continues to develop his research convinced that the procedure that he had invented could be exclusively applied to the electromagnetic field, because as he had the opportunity to explain in '32, *"we cannot imagine that the field is a dynamic system of the same type as material particles."*

3 – Two papers by Heisenberg and Pauli [11] illustrate the following step in which the two authors try to give a paradigmatic formulation to the theory. They start with a Hamiltonian of the electromagnetic field interacting with material particles which satisfies the principles of relativity, treating the first with the Dirac method and the second with that of Jordan and they both propose a method of solving concrete problems *involving phenomena concerning this area of physics.* They immediately find themselves in various kinds of difficulty. The first derives from the necessity of describing the electromagnetic field through a scalar and a vector potential (which are not completely determined by the electrical and magnetic components of the field). The Lorentz condition which binds them is in fact incompatible with the rules of commutation which the corresponding quantum variables must satisfy. This is a difficulty to which Fermi gives an original solution in his work considering the Lorentz condition as a constraint on the state of the system rather than a relationship between the variables. A second difficulty arises from the infinite value of energy radiation in the fundamental state and from infinite coulombian energy of point charges. The authors do not seem very worried about this because it is about infinite constants which disappear during the calculation of the difference of energy between two levels of the system by means of the perturbation theory. Neither do they seem worried about what could happen in the higher order of approximation. Finally they mention the difficulties that arise from the negative energy in Dirac's equation used to describe the particles in agreement with the theory of relativity. On the whole even though they admit that *"there is a long way to go to reach a definite theory"* they declare that they are convinced that substantial progress will be possible following the lines suggested. Jordan is not of the same opinion. In fact in a paper [12] immediately following theirs, he clearly judges the theory they propose as inadequate. *"It seems clear* —one reads— *that this Maxwell quantum theory contains fundamental difficulties and defects."* In fact —he rightly observes— *"the self-energy of the electron is an infinite variable and represents such a serious and worrying obstacle that it makes the practical application of the theory almost impossible."* In short —continues Jordan— *"the difficulties which we have to face today are of a completely different nature from those of the pre-quantum period, when no solid basis existed to answer the question "why do electrons exist?" It is not about the existence of electrons anymore but now it is the manner and the nature* *of their interaction with the radiation field that still remain without solution and present us with difficult enigmas"*.

4 – At almost the same time Enrico Fermi presents a series of notes on the formulation of quantum electrodynamics to the Accademia dei Lincei [13-15] (the date of receipt of Heinsenberg's first paper is 19/3/29 while that of Fermi's first note is 5/5/29). According to Amaldi [16] Fermi started to work on this problem in the winter of 1928-29 and did not thoroughly study Heinsenberg and Pauli's work. His theory differs both from Dirac's and from Jordan's even if his approach is through Hamiltonian formalism.

His objective is to *"face the particular problem of the construction of quantum electrodynamics for this case which is practically the most important in which there are also electric currents in the field"*. *"A partial solution to this problem* —continues Fermi— *has been given by Dirac in his fundamental work on the irradiation theory. Although Dirac's electrodynamics is incomplete for the fact that it only takes into consideration the electromagnetic radiation field that is, a field that can show itself only as a sum of electromagnetic plane waves. In fact the electromagnetic field which is produced near a moving electric corpuscle is rather more complicated, in a way that Dirac's theory is a correct description of reality only for regions of space quite far from the electric corpuscles that generate the field. For example it could not be applied to the problem of interaction between two electrons of the same atom or of nearby atoms taking into account that the field is propagated from one to the other because of retarded potentials; similarly his theory cannot be applied to the problem of the quantum theory of electromagnetic mass etc."*

To write quantum electrodynamic equations Fermi [13] first tries to set out the equation of classical electrodynamics in such a way as to render the translation in quantum form possible. To do this, instead of taking the scalar and vector potentials as variables, which describe the electromagnetic field in every point of space, he takes their Fourier expansion coefficients which are functions only of time. As far as the particles are concerned, they are described according to their position in space. Next to the coordinates, their conjugated momenta are introduced into the theory.

"Having written the equations of electrodynamics and of the charges motion in canonical form —concludes Fermi at the end of the first note— *their translation into quantum-mechanical equations is straightforward. For this reason it is enough to consider the expression of the Hamiltonian as an operator, in which the momenta, according to the usual rule, are equivalent to the operation of derivation with respect to the corresponding coordinate and multiplication by* $(-h/2\pi i)$*."*

The simplicity, the clarity and the rigour of this exposition almost seem to hide the underlying difficulties that stand in the way of further development of the theory.

5 – In the note that immediately follows [14] "Sulla teoria quantistica delle frange di diffrazione" (On the quantum theory of diffraction fringes) Fermi returns to Dirac's original theory —disregarding a "refining of the theory" which he had proposed in the preceding note— to demonstrate that it correctly takes into consideration the phenomena of interference (typically wave-like) and that *"one can well say that it includes all the properties of radiation."* Even here the simplicity of the formulation is striking.

"The elements of the problem —we read— *are: an emitting atom A, that works as a source of light, an interferometer, which we can schematically think of as made up of a set of opaque screens, or made up of reflecting surfaces which delimit media of different refraction index, finally another atom B which, absorbing more or less intensely the light emitted from A, tells us in which regions the light emitted from A arrives with greater intensity (bright fringes) and in which it arrives with minor intensity (dark fringes). Naturally, according to the fundamental criterion of Dirac's theory, the two atoms A and B and the radiation field must be considered as a unique system, which must be treated with quantum-mechanical methods."* As one can see photons are not mentioned, but only quantum states of the different harmonic components of radiation. Atom A is initially excited and atom B is in the fundamental state of minimum energy. The calculation shows that the probability of the excitement of B as a function of time depends on its position according to the classical theory of stationary waves. *"In this way* —he concludes— *we rediscover the results of the classical theory of wave interference."*

The same formal simplicity distinguishes the following work [15] dedicated to the generalization of the theory in the case of relativistic motion. *"It is well known* —comments Fermi— *that recently even Heisenberg and Pauli have dealt with the problem of quantum electrodynamics. However, since the methods followed by these authors are essentially different from mine, I believe it is not useless to publish my own results too."* It was not only modesty. To understand how much the publication was not "useless" it is enough to compare the eight pages of Fermi's two notes with the more than eighty pages of the two papers by Heisenberg and Pauli.

His conclusions are stamped with pragmatic optimism. *"Naturally* —we read— *as we have already hinted, even this theory holds two fundamental defects; which rather than of electromagnetic origin, can be considered to derive from the incomplete knowledge of electronic structure. They are the possibility Dirac's electron has of falling into energy levels with negative energy, and the fact that intrinsic energy has an infinite value if one has to accept that the electron is exactly a point charge."*

Fermi dedicates his following publication in "Nuovo Cimento" at the beginning of '31 [17] to the latter problem. It is useful to examine some of the methodological aspects even though they are no longer current, because they demonstrate once more how Fermi's approach to the problem is marked by physical observation, more than by formal reasoning.

After having briefly drawn our attention to the well-known fact that *"the energy of the electrostatic field produced by a point charge is infinite"*, and having observed that *"it is not possible to eliminate this difficulty by modifying the value of the energy using an additive infinite constant, because the infinite term is not constant, but varies with the variation of quantum states"*, Fermi reminds us again that *"the expedient that appears first to avoid these setbacks, is to try to introduce something equivalent to a finite radius of the electron into quantum electrodynamics."*

"If you try to do this, though, —continues Fermi— *you find yourself confronted with difficulties greater than those you find in the corresponding classical problem; so that it does not seem possible, at least within the current framework of quantum electrodynamics,*

to construct a theory foreseeing the finite radius of the electron and preserving relativistic invariance. It is nevertheless rather interesting to study to what extent a theory of electromagnetic masses can be constructed, since it presents several characteristic differences from the classical theory". It is at this point that one appreciates the originality of Fermi's approach.

"To introduce in a simple analytical way the hypothesis that the electron has finite extension —he explains— *let us observe, if it is so, that the different parts of the electron will find themselves in points of equal phase, with respect to waves which have a greater wave length compared to the size of the electron, whereas waves whose wavelengths are of the same order of the size of the electron, or even smaller, will have different phases in the different points of the electron. Now we can easily understand and immediate calculations confirm this, that the electron will interact much less intensely with the different waves, the smaller their wavelength is since, for very short waves, the different parts of the electron interfere destructively."*

From this physically clear formulation it follows that *"the electron interacts with the high-frequency harmonic components of the field as if it had an effective electric charge smaller than that which determines the interaction with the harmonic components at low frequency, and which naturally coincides with the ordinary value of the electronic charge which is supplied from static or quasi static measurements."* There is an interesting anticipation of the "renormalization of the electric charge" in this observation which will later be introduced into the formulation of Feynman, Schwinger and Tomonaga's quantum electrodynamics which is now accepted as "definitive".

After having published in '32 an exhaustive monograph in "Review of Modern Physics" [18] in which the theory is presented in a systematic way —a monograph from which generations of young researches (myself included) learned quantum electrodynamics immediately after the war— Fermi dedicates his last paper [19] on the subject in collaboration with Hans Bethe, in 1933. It deals, not accidentally, with an applicative calculation in which, using his own methods, he retraces Bethe and Møller's formulae for the scattering of two electrons and compares them.

6 – It now remains for us to briefly comment (we will go more deeply into the subject later with a further contribution) on the connection between Fermi's quantum-electrodynamic research and his theory on beta emission in '33. Rasetti [20] tells us that, *"To understand the fundamental contribution brought to us by Fermi with his theory of beta decay, we must emphasize that the idea of the neutrino (brought forward by Pauli in December 1930) had been up to that point a rather vague hypothesis, and that the construction of a formal theory had never been attempted. When Pauli had advanced his suggestion it was thought that electrons already existed in the nucleus, and the hypothetical neutral particle was by analogy imagined as another component of the nucleus endowed with a small but finite mass. Only vague hints at the possibility that the electron was created with the neutrino, in the process of beta decay, can be found in the Proceedings of the Solvay Conference of '33."*

But let us see what Fermi himself has to say about it [7]. After having declared that the hypothesis of the neutrino's existence is at the root of the theory that he intends to present, and having reminded us of how difficult it is to explain satisfactorily how light particles can be stably bound inside the nucleus, Fermi proceeds in this way:

"The simplest way to construct a theory which allows a quantitative discussion of the phenomena in which nuclear electrons are involved, seems consequently to reside in the hypothesis that electrons do not exist as such in the nucleus before beta emission, but they acquire existence, so to speak, the instant they are emitted; in the same way as a light quantum emitted from an atom during a quantum jump can in no way be considered pre-existent in the atom before the process of emission. *Therefore in this theory the total number of electrons and neutrinos* (like the total number of light quanta in the theory of radiation) *will not necessarily be constant, because there might be processes of creation and destruction of light particles* (1)."

And again: *"We will then look for an expression of the interaction energy among the light particles and the heavy ones which allows the transformation of a neutron into a proton and vice-versa, in such a way that the transformation from neutron to proton has to be connected to the creation of an electron, which is observed as a beta particle and a neutrino; while the opposite transformation from proton to neutron is connected to the disappearance of an electron or a neutrino;* as in the radiation theory a light quantum is connected to a certain quantum jump in the atom, while the opposite quantum jump is connected to the absorption of a quantum (1)."

The link with all his previous work could not be more explicit. Anyway it is worth adding, without going into detail, that history has shown fifty years later, with the unification of quantum electrodynamics and of weak interactions by Weinberg and Salam and with experimental confirmation by Rubbia, how deep the link between the two theories really was. But this is another story.

7 – We are not interested in going further into the subsequent events that led to the development of quantum electrodynamics. It seems to me more useful to conclude with some remarks which result from an examination through the eyes of a historian, of the different elements of its initial phase which we have schematically reconstructed. We have in fact shown how the approaches chosen by the physicists —Jordan, Dirac, Heisenberg, Pauli and Fermi— who have contributed most to the problem of formulating a satisfactory quantum-electrodynamic theory have been characterized by substantial methodological and epistemological differences. A spontaneous question comes to mind: Are these differences due only to characterial and psychological factors or philosophical prejudice which is purely individual, or is it possible to try and trace the origin, at least in part, in the different cultural and social context in which they were working?

I am well aware that for the majority of scientists carrying out research in their own particular field this is a meaningless question, because either it is not verifiable, or, at

---

(1) The sentences in roman characters in the quotations belong to the author.

most, a question without interest. However, it is not so for those who are convinced that history is neither the carrying out of a preordained plan nor the pure fruit of circumstance. If we accept this premise, in fact, it becomes the duty of historical doctrine —from the most general, which studies the evolution of the society, to the more specific which studies the events which have characterized the development of science— to try to reconstruct, as accurately as possible, the context which creates the ties into which, even though there is a wide margin of risk and unpredictability, the distinct events which mark the stages in history are channelled.

In particular as far as science is concerned it is worth keeping in mind that the creativity and brilliance of each scientist is only one aspect, even though fundamental, of its growth process, the other is made up of the acceptance by the community —founded on the principle of *peer evaluation*— of every radical proposal of innovation in the shared patrimony of knowledge which characterizes the identity of the subject. And it is in judging if the proposal should be accepted or rejected that, besides the technical checks of compatibility with that patrimony, general criteria of opportunity, borrowed, more or less explicitly, from the social and cultural context, come into play.

From this point of view it is useful to remember how differently the new quantum mechanics was received in England and in Germany, linked to —in Paul Forman's opinion [21]— the profound differences that existed then between the environment in which the English physicists were formed and that of the community of German physicists. This author remarks that *"only in Germany the indeterminism of the theory was immediately recognized and well received by a significant number of physicists, in most cases as a desire to obtain more favourable public recognition of their subject. In England, on the contrary, where the physicists had not been criticized by the intellectual environment, and therefore the theme of causality had not become a controversial question, the epistemical consequences of the new theory were not noticed and its formalism was adopted without criticism under its most congenial aspects."*

In other words, one can say that this difference between cultural environments favoured an approach to the new mechanics that was characterized by a search for continuity with the classical tradition in England, whereas in Germany the elements of a break away from it was emphasized [22]. In particular for Dirac the classical theory remained, as it is evident from his research during the following years, the only model that is able to provide directions of how to construct the new theory, while for the German physicists the only link between classical physics and the new physics was Bohr's principle of correspondence, which simply guaranteed the *approximate* validity of the first, through the description of the properties of macroscopic bodies. The starting point for the construction of the new quantum electrodynamics therefore could not have been more distant.

8 – It is therefore easy to retrace the origin of the different approaches chosen by the above-mentioned authors, in the variety of their cultural backgrounds [23]. The first question concerns the contrast between Dirac and Jordan on the problem of the existence of material particles. The second, the description of the state or a system of N particles in an abstract space at $3N$ dimensions rather than in the ordinary three-dimensional space of our daily experience. For Dirac these problems do not exist. Educated in the tradition of Newton and Hamilton, he lived in Cambridge at a time when Whittaker taught analytical mechanics, his starting point was material particles. Therefore it was not a scientific problem to explain their existence. Neither was it a problem to have to describe the state of a system by means of the Lagrangian coordinates in configuration space.

For Jordan instead, the existence of electrons was a fundamental problem which was deeply rooted in the philosophical question of the "liquidation of materialism". In fact Joardan wrote on this subject: *"The new concepts which arise from the experiment on quantum physics and from their theoretical elaboration lead to the liquidation of the materialistic image of the world developed by classical western science, which in its turn derives from the materialistic philosophy of the Greeks."*

And again: *"Before now we believed that wave radiation and corpuscular radiation existed in Nature, and our classical intuition led us to represent them as different incompatible things. Now we see that in reality Nature knows only one type of radiation, which, on the one hand, demonstrates properties that correspond to the classical wave representation and, on the other, all the properties that we attribute to corpuscular rays."* Both the corpuscular and wave nature of radiation are therefore consequences of a theory which represents them in ordinary space. The conclusion is that *"thanks to observations, one of the most visible aspects of the materialistic representation of the world is liquidated once and for all, while the positivist theory emerges decisively justified and confirmed."* Be careful this is not idle philosophical chat. These statements demonstrate in fact that both the problem of the existence of particles, and that of their representation in ordinary three-dimensional space, problems that are inexistent in the British culture, are important problems for the representatives of the German School, which have to be resolved through a unitary description of reality. This is not the only difference between the two formulations of the theory which stems from the cultural difference in which the scientists of the two schools work. There is at least one other which is also fundamental. It is the importance attributed to the theory of relativity. It was obvious to all that the interaction between charged particles and the electromagnetic field should have been invariant with respect to Lorentz's transformations. Nevertheless it appears clear that for Dirac —after the first paper in which the choice of a non-relativistic approximation is justified by the necessity not to tackle too many difficulties at the same time— the construction of an explicitly invariant theory to include into quantum mechanics Lorentz's transformations becomes, coherently with the formal strategy of classical mechanics, a priority motive. The faithfulness to this euristic principle is on the other hand rewarded by the success of his relativistic equation for the electron, a success which demonstrated that even an apparently non-relativistic property like the spin, on reality derived from a rigorous application of the constraints of special relativity.

The direction chosen by the physicists of the German School for the construction of the theory is, even in this case radically different. If the quantum world has nothing to do with the classical one, relativity becomes only a formalism to be used when necessary, that is when the speeds at stake are comparable with the speed of light. On the contrary, the complications that its introduction can produce risk obscuring the simplicity of the quantum phenomena, which are considered more essential. Be that as it may, it is not possible not to emphasize that this methodological attitude hindered the German physicists during the following years, from taking significant steps ahead to unblock the stalemate which had formed toward the middle of the thirties. Only ten years later, when Dirac's two papers of 1932, which had formulated electrodynamics in a strictly relativistic way, were taken up again, the solution to the problem came out, as we know, independently found by Tomonaga in Japan and by Feynman and Schwinger [24] in the United States.

9 – In Italy as we have seen, it was Fermi who imported and diffused the new quantum mechanics. It is nevertheless important to emphasize, in order to reconstruct the cultural environment in which this activity was carried out, that the subject area mainly involved in international research during that period was that of mathematical physics, with figures like Volterra V. and Levi-Civita T., closely linked to Einstein, Planck and Schrödinger's Berlin group. Perhaps this also explains why, as we have seen, it was through the study of Schrödinger's memoirs that Fermi mastered the formalism and physical content of quantum mechanics.

It was probably the absence of a cultural hinterland involving a large group of physicists in the study of quantum problems which allowed Fermi to take on the role of leader, imprinting it with his personality and pragmatic attitude, free from epistemological interests, directed essentially towards the use and application of the new theoretical instruments to study a variety of physical phenomena.

Fermi's approach to the formulation of quantum electrodynamics strikes us, as we have seen, with its simplicity, its pragmatic character and the evident lack of interest he shows for the abstract conceptual questions which, like wave-corpuscle dualism, worried Jordan and Dirac. Four pages are enough for him in his first note, to formulate the problem in a strictly Hamiltonian description according to canonical rules. Those that followed are applications of this canonical formalism to the cases in which it works. When the formalism gives senseless results Fermi does not worry too much. One day we will have a better theory.

And this, to sum up, is the thesis with which he concludes his monograph in the "Review of Modern Physics" [18]: *"We can conclude* —he writes— *that practically all the problems of radiation theory which do not involve the structure of the electron have a satisfactory explanation; while the problems connected with the intrinsic properties of the electron are still very far from being resolved."* As if to say: let us be content for the moment with what the theory can give us, and let us use it to resolve the practical problems that we are interested in.

In this way Fermi's interest in electrodynamics ends, and it moves on to problems newer from a physical point of view and more intellectually stimulating. The future of quantum physics was in the study of the atomic nucleus.

REFERENCES

[1] *Electrons et photons,* Rapports du 5e Conseil de Physique, Paris, 1928.
[2] Born M., Heisenberg W. and Jordan P., *Z. Phys.*, **35** (1925) 557.
[3] Pauli W., *Handbuch der Physik,* Vol. **24** (Springer Verlag, Berlin) 1933.
[4] Kuhn T., *The Structure of Scientific Revolutions* (The University of Chicago Press) 2nd extended edition, 1970.
[5] Segrè E., "Biographical Introduction", in *E. Fermi, Note e Memorie (Collected Papers),* Vol. **I** (Accademia Nazionale dei Lincei, The University of Chicago Press) 1962, p. XXVI.
[6] Segrè E., op. cit., p. XXXIII.
[7] Fermi E., *Ric. Sci.*, **4** (2) (1933) 491-495.
[8] Dirac P. A. M., *Proc. R. Soc. A*, **114** (1927) 243.
[9] Jordan P., last chapter in Born M., Heisenberg W. and Jordan P., *Z. Phys.*, **35** (1925) 557.
[10] Jordan P., *Z. Phys.*, **44** (1927) 1.
[11] Heisenberg W. and Pauli W., *Z. Phys.*, **56** (1) 1929; **59** (160) 1930.
[12] Jordan P., *Phys. Z.*, **30** (1929) 700.
[13] Fermi E., *Rend. Acc. Lincei,* **5** (881) 1929.
[14] Fermi E., *Rend. Acc. Lincei,* **7** (1930) 153.
[15] Fermi E., *Rend. Acc. Lincei,* **12** (1930) 431.
[16] Amaldi E., in *E. Fermi, Note e Memorie (Collected Papers),* op. cit., p. 305.
[17] Fermi E., *Nuovo Cimento,* **8** (1931) 121.
[18] Fermi E., *Rev. Mod. Phys.*, **4** (1932) 87.
[19] Bethe H. and Fermi E., *Z. Phys.*, **77** (1932) 296.
[20] Rasetti F., in *E. Fermi, Note e Memorie (Collected Papers),* op. cit., p. 538.
[21] Forman P., in *The Reception of Unconventional Science,* edited by S. H. Mauskopf (AAS, Westview, Boulder) 1978.
[22] De Maria M. and La Teana F., *Fundam. Sci.*, **3** (1973) 2.
[23] Cini M., *Fundam. Sci.*, **3** (1982) 229.
[24] Tomonaga S. I. and Schwinger J. in *The Physicist's Conception of Nature,* edited by J. Mehra (Reidel, Dordrecht) 1973, p. 405.

About the Author

Marcello Cini has been Full Professor of Theoretical Physics and then Quantum Theories and today is Professor Emeritus at "La Sapienza" University in Rome. He has been vice-president of the Società Italiana di Fisica, deputy editor of the international journal of physics "Il Nuovo Cimento", director of CIRMS. He has published over a hundred papers in international journals of physics and of the history of science, several books among which "L'ape e l'architetto" (The Bee and The Architect) (with G. Ciccotti, G. Jona-Lasinio and M. De Maria (Feltrinelli, 1976)); "Un paradiso perduto" (A Lost Paradise) (Feltrinelli, 1994); "Quantum Theory without Reduction" (with J.M. Lévy-Leblond, Adam Hilger, 1991); "Dialoghi di un cattivo maestro" (Dialogues of a Bad Teacher) (Bollati Boringhieri, Torino, 2001).

Weak interactions

Nicola Cabibbo

1. – Introduction

What are weak interactions? What is their role in the world around us? Why talk about them in a book dedicated to the memory of Enrico Fermi? To get our bearings let us start by giving some simple answers to these questions, starting from the third, perhaps the simplest.

In December 1933 an article entitled "Tentativo di una teoria dell'emissione dei raggi beta" (Tentative theory of beta rays) appeared in "La Ricerca Scientifica", the journal of the "Consiglio Nazionale delle Ricerche" (National Research Council). The title could be deceiving: the theory presented in that article is much more than an attempt and with a few improvements which we will talk about later it is completely valid today after nearly seventy years.

In spite of its great importance, Enrico Fermi's work did not initially have an easy life. It was refused by the English magazine "Nature", because it was "too abstract", and Fermi had to turn to "La Ricerca Scientifica". The Italian route to rapid publication proved to be precious the following year when the results of radioactive transmutation produced by neutrons were released in ten brief articles published in the course of a few months.

In his theory of beta radioactivity Fermi introduced a new type of interactions among elementary particles, which today we call "weak interactions". Many new manifestations of the weak interactions, which could be interpreted using Fermi's 1933 theory, were found in the following decades. The study of weak interactions has led to surprising discoveries, among which the violation of specular symmetry (known as parity symmetry or P symmetry), and the violation of time reversal symmetry (T symmetry) and of the symmetry between matter and antimatter (CP symmetry).

Weak interactions are very interesting because they lead to the transformation of an elementary particle into different ones, and they lay bare the relationships among the different types of particles.

At the end of the sixties the study of Fermi's theory and some of its most problematic aspects led to the formulation of a unified theory of fundamental interactions, in which the weak interactions and electromagnetism appear as different expressions of a single force which is now called the electroweak force. The unified theory contains as particular limits both Maxwell's theory of electromagnetic forces and Fermi's theory of weak interactions.

2. – Beta decay and the neutrino hypothesis

In the first chapters of his book "Inward Bound", Abraham Pais tells the fascinating story of the discovery of radioactivity, and of the studies which led to the explanation of the existence of three different types of radioactivity, differentiated by the Greek letters α, β. γ. Let us briefly outline the state of knowledge just before Fermi's work, concentrating in particular on the problems posed by beta radioactivity.

The substances endowed with alpha radioactivity emit helium nuclei (α particles); for example radium (Ra), the element discovered by Marie Curie, decays into radon (Rd, a noble gas) according to the reaction:

$$\mathrm{Ra}_{88}^{226} \to \mathrm{Rd}_{86}^{222} + \mathrm{He}_2^4 \,. \tag{1}$$

For each nucleus we have indicated the "atomic number" (lower index), which represents the electric charge of the nucleus in unit e, and is also equal to the number of protons, and the "atomic weight" (upper index), equal to the total number of protons and neutrons. Alpha radioactivity leaves the total number of protons and neutrons unchanged, and corresponds to a rearrangement of the components of the initial nucleus. This process reflects a structural instability of the "father" nucleus, a phenomenon which was clearly explained by George Gamow, at the beginning of the thirties, in the frame of quantum mechanics.

In gamma type radioactivity, photons (light quanta) of very high energy are emitted; we are dealing with a phenomenon which is very similar to the emission of light by an atom, thus its existence did not create, in principle, particular problems.

The origin of beta radioactivity remained problematic up to the time of Fermi's work. Beta rays are nothing more than electrons, but where do these electrons come from? Before the discovery of the neutron (Chadwick, 1932) it was believed that the nucleus was made up of protons and electrons. As an example a helium nucleus would be made up of 4 protons and 2 electrons. This allowed an explanation of its "atomic weight", equal to 4 (the mass of electrons is negligible compared to that of protons), and its electric charge, equal to that of two protons (the charge of two of the four protons would be neutralized by the opposite charge of the two electrons). In this scheme beta decay would arise from the emission of one of the electrons in the nucleus; as in the case of alpha decay one would thus obtain a true disintegration, a rearrangement of the components already present in the initial nucleus.

The discovery of the neutron rapidly led physicists to drop the hypothesis that the nucleus contained electrons, in favour of that which is substantially the modern view of the structure of the nucleus, made up of protons and neutrons. The transition to the modern point of view was very rapid. Several measurements on the spectroscopy of nitrogen molecules, carried out by Franco Rasetti in 1929, had already put the hypothesis of the presence of nuclear electrons in a difficult position. Rasetti's results are for example mentioned in Heisemberg's 1932 paper on the new theory of nuclear structure. A question comes naturally to the mind, as soon as we accept the fact that the atomic nucleus does not contain electrons: where do beta decay electrons come from?

Beta radioactivity posed an even greater problem: electrons are not emitted with one single energy but with an energy spectrum which varies with continuity. The situation is very different from that encountered in alpha or gamma radioactive decay, in which the energy of the emitted particle equals the energy difference between the initial nucleus and the final one, and is therefore always the same for a given type of decay. For example in the disintegration of radium (eq. (1)) alpha particles are emitted with an energy of 4.88 MeV. How is it that this simple deduction does not work in the case of beta decay? Niels Bohr, the father of atomic physics, played with the idea that in the case of beta decay the energy was not exactly preserved, and Bohr's ideas enjoyed undisputed prestige.

The solution to this second problem was found by Wolfgang Pauli: in beta decay a second particle is emitted together with the electron, but it escapes detection from our instruments. The two particles can share the available energy in different proportions, in such a way that the energy given to the electron is not univocally determined. The "second particle" must be neutral, otherwise it would easily be revealed through its ionizing power, and could not be a photon, a possibility that experimental data seemed to exclude. It must be an entirely new particle never seen before.

The hypothesis of the "second particle" appeared so fanciful that Pauli himself, even perhaps not to oppose Niels Bohr, decided not to publish it. He wrote a letter to his closest colleagues, "Dear radioactive ladies and gentlemen", and called these particles neutrons. He talked about the new particle in the corridors of physics congresses but never officially. He spoke to Fermi during the Nuclear Physics Congress held in Rome in 1931, and it was on this occasion that Fermi jokingly proposed that the most appropriate name should not be neutron but neutrino, given that its mass must be very small. There was not to be any further contact between them on the subject: while Pauli adopted the name proposed by Fermi without reservation, Fermi himself preferred to use the name initially proposed by Pauli. At an international conference, held in Paris in 1932, Fermi said:

"... One could think for example, following a suggestion by Pauli, that in the atomic nucleus you can find neutrons which should be emitted at the same time as beta particles."

Following the question of a participant at a congress, Fermi made clear that these neutrons could not be the same neutrons recently discovered by Chadwick, but much lighter particles.

This brief citation demonstrates how in 1932 Fermi still talked about beta radioactivity in terms of the emission of particles already present in the nucleus. The solution introduced in the article of 1933 was to be radically different.

The modern point of view on the structure of the nucleus, made up of protons and neutrons, had its official approval during the Solvay Conference held in October 1933 in Brussels. In the discussion which followed Heisenberg's report on nuclear forces, Pauli finally came out into the open with several comments on the hypothesis of the neutrino.

3. – Fermi's Theory

The introduction of Chadwick's neutron in the structure of the nucleus disposed of the electron and left very little room for Pauli's neutrino. A new idea was needed, and this was supplied by Fermi's work which appeared in "La Ricerca Scientifica" in the December edition of the same year.

Let us briefly examine Fermi's work, considering the beta radioactive decay of a nucleus A in a nucleus B with the emission of an electron and a neutrino,

$$\mathrm{A} \rightarrow \mathrm{B} + \mathrm{e}^- + \bar{\nu}\,. \tag{2}$$

If the electron and neutrino are not present in A, they must be created during the act of transition. This conclusion was hard to accept, given that one was used to thinking of the electron as a material particle, having its own solidity and persistence.

The possibility of creating (and destroying) particles had a well-known precedent in the case of photons. Light is made up of particles, photons, which are created when light is emitted, and destroyed when it is absorbed. An atom can emit a photon when it passes from a higher level of energy to a lower one. In the reverse process, an atom can absorb a photon going from a lower energy level to a higher one.

The processes of creation and destruction of photons are described in the quantum theory of the electromagnetic field, developed by Dirac immediately after the birth of Heinsemberg's new quantum mechanics. In 1927 Jordan and Klein showed that the theory of quantum fields can be applied to any particle. In other words, electrons can be seen as particles —the classical point of view —but also as a wave phenomenon. This strange situation is well espressed in the poem in which Enrico Persico jokingly described his missionary work at the Turin University:

...Credon poi con fe' profonda	...They then believe with deep faith
Cui s'inchina la ragion	To which reason bows low
Che la luce è corpo e onda	That light is body and wave
Corpo ed onda è l'elettron	Body and wave is the electron
Sono questi i dogmi santi	These are holy dogmas
Ch'egli insegna agli infedel	That he teaches the unfaithful
Con esempi edificanti	With edifying examples
Appoggianosi al Vangel	Relying on the Gospel[1]

---

[1] The "Vangelo" (Gospel) was Enrico Persico's duplicated lecture notes, one of the first textbooks on quantum mechanics that appeared in Italy, still in use in Italian universities during the nineteen fifties.

The first traces of field-particle duality can be found in the 1905 paper on the photoelectric effect, where Einstein introduces the concept of the photon. An important consequence of the duality between field and particles is the perfect identity of particles which correspond to a given field. Two photons are "identical" in a much stronger sense than that in which this term is used in everyday language. The mother of two "identical" twins, let us call them Ida and Ada, is almost always able to distinguish them. The situation in which Ida is on the left of Ada is different from that in which Ida is on the right. To exchange two photons instead does not change the situation in any way —as one says in technical terms, the state of the system does not change.

In quantum physics the concept of particle and that of field are totally interchangeable. To every field a kind of identical particles corresponds, and vice versa: to every kind of identical particles a field corresponds. We can consider Jordan and Klein's work as a dictionary which allows us to translate from the language of fields to that of particles and vice versa.

The language of fields allowed the description of phenomena in which particles are created or destroyed, but Fermi's work on beta radioactivity is the first in which this possibility was used outside the photon theory.

At the basis of Fermi's theory is the hypothesis that the beta decay of a nucleus (see eq. (2)) is due to a new type of interaction among particles which causes the transition of a neutron into a proton with the simultaneous creation of an electron and a neutrino,

$$\mathrm{N} \to \mathrm{P} + \mathrm{e}^- + \bar{\nu}\,. \tag{3}$$

Given that proton and electron have an electric charge respectively equal to +e, −e, while neutron and neutrino are electrically neutral, the value of the total electric charge is preserved. Fermi constructed his theory starting with the hypothesis that an analogy —as close as possible— existed between this process and that which is the basis of the emission of gamma rays:

$$\mathrm{P} \to \mathrm{P} + \gamma\,, \tag{4}$$

in which one of the protons of the nucleus passes from a higher energetic state to a lower one, emitting a photon. It can be instructive to compare process (4) with the emission of radio waves where the source of radiation is the oscillating electric current in the antenna. In process (4) the photon (which is the quantum manifestation of electromagnetic radiation) is produced by the electric current which manifests itself at the moment when the proton moves from one energetic state to another.

The couple electron-neutrino emitted from process (3) is, in Fermi's theory, the analogue of the photon emission in process (4), and the mechanism of the emission is strictly analogous. Fermi proposed the existence of a new type of current, today called weak current, which manifests itself at the instant of the transition of a neutron into a proton leading to the creation of the pair e-ν.

In his 1933 papers Fermi presented the mathematical structure of the new theory, and its application to the study of radioactive beta decays. These can be divided into two classes: the allowed decays, which according to the theory could occur even if the nucleons (protons and neutrons) were stationary inside the nucleus, and the forbidden ones which are only made possible by the fact that the nucleons are in motion. Forbidden decays proceed more slowly, and their mean life is about one hundred times longer than that of permitted decays. It was known that some types of beta decay proceed more rapidly than others, but it was only with Fermi's work that this phenomenon found a quantitative explanation.

A second important result of Fermi's work is in the determination of the energetic distribution of the emitted electrons. Fermi demonstrates that a measurement of this distribution allows the determination of the neutrino's mass, and that the existing data seemed to favour a very small mass, and were compatible with a mass exactly equal to zero.

Fermis's theory contains only one unknown parameter which can be determined by measuring the mean life of "allowed" beta decays. This parameter, G, today known as "Fermi's constant" determines the intensity of the new interactions. The very small value of G earned the new interactions the name of "weak interactions". How weak are weak interactions? If their intensity was comparable to that of the forces which bind the nucleons in the nucleus, beta disintegration would be ten billion times faster, the mean life of radioactive beta matter would be ten billion times shorter than we can observe.

Before closing this section, let us give some further details on the structure of Fermi's theory. We will use for this purpose a term which came into use at a time later than that of Fermi's work, using "nucleon" to designate indifferently protons or neutrons, and the term "lepton" to designate electrons or neutrinos. The term "lepton" has in fact a wider meaning, because it is also used for two particles which are essentially "heavy electrons", the muon whose mass is about two hundred times that of the electron, and the tau, which has a mass which is three thousand five hundred times that of the electron. We now know that there are three kinds of neutrinos, associated to the three leptons (electron, muon, tau) endowed with electric charge.

We have already said that the transition between neutron and proton in process (3) generates a weak current which leads to the creation of the electron-neutrino pair. The mechanism of this phenomenon is similar to that of magnetic induction, in which a variable current in an electric circuit generates a magnetic field which in its turn can generate a current in a physically separate circuit. In Fermi's theory there is a sort of short circuit between the weak current of the nucleons, activated by the transition from neutron to proton, and a corresponding current of leptons (electron and neutrino) the activation of which leads to the electron-neutrino pair. The weak interaction is therefore, according to Fermi, a direct interaction between weak currents, without the action of an intermediate field, as in the case of magnetic induction.

At the end of the nineteen sixties a new theoretical framework was developed, in which weak interactions are much more similar to electromagnetic interactions than Fermi suspected. In the modern theory, in fact, the interaction among weak currents is, as in magnetic induction, mediated by an intermediate field, whose quanta, the W bosons, were effectively identified at the beginning of the eighties. Nevertheless Fermi's theory remains an excellent approximation to describe processes in which the energy available is much less than the mass energy (Mc^2) of the W bosons, a condition which is well satisfied by beta radioactivity, but also in all decays of elementary particles([2]).

4. – Weak interactions in the economy of the Universe

In ordinary matter, that which forms our bodies and the objects which surround us, weak interactions do not seem to play a significant role. Beta radioactivity —all radioactivity— is rightly considered a damaging phenomenon to be wary of and of which we could readily do without. Radioactive materials are relatively rare in nature, and the artificial ones, produced, for example in nuclear reactors, are accurately segregated in safe deposits.

Ignoring for a moment the practical applications of beta radioactivity, for example in medical diagnosis and for dating archaeological finds, one could ask oneself if weak interactions are not a mere curiosity in the world of elementary particles, put there only for the joy of a few physicists. The reality is very different: these interactions have a fundamental role in the economy of the universe. Weak interactions are the basis of solar energy, and they act as cosmic road sweepers, being responsible for the decay of very many particles produced during the "Big Bang" or in cosmic radiation: our lives essentially depend on the existence of weak interactions.

4·1. *Solar energy.* – The nuclear reaction on which the production of energy by the sun is based consists in the transformation of hydrogen into helium. The hydrogen nucleus is made up of a single proton (P), while that of helium contains two protons and two neutrons. During the reaction some protons must be transformed into neutrons, and this requires the action of a weak interaction. The reaction takes place in a series of steps of which the first is a weak interaction,

$$\mathrm{P} + \mathrm{P} \rightarrow \mathrm{D} + \mathrm{e}^+ + \nu\,. \tag{5}$$

This reaction transforms two protons in a deuton (D), the nucleus of the heavy hydrogen isotope, that contains a proton and a neutron. In the second step deuterium is burned up to form helium. This is a complex chain of reactions during which nuclei that are relatively heavier than helium, for example lithium and beryllium are also constructed. For the present discussion we can express this second step in a simplified way as

$$\mathrm{D} + \mathrm{D} \rightarrow \mathrm{He}_2^4 + \text{ energy} \tag{6}$$

The main part of the energy is produced during the second step which does not involve weak interactions: once they are formed the deutons are rapidly burned. The

([2]) The only exception is the decay of extremely heavy particles such as the t quark, whose mass energy is twice that of the W boson.

first reaction produces relatively little energy, but its role is fundamental, because it is there that the neutrons necessary for the energetic second step are formed. In very simple terms, without weak interactions the sun could not produce its energy.

The first reaction is also essential as a regulator of solar activity: the fact that it takes place through a weak interaction guarantees that the hydrogen is "burned" slowly, so that the sun's life can be extended over billions of years. The reaction entails extremely high temperatures and pressures, and takes place in the core of the sun.

The sun, apart from being a source of light and heat, is also a very powerful source of neutrinos. In recent decades many experiments carried out in underground laboratories have enabled the detection of neutrinos emitted from the inner regions of the sun, which reach us after having crossed the entire mass of the sun practically without being disturbed. The neutrinos produced by the PP reaction (see eq. (5)) are those with the lowest energy, and were observed for the first time in 1991 in the "Gallex" experiment which operates in the Gran Sasso laboratory. The results of this and other experiments have presented a puzzle: the flux of neutrinos which is observed is definitely smaller than what is expected. The most recent results have demonstrated that the solution of this puzzle is given by the phenomenon of neutrino oscillations, proposed in the sixties by Bruno Pontecorvo and which we will discuss in the following.

4'2. *The cosmic sweeper.* – New types of particles were discovered in the years following Fermi's work; the first were revealed as far back as the thirties during studies of the cosmic radiation, but the rhythm of the discoveries increased in the second half of the last century, with the emergence of powerful particle accelerators.

The new particles are short-lived; once they have been produced, for example during cosmic radiation, they disappear in a more or less brief period of time changing themselves into stable particles: protons, electrons, photons, neutrinos.

In the absence of weak interactions, some of these particles would be stable. For example hyperons would be stable; these are particles which are very similar to the nucleons, but where one or more u quarks are substituted with s quarks. Consequently new types of stable nuclei would exist. The stability of muons would be even more worrying, since these particles can displace electrons and form new exotic species with unusual chemical properties. The presence of weak interactions guarantees that these particles, and many others, are sweeped away to leave the relatively simple structure of ordinary matter.

The exploration of hypothetical consequences of the absence of weak interactions could appear as an idle curiosity and we will not pursue it further. Although it is clear that weak interactions have, even with their work of "ecological operator" an important role in the structure of our world.

5. – Weak interactions after Fermi

The story of weak interactions can be roughly subdivided into two periods. The first period starts with Fermi's 1933 work and finishes with the discovery of the violation of parity (T. D. Lee and C. N. Yang, 1956) and with the first synthesis represented by

"$V - A$ Theory" (Feynman and Gell-Mann, Marshak and Sudarshan, 1958). The second period goes up to the present day, and has seen the emergence of a unified theory of weak and electromagnetic interactions which has fully confirms Fermi's hypotheses of a strong similarity between two types of interactions which are apparently so different.

A characteristic which is common to all of Enrico Fermi's scientific work is his search for the simplest and most effective solution, achieved with his own legendary "sense of physics". This characteristic is fully illustrated in his "Tentativo di una teoria dell'emissione dei raggi beta" (Tentative theory of beta rays): Fermi did not try to determine the most general rule for weak interactions, but he concentrated on the more promising one starting from the hypotheses of a strict analogy between weak and electromagnetic interactions. If one abandons this hypothesis one can obtain a more general — but more complex — theory, characterized by five arbitrary parameters, against the single G parameter of Fermi's theory.

The existence of theories more general than Fermi's stimulated the experiments which finally allowed the selection of the correct theory. This research was only concluded between 1956 and 1958 after the discovery of the violation of parity, and led to the theory known as $V - A$ which, apart from the violation of parity, is substantially identical to the one initially proposed by Fermi.

This first period witnessed several important discoveries, first of all that of muons (Anderson 1936), charged particles with a mass equal to two hundred times the mass of the electron. Initially these were identified with Yukawa's mesons, but the experiments carried out by Conversi, Pancini and Piccioni (1946) showed that the muon is essentially a "heavy electron", a completely unexpected new particle.

The following year saw the discovery of the π meson (Lattes, Occhialini, Powell, 1947), a particle endowed with strong interactions which could be identified with Yukawa's meson, the hypothetical quantum of the nuclear forces. Also in 1947 the first examples of "strange" particles were discovered in the cosmic radiation, those which we now call K mesons and hyperons (Λ, Σ, Ξ). What is *strange* about these particles is that they are capable of strong interactions, but the intensity of their decay, for example

$$\mathrm{K}^+ \to \pi^+ + \pi^+ + \pi^- , \tag{7}$$

is characteristic of weak interactions. This decay is the first example of a new type of weak interaction which does not involve the emission of electrons. The study of K mesons led in the fifties to the first clues to the breakdown of specular symmetry in weak interactions.

In 1956 Cowans and Reines managed to directly reveal the neutrinos produced from the intense concentration of radioactive material in a nuclear reactor. For a quarter of a century the neutrino had remained a hypothesis, certainly a convincing one, in the light of Fermi's successful theory, but still a hypothesis. The experimental discovery of Pauli's neutrino, that of the violation of parity and the achievement of the $A-V$ theory, so similar to Fermi's original theory, close this first period of the history of weak interactions.

5·1. *Quark mixing.* – Fermi had built up his beta decay theory starting from the similarity between this phenomenon and electromagnetic interactions but was it really a close analogy? The intensity of electromagnetic interactions among elementary particles is determined by their electrical charge, which is always a multiple of the electron's charge([3]), e. The electron charge therefore represents a universal unit of measure for the intensity of electromagnetic interactions. Is there a similar universal unit of strength for weak interactions? The question emerged with the discovery of new types of particles and new types of weak interaction and was for the first time clearly posed by G. Puppi, who considered three processes of weak interaction: beta radiation of the neutron (eq. 3), the decay of the muon into an electron and two neutrinos,

$$\mu^- \to \mathrm{e}^- + \nu + \bar{\nu}\,, \tag{8}$$

and its capture by a nucleus with the emission of a neutrino,

$$\mu^- + \mathrm{A} \to \mathrm{B} + \nu\,, \tag{9}$$

a process identified for the first time in the famous experiment by Conversi, Pancini and Piccioni.

The three processes studied by Puppi can be described in a Fermi-like theory, and with the data available in the fifties seemed to be characterized by the same intensity: the first indication that weak interactions have, like electromagnetic ones, a universal unit of strength.

At the beginning of the sixties the situation became more complicated when the first data on the decay of strange particles with the emission of an electron-neutrino pair became available, for example the decay of the hyperon Λ,

$$\Lambda \to \mathrm{P} + \mathrm{e}^- + \bar{\nu}. \tag{10}$$

This process is very similar to the decay of the neutron, see eq. (3), but the experimental results indicated that in this case the weak interactions act with an intensity which is almost five times smaller than in beta radioactivity. At the same time more precise data and a more accurate theoretical analysis showed that in beta radioactivity and in muon decay the weak interactions show a different intensity, if only by a few percent.

The solution to these problems was found by this author in 1963. I will try to explain it in simple terms, even if the use of some elementary notions of quantum mechanics is inevitable. The reader who finds it difficult to understand the following sentences could by all means skip a few lines.

Neutron and Λ decay (see eq. (3) and (10)) lead to the same final state and can therefore interfere among themselves (the photon is corpuscle and wave...). A certain

---

([3]) In the case of quarks the charge is a multiple of $e/3$, where e is the electron charge. More exactly we could say that all the particles bear a charge which is a multiple of $e/3$.

combination of N and Λ will exist in which the interference is destructive and another in which the interference it highly constructive, leading thus to a situation where the weak interaction acts with maximum intensity. There is no sense in comparing separately the intensity of muon decay (see eq. (8)) with that of reactions (3) and (10), but rather, we need to compare it with the decay of that particular combination of N and Λ which displays the maximum constructive interference. This combination, N_θ is determined by a "mixing angle" (the Cabibbo angle) between N and Λ,

$$N_\theta = N\cos(\theta) + \Lambda\sin(\theta). \tag{11}$$

The experimental data allows us to reconstruct what the intensity of weak interactions would be in the imaginary process

$$N_\theta \rightarrow \mathrm{P} + \mathrm{e} + \nu. \tag{12}$$

The intensity of this process coincides with that which manifests itself in muon decay; weak interactions are therefore, like electromagnetic ones, characterized by a universal intensity. This conclusion has opened the way to the modern unified theories.

The angle θ (the Cabibbo angle) which appears in eq.(11) reflects a new property of weak interactions: they do not act directly on the common elementary particles, for example N and Λ, but on certain mixtures between them. For a more correct discussion than what is presented here we should take into account the fact that N and Λ are particles made up of quarks, and that the phenomenon of mixing which we have discussed really comes about at the level of d and s quarks.

The discovery of new types of quark (charm, beauty, top) has greatly enriched the picture of the effects which arise from the mixing among quarks. The first effects of mixing dealt with the relationship between the decays of "strange" and "common" particles, but today we can also examine these effects in the decay of particles endowed with "beauty" (quark b) or "charm" (quark c). A complete description of quark mixing requires four parameters, one of which coincides with the original angle θ. Mixing among quarks is very probably at the root, as proposed by Kobayashi and Maskawa, of symmetry violation between matter and antimatter, first observed in decays of the K^0 mesons. These ideas have recently (2001) found a brilliant confirmation in the experiments on the decays of B mesons (particles which contain a b quark), carried out at the Stanford laboratory in California and at Tsukuba in Japan.

5·2. *Neutrino oscillations.* – The effects of interference among different kinds of decay, with which we have illustrated the mixing phenomenon, cannot be directly observed but only reconstructed by theoretical analysis. The reason is very simple: to be able to interfere, two waves must be able to maintain the relative phase, therefore have the same frequency. Given that the frequency ν of the wave associated to a particle is proportional to its energy E (according to Planck one has $E = h\nu$), interference is only possible among particles which have the same energy. If the energies (therefore the frequencies) of two particles are very near, one could observe beats, but for very different energies, like those of a neutron and a Λ hyperon, which have widely different masses, the interference is effectively washed out.

The situation is very different in the case of neutrinos: from the time of Fermi's work it was known that neutrinos have very small masses, and for a long time it was thought that their mass could be exactly zero. Bruno Pontecorvo, one of Enrico Fermi's collaborators at the time of "via Panisperna", had the idea that a phenomenon of mixing, similar to the one that I have proposed, can lead to observable interference phenomena which take the name of "neutrino oscillations". The reason is that since the masses of neutrinos are very small their differences are in absolute terms extremely small. This opens the way for the observation of interference between different neutrino types. The effects predicted by Pontecorvo would in fact vanish if the neutrino masses were exactly equal.

After years of research with uncertain or negative results, a positive result was reached few years ago by researchers from the Japanese laboratory "Kamioka", which studied neutrinos produced by cosmic radiation in the atmosphere. These results have been confirmed by the "Macro" experiment in the Gran Sasso laboratory. A new and more accurate experiment, in preparation at the moment, will use neutrinos launched from the CERN laboratories in Geneva toward the underground laboratories of Gran Sasso.

A separate piece of evidence for the existence of neutrino oscillation is supplied by the discrepancies between the measurements of solar neutrinos and the theoretical expectation. Given that the mixing of neutrinos, like that of quarks, can be described by four parameters, one of which is associated to a possible violation of the symmetry between matter and antimatter, a complete clarification of the situation will keep physicists busy for several decades to come.

6. – Conclusions

With his theory Fermi demonstrated his capacity to give a simple and accurate solution to a problem of physics, but at the same time his capacity to recognize in a physical phenomenon the mark of a mathematical idea which up to then could have appeared to be a purely abstract construction.

Two facts allow us to evaluate the importance of the field of research opened up by Enrico Fermi's paper on weak interactions. The first concerns Nobel prizes, a rough but efficient way to measure the importance of a specific area of scientific research and the progress attained in it: more than ten Nobels for physics have been attributed to discoveries relevant to weak interactions. If Fermi had not been given a Nobel for his research on neutrons he would certainly have deserved one for the discovery of weak interactions.

A second evaluation of the importance of Enrico Fermi's discovery can be deduced from the fact that over half the experiments in progress at the moment or in preparation with particle accelerators —at CERN in Geneva, at Fermilab in Chicago, at Stanford or Frascati, or Tsukuba in Japan, or Novosibirsk in Russia— are dedicated to the study of different aspects of weak interactions. The same prevalence in studies on weak interactions can be noticed in the experimental programmes carried out in large underground laboratories such as the Italian one at Gran Sasso, the Japanese one at Kamioka, and others still in Canada and the United States.

Fermi's theory of weak interactions has become an essential component of the more general theory of elementary particles which goes under the name of "Standard Model". These developments are described in the contribution by M. Jacob and L. Maiani in this volume. It is important to remember though that Fermi's theory still maintains its value today, both for the validity of the solutions proposed and as a stimulus for research which has kept physicists busy for almost seventy years, and that will still do so for decades to come. In this theory Fermi's greatness is reflected, the signature of a great physicist.

About the Author

NICOLA CABIBBO teaches elementary particle physics at Rome University "La Sapienza". He chairs the Pontifical Academy of Sciences. He has been president of INFN and ENEA. He discovered "quark mixing" and he is interested in various problems which range from the interaction between radiation and crystals to the magnetism in superconductors. He directed the realization of APE parallel supercomputers.

Nuclear physics from the nineteen thirties to the present day

Ugo Amaldi

A systematic presentation of seventy years of development in nuclear physics would need much more space than what is available here; for this reason only three important moments of this fascinating intellectual adventure are focused on.

The first part is about the birth, in the thirties, of this type of physics and its first developments emphasizing in particular the scientific work of Enrico Fermi and his collaborators. The second chapter is dedicated to the state of nuclear physics half-way through the sixties, when the study of what was then called "particle physics" had not long been separated from the experimentation and interpretation of phenomena which occurr in atomic nuclei.

To pay homage to Enrico Fermi on the centenary of his birth, during the discussion of these two themes ample references are made to the writings of his first pupils: Edoardo Amaldi and Emilio Segrè.

The last chapter deals with three of the themes on which nuclear physical research is concentrated today; such topics have been chosen for their importance and representation of new lines of development, but without any pretence of completeness.

1. – The thirties: birth and first development

1·1. *The "Physics Report" by Edoardo Amaldi.* – As a guide to this extraordinary period of growth in modern physics we use the essay published in 1984 by Edoardo Amaldi for "Physics Reports", a publication of monographies from the prestigious scientific journal "Physics Letters".

The large volume (331 pages with 924 bibliographical references and notes) is entitled "From the Discovery of the Neutron to the Discovery of Nuclear Fission" [1]. The index gives a precise idea both of the contents and of the tone, which is at times conversational and anecdotical but always remaining rigorously scientific:

1. *The discovery of the neutron opens the great season of nuclear physics (1932-1933).*
2. *Other highlights of the same period of nuclear physics (1932-1933).*
3. *Beta decay discloses the existence of a new particle and a new interaction (1930-1939).*
4. *Artificial radioactivity, the road to thousands of new nuclear species for research and applications. The photoneutron and (n, 2n) reactions (1934-1937).*
5. *Low velocity neutrons reveal new quantum effects in nuclei (1934-1935).*
6. *Long-lived systems: Neutron resonances and isomeric states (1935-1936).*
7. *Slowing down of fast neutrons; diffusion and diffraction of slow neutrons.*
8. *Other fundamental discoveries and results.*
9. *An intriguing puzzle: What happens to Thorium and Uranium irradiated with neutrons?*

1·2. *The discovery of the neutron.* – Nuclear physics was born with the discovery of the neutron, which was preceded in 1930 by the observation carried out by W. Bothe and H. Becker of penetrating radiation emitted from several light nuclei, including beryllium, when it was irradiated with alpha particles emitted from polonium. In the first chapter Amaldi writes [2]: "*In a general review of transmutation and excitation of nuclei by* (α) *particles, presented by Bothe to the International Conference on Nuclear Physics, held in Rome in October 1931, he* [Bothe] *reported on new measurements of the absorption in Pb and Fe of the secondary radiation emitted from Be. The result of this new experiment, made under better geometrical conditions, was that the observed penetration was so high to require photons of energy about twice the energy of the incident* (α) *particles.*"

Several experiments followed, in particular one by the couple Joliot-Curie, which were very difficult to explain with the hypothesis that during the reaction high energy photons were emitted. Less than six months later, in February 1932, James Chadwick measured the ionization produced by single particles of the secondary radiation of beryllium. As written in "Physics Report" [3], these "*results were summarized in a Letter to the Editor of Nature, dated 17 February, 1932, and entitled "Possible existence of the neutron" which is universally considered the birth-certificate of the neutron*" and of nuclear physics. Since then a nucleus is thought of as a whole of N neutrons and Z protons bound together, so that the mass number A is equal to $A = N + Z$. Amaldi continues [4]: "*Heisenberg, Majorana and Wigner were the first to appreciate fully the importance of the new model describing the nucleus as a system composed only of protons and neutrons. The formalism of nonrelativistic quantum mechanics could be applied to it for trying to explain qualitatively, and in part quantitatively, a few nuclear properties. Such a programme clearly involved the introduction in the Schrödinger equation of a 'potential' describing the new forces acting between the nuclear constituents.* [that is the nucleons]. *The choice of this potential were made by the three authors mentioned above, in different ways. Each of them had its grounds and merits and remained as a useful (or even necessary) ingredient of many successsive developments.*"

Different types of forces were introduced in this way (exchange or ordinary) which are still called after Heisenberg, Majorana and Wigner. With them it was possible to explain nuclear saturation, which is the fact that the energy binding the nuclei is approximately proportional to the number A of nucleons, so that the binding energy per nucleon is practically independent from A and for medium and heavy nuclei equals approximately 8 MeV.

Between 1932 and the beginning of 1934 a large number of discoveries was made. Limiting ourselves only to nuclear physics, we need to remember: the first reactions of nuclear disintegration produced by protons artificially accelerated with an electrostatic accelerator (Cockcroft and Walton, 1932) the discovery of deuterium —which is the isotope of hydrogen with mass number $A = 2$ (Urey *et al.*, 1932), the acceleration of deuterons (nuclei of deuteron) and of protons at energy greater than one MeV in the new cyclotron measuring 27 inches (Lawrence and Livingston, 1933), the discovery of tritium —one isotope of deuterium with $A = 3$ (Oliphant, Kinsey and Rutherford, 1933), the theory of beta decay, in which the emission of electrons is interpreted as the transformation of a neutron into a proton with the contemporary emission of an (anti)neutrino (Fermi, 1933), the discovery of the photo-disintegration of the deuteron (Chadwick and Goldhaber, 1934) and, lastly, the discovery of a new type of radioactivity induced in an aluminum target irradiated with alpha particles (nuclei of ^{4}He) in which anti-electrons (positrons) are emitted instead of the electrons emitted in natural beta radioactivity (Joliot and Curie, 1934).

1·3. *Radioactivity induced by neutrons.* – These discoveries are discussed in chapters 2 and 3 of "Physics Report" by Amaldi who, in chapter 4, writes [5]: "*After the papers of Joliot and Curie were read in Rome, Fermi, at the beginning of March 1934, suggested to Rasetti that they should try to observe similar effects with neutrons* [instead of alpha particles] *by using the Po(α) + Be source prepared by Rasetti. About two weeks later several elements were irradiated and tested for activity by means of a thin-walled Geiger-Müller counter but the results were negative due to lack of intensity. Then Rasetti left for Morocco for a vacation while Fermi continued the experiments. The idea then occurred to Fermi that in order to observe a neutron induced activity it was not necessary to use a Po(α) + Be source. A much stronger Rn(α) + Be source could be employed, since its beta and gamma radiations (absent in Po(α) + Be sources) were no objection to the observation of a delayed effect.* [...] *When Fermi had his stronger neutron source (about 30 millicurie of Rn), he systematically bombarded the elements in order of increasing atomic number, starting from hydrogen.* [...] *Finally, he was successful in obtaining a few counts on his Geiger-Müller counter when he bombarded fluorine and aluminum.*"

The discovery of artificial radioactivity induced by neutrons in these light elements, in which usual electrons instead of positrons are emitted, was correctly interpreted by Fermi as being due to a new type of nuclear reaction —indicated with the symbol (n, α)— in which a nucleus absorbs a neutron and emits an alpha particle. In particular for aluminum he wrote the nuclear reaction in the form of

$$^{1}\mathrm{n} + {}^{27}\mathrm{Al} \longrightarrow {}^{24}\mathrm{Na} + {}^{4}\mathrm{He}$$

and stated that "*^{24}Na formed in this way would be a new radioactive element and would be transformed into ^{24}Ca with the emission of a beta particle (an electron)*" [6].

"*Fermi wanted to proceed with the work as quick as possible and therefore asked Segrè and me* —says Amaldi [5]— *to help him with his experiments.* [...] *A cable was sent to Rasetti, asking him to come back from his vacation. The work immediately was organized in a very efficient way. Fermi, helped a few days later by Rasetti, did a good part of the measurements and calculations, Segrè secured the substances to be irradiated and the necessary equipment and later became involved in most of the chemical work. I took care of the construction of the Geiger-Müller counters and of what we now call electronics. This division of the activities, however, was not rigid at all and each of us participated in all phases of the work.*" In a few months about sixty elements were irradiated and forty-four new nuclides were discovered; in as many as sixteen cases the chemical separation of the radioactive element was obtained using the technique of chemical transport, which the chemist Oscar D'Agostino, who had been invited to join the group of physicists in Via Panisperna, had mainly worked on. In all the cases it was found that in radioactivity artificially induced by neutrons, electrons were always produced and not positrons, as should be expected for nuclides rich in neutrons produced in nuclear reactions of the type (n, α), (n, p) and (n, γ).

In the final work published in "Proceedings" by the Royal Society [7] the case in which the radioactive nuclide produced by neutron capture is the isotope of the original nuclide, was discussed in detail. Two possible reactions were considered, the reaction (n, 2n), with the emission of two neutrons, and the reaction (n, γ), in which after the capture a gamma quantum is emitted. Since the capture of a neutron frees its binding energy, about 8 MeV, and the emission of two neutrons requires twice this binding energy, the reaction (n, 2n) is endoenergetic while the (n, γ) reaction is esoenergetic.

The second was therefore the interpretation to favour, but it "*...gives rise to serious theoretical difficulties when one tries to explain how a neutron can be captured by a nucleus in a stable or quasi-stable state. It is generally admitted that a neutron is attracted by a nucleus only when its distance from the centre of the nucleus is 10^{-12} cm in size. It follows that a neutron of a few million volts' energy* [which is the energy possessed by neutrons emitted in (α, n) reactions typical of radon-beryllium sources] *can remain in the nucleus (i.e. have a strong interaction with the constituent particles of the nucleus) only for a time of about 10^{-21} s, that is of the order of the classical time needed to cross the nucleus. The neutron is captured if, during this time, it is able to lose its excess energy, (e.g. by emission of a (γ) quantum). If one evaluates the probability of this emission process by the ordinary methods, one finds a value much too small to account* *for the observed cross-sections. In order to maintain the capture hypothesis, one must then either admit that the probability of emission of a* (γ) *quantum (or of an equivalent process as, for example, the formation of an electron-positron pair) should be much larger than is generally assumed; or that, for reasons that cannot be understood in the present theory, a nucleus could remain for at least* 10^{-16} *s in an energy state high enough* [that is over 8 MeV] *to permit the emission of a neutron*" [8]. Amaldi continues "... [...] *the proof that neutrons can undergo radiative capture with an appreciable cross-section was one of the first experimental evidences that the "one-particle model" of the single nuclear particle is inadequate for describing many important properties of nuclei.*" The explanation of this important phenomenon had to wait for the discovery of the great probability of absorption of slow neutrons, certainly the most important contribution to nuclear physics that is owed to Fermi's group.

1·4. *Slow neutrons*. – The history of this discovery began with a series of mysteries that put the young men who worked in the institute of Via Panisperna in a difficult position.

In September 1934 Amaldi and Segrè repeated the measurements of radioactivity induced in aluminium with the intention of verifying if this was due to an (n, γ) reaction and immediately sent the interesting result by telegram to Fermi, who had stopped off, in London for an international congress on his way back from South America. Fermi mentioned this important observation in his speech, but when he returned to Rome he had reason to be worried. In fact a few days later the measurement had been repeated by Amaldi alone, because Segrè was in bed with a cold, and a different half life in the induced radioactivity was found. Segrè tells how much the two young researchers, who were sure of their first results, regretted Fermi's irritation and adds: "...*I could not imagine what was happening. What is more, Edoardo was finding several new inexplicable phenomena; in a few weeks the mysteries were multiplied*" [9].

In "Physics Report" the events are told as follows [10] "[...] *we decided to try to establish a quantitative scale of activities* [...] *This work was assigned to me and B. Pontecorvo (b. 1913), one of our best students, who had taken the degree (laurea) in July 1934 and after the summer vacations had joined the group. We started by studying the conditions of irradiation most convenient for obtaining well reproducible results* [...] *We immediately found, however, some difficulty because it became apparent that the activation depended on the conditions of irradiation. In particular in the dark-room, where usually we carried out the neutron irradiation, there were certain wooden tables near a spectroscope which had miraculous properties. As Pontecorvo noticed accidentally silver irradiated on those tables gained more activity than when it was irradiated on the usual marble table in the same room. These results, daily reported to Fermi and the others, were friendly, but at the same time strongly, criticized by Rasetti, who, in a teasing mood, insinuated that I and Pontecorvo were unable to perform 'clean and reproducible measurements.'* "

On the morning of 22 October most of the young men from Via Panisperna were busy examining students and Fermi, finding himself alone in the laboratory, decided to continue the measurements but, instead of using a lead absorber between the source of neutrons and the silver to be activated, he decided to use a piece of paraffin wax. Enrico

Persico from Turin University was present in the laboratory; he was in Rome by chance and helped Fermi to write the first results of the measurements in the log-book of the laboratory. At the end of the morning the increase in induced radioactivity caused by the presence of the paraffin was shown to Bruno Rossi of Padua University, who was also visiting that day and to the other members of the group who had finished the examinations.

Many years later Fermi himself told Chandrasekhar how it had happened [11]. "*We were working very intensely on radioactivity induced by neutrons and the results did not make sense at all. One day while I was going to the laboratory, it occurred to me to study what would happen if I placed some lead in front of the source of neutrons. I took a long time to work the piece of lead very carefully on the lathe, which was unusual for me; I was clearly dissatisfied with something and was looking for every possible excuse for delaying the moment for putting the lead in place. At a certain point I said to myself: 'No, I do not want a piece of lead here: what I want is a piece of paraffin.' And that is how it was, without prior warning or conscious reasoning. I immediately took any old piece of paraffin and put it there where I should have put the piece of lead.*"

The events of October 22, 1934 have been told by many who were present with additions and different tinges. Bruno Pontecorvo for example, reports several illuminating sentences spoken by Fermi on that occasion [12]:

"*The result was clear: the paraffin 'absorber' did not diminish the activity but (even if by very little) it increased it. Fermi called us all and said: 'This fact presumably comes about thanks to the hydrogen contained in the paraffin; if a small quantity of paraffin gives an evident result in any case, let us try and see what happens with a larger quantity.' The experiment was immediately carried out first with paraffin and then with water. The results were astounding: The activity of the silver was hundreds of times greater than that which we had achieved before! Fermi put an end to the noise and agitation of his collaborators by saying a famous sentence which, they say, he repeated eight years later at the moment when the first reactor was started up: 'Let's go to lunch'...* [On the discovery of slow neutrons] *several casual situations and, the depth and intuition of a great intellect have both played a substantial role. When we asked Fermi why he had used a paraffin wedge and not a lead one, he smiled and with a mocking air articulated: 'C.I.F.' (Con Intuito Fenomenale —With Phenomenal Intuition). If the reader of this statement gets the idea that Fermi was immodest, he would be completely wrong. He was a direct man, very simple and modest, yet conscious of his own capacities. On this subject, when he returned to the Institute of Physics after lunch that famous day, and with incredible clarity explained the effect of the paraffin, introducing in this way the concept of the slowing down of neutrons, he said with absolute sincerity: 'What a stupid thing to have discovered this phenomenon by chance without having known how to foresee it.'* "

The explanation found by Fermi during the lunch break is well known: neutrons of several MeV —emitted from a radon-beryllium source— lose practically all this kinetic energy in a great number of successive collisions with protons, *i.e.* with the nuclei of hydrogen atoms contained in paraffin. Slowed down to the same speed as molecules, the neutrons stay longer near the silver nuclei in the target and are therefore much more efficient in inducing radioactivity.

The interpretation requires that, contrary to what was expected at that time, the probability by a nucleus to capture a neutron greatly *increases* with the *decreasing* speed of the neutron; this is the essential core of the discovery made in Rome on October 22, 1934, confirmed in the afternoon for copper and iodine (but not for silicon, zinc and phosphorous) and described in a short memorandum) [13], which was prepared in a meeting full of excitement held at Amaldi's house the same evening. The next day the letter (signed by E. Fermi, E. Amaldi, B. Pontecorvo, F. Rasetti and E. Segrè) was taken to the "Ricerca Scientifica" by Ginestra Amaldi who worked there as editor assistant.

It was immediately clear to all the members of the group that the discovery was also important for practical purposes because the rate of production of artificial radioactive isotopes increases greatly when slow neutrons are produced by any source immersed in speed "moderating" substances in that they are rich in hydrogen. In particular the researchers from Via Panisperna immediately thought of uses for these radioactive isotopes for medical purposes and as tracers with —physical, chemical and biological— aims. Orso Mario Corbino, the director of the Institute of Physics, immediately suggested taking out a patent, which was done on 26 October by the authors of the work and O. D'Agostino and G. C. Trabacchi, who were added among the inventors considering the help they had given to neutron research in the month preceding the discovery.

As Amaldi writes [14]: "*We were extremely pleased and amused, but not so much because a patent could result, sometime in the future, in a financial benefit for the 'inventors', but rather because a work, carried out with great energy and dedication, only for its intrinsic scientific merits, had, unexpectedly, brought us to applications which, in addition, would be mainly of a scientific and medical nature.*" In fact, because of a series of circumstances linked first to the Second World War and then to the Cold War, the economic benefit was small but the satisfaction remained great. It is enough to think that seventy years later about 80% of all diagnostic examinations in nuclear medicine use a technetium isotope, which is produced from thermal neutrons produced in nuclear reactors. It is interesting to recall that the element technetium was discovered in Palermo a few years later by Emilio Segrè and Perrier in a piece of metal irradiated in the cyclotron at Berkeley.

Going back to the research carried out in Via Panisperna, in the following months, sixty new radioactive isotopes were discovered and it was confirmed that the probability of neutron capture varies approximately with the $1/v$ law, *i.e.* it increases with the inverse of the speed. But soon it was observed that this law was not valid for all nuclei; in 1935 Bjerge and Westcott, at the Rutherford laboratory near Oxford, and Moon and Tillman at the Imperial College London found several exceptions so that, after the Summer of 1935, Amaldi and Fermi undertook a systematic study of the phenomena of the absorption of neutrons filtered by different substances. The other members of the group were no longer in Rome since Rasetti was in the United States, Segrè had been nominated professor in Palermo and Pontecorvo was a theoretical physicist and was preparing to go to work with the Joliots in Paris.

Amaldi and Fermi worked with enormous energy and concentration for many months "...[...] *as if by our own more intensive efforts we wanted to compensate for the loss of manpower in our group*" [15], and within a few months they had published six papers

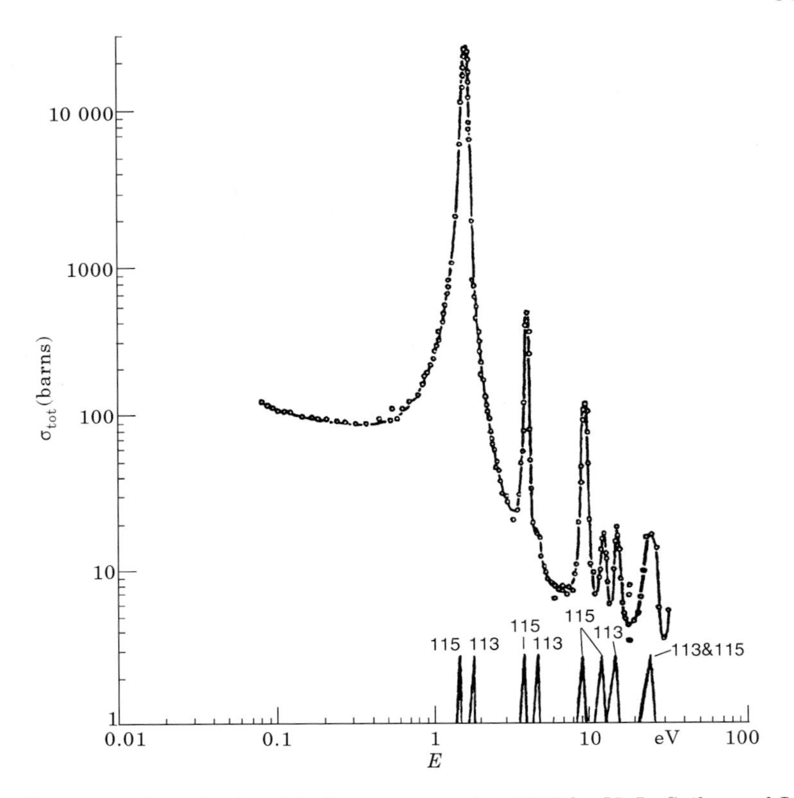

Fig. 1. – Total cross-section of natural indium measured in 1952 by V. L. Sailor and L. H. Boost as a function of the energy of the captured neutron expressed in eV; 1 barn is 10^{-24} cm^2.

in the "Ricerca Scientifica". The conclusion of this systematic approach, combined with results obtained by Leo Szilard in Oxford, was that in many elements, the neutrons, whose kinetic energy is within several bands characteristic of the nuclear target, are exceedingly well absorbed.

This new phenomenon completed the series of discoveries made in Rome between 22 October, 1934 and 29 May, 1936, the day on which Amaldi and Fermi sent their ponderous work which concluded the systematic study of absorption and diffusion of neutrons to the "Ricerca Scientifica" and the "Physical Review" [16]. Summing up they are:

1. Radioactivity produced by neutrons slowed down by impact with hydrogen and other light nuclei.

2. The $1/v$ law according to which the slower the speed v of neutrons, the more they are absorbed by nuclei.

3. The huge cross-section of cadmium and the existence, in cadmium, and in many other nuclides, of bands of absorption of slow neutrons.

4. The effect of the chemical bond on the absorption of neutrons.

The discovery of absorption lines was particularly important for the future of nuclear physics. These bands were immediately studied with ingenious methods but they only appeared in a spectacular way after the war when thanks to nuclear reactors accurate measurements could be made. As an illustration of the phenomenon of selective absorption, the "Physics Report" quotes the cross-section of natural indium which has two isotopes with mass numbers 113 and 115 (see fig. 1).

1·5. *The model of the compound nucleus.* – In less than a year and a half nuclear physics had changed its nature. In fact if, as we have seen, nuclear theory at the beginning of 1934 was not able to explain the cross-section of (n, γ) reactions, observed when neutrons are endowed with an energy of a few MeV, it could interpret even less the absorption bands observed for epithermal and thermal neutrons, those which have energy lower than about ten electronvolts. At the beginning Niels Bohr could not believe these results, so that in June 1934 he wrote to Rutherford [9]: "*I am full of doubts about Fermi's idea that in certain cases neutrons attach themselves directly to the nuclei with emission of radiation* [that is producing reactions (n, γ)]; *it seems much more probable to me that the result is the expulsion of two neutrons* [reaction $(n, 2n)$]." But the confirmation of the majority of the reactions (n, γ) and the discovery of nuclear resonance obliged him to reconsider his ideas. So at the beginning of 1936 Bohr published an important work [17], which was followed shortly after by another that was just as important by Breit and Wigner [18].

The penetration of a neutron in a medium or heavy nucleus frees its binding energy, about 8 MeV, which has to be added to the kinetic energy to calculate the total energy acquired by the nucleus. With the acquisition of this energy as Bohr writes [17], "*...a collision between* [...] *a neutron and a heavy nucleus first of all gives rise to the formation of an extraordinarily stable system. The next possible breakdown phenomenon in this intermediate system with ejection of a material particle, or its transition to a final state with the emission [of a quantum of] radiation, must be considered as different processes in competition between them and which are in no way connected with the first stage of the encounter.*"

In chapter 6 Amaldi reproduces the figure used by Bohr to explain his model (see fig. 2) and he explains it like this [19]: "*The extraordinary stability of the intermediate state is explained by Bohr by noticing that as soon as the incident neutron has entered the nucleus it starts to collide with its 'constituents'. As a consequence its energy* [always greater than 8 MeV] *is rapidly shared among many particles none of which acquires an energy large enough to leave the nucleus* [...]. *According to this model a nuclear process can be described as taking place in two independent steps; the first is the formation on an 'excited compound system' as a consequence of the capture of the incident neutron, the second is the decay of the compound system or 'intermediate state' either by emission of a particle or by irradiating a photon.*"

In the same work in which Bohr proposes this '*model of the compound nucleus*' he also explains the selective absorption of neutrons. The presence of absorption lines in the case of slow neutrons is due to the fact that in medium and heavy nuclei, when the nuclear excitation energy is smaller than about ten MeV (typical in the capture of

Fig. 2. – Model made in the workshops at the Physics Institute of Copenhagen to illustrate Bohr's idea on the formation of the excited compound system following the capture of a neutron.

thermal and epithermal neutrons), the states of the composite nucleus are well separated from one another; in correspondence to each of them the neutrons are more likely to be captured and the cross-section has a "peak". Several of these peaks are shown in fig. 1. For excitation energy in the order of 15 MeV —typical in the capture of fast neutrons which have several MeV of kinetic energy— the energy levels of the compound nucleus are very near one another and tend to overlap; there are therefore no absorption bands and the cross-section varies following approximately the $1/v$ law.

While in the brief letter he sent to "Nature" Bohr did not write a single formula , Breit and Wigner's paper contained a lot of them and, in particular, for the first time the very famous "Breit and Wigner formula" . This formula expresses the behaviour of the cross-sections for neutron capture and for neutron scattering in the neighbourhood of an isolated nuclear resonance, and reproduces, for instance, the shape of the peaks of fig. 1. These two complementary papers, published at the beginning of 1936, represent a milestone in the construction of nuclear physics.

1·6. *The nucleon-nucleon forces.* – Since the thirties the experimental and theoretical study of nuclei has followed two parallel and complementary approaches: the construction of models (which permit the explanation of energy levels, the probability of transition between levels and nuclear reactions) and the justification of these models starting from the nucleon-nucleon forces. Physicists have always hoped to be able to derive the properties of the nucleus from the knowledge of the interactions between their components, but today this goal still has not been reached (apart from some light nuclei) and indeed many think that it never will be.

The experimental study of proton-proton scattering, begun at Carnegie Institute in Washington in 1926, in 1934 continued at the Berkeley cyclotron. The following year in Rome Amaldi and Fermi proved that at low energy the neutron-proton force is equal to the proton-proton one [9]. This and other results allowed Breit, Condon and Present to state in a famous paper, also published in 1936 [20], that "*the agreement between the experimental values of proton-proton and proton-neutron interactions in the* ^1S *state suggests that without considering Coulomb effects* [that is electric] *and spin, the interactions among heavy particles* [nucleons] *are independent from their charge.*" We know today that this very important property (charge independence) of nuclear forces which act between couples of nucleons is a consequence of the composite nature of nucleons, made up of three quarks with a mass much smaller than that of a nucleon.

In the field of internucleonic force the most important development was made by Hideki Yukawa in a paper that, published at the beginning of 1935 [21], remained unknown in the West for more than two years. As Amaldi explains [22], this paper —taking up a proposal previously made by others according to which the neutron-proton force was due to the exchange of an electron-neutrino pair— "*...it begins with two observations and a proposal. The first observation concerns the importance of the exchange forces, the second refers to the unfruitful attempts mentioned above. They are contained in the first ten lines of Yukawa's paper, which presents an extraordinary new proposal in the following ten lines: if Heisenberg's exchange forces cannot be interpreted as an effect of the exchange of an electron-neutrino pair, why not try to interpret them as being due to the exchange of a new particle of intermediate mass* [between those of an electron and a nucleon]? *Yukawa writes: 'Now this interaction among elementary particles can be described through a force field, exactly as the interaction among charged particles is described by the electromagnetic field.* [...] *In the quantum theory this field* [called U field by the author] *should be accompanied by a new quantum, exactly as the electromagnetic field is accompanied by the photon.'* "

Since nuclear forces only act up to a distance of about 10^{-13} cm, Yukawa calculated that their mediator should be about 200 times more massive than an electron. This expectation was verified when this new particle, now called "pion", was discovered about ten years later in cosmic rays. It was only one of the many integer spin particles (bosons) whose the exchange was used in the sixties to try to quantitatively explain the nucleon-nucleon force. These attempts have been abandoned almost completely, but in nuclear calculations the nucleon-nucleon force at a distance greater than 10^{-13} cm is still described through the exchange of a pion, using the scheme which is known as OPEM (One Pion Exchange Model).

1·7. *Transuranics and fission.* – Chapter 9 —entitled "An intriguing puzzle: What happens to Thorium and Uranium irradiated with neutrons?"— opens with the description of the many activities discovered at the time when the group from Via Panisperna irradiated thorium and uranium. Subsequently there is a description of the history of the innumerable experimental and theoretical papers in which the group from Rome was involved —together with the American, Austrian, French and German physicists, Angruss, Irène Curie, Grosse, Lise Meitner, Ida Noddack, Hahn and others in the attempt to explain these radioactive nuclides as "transuranic". The confusion lasted four years and, looking back, the only valid intuition during this period was that of Ida Noddack [23] who in 1934 suggested that due to neutron bombardment heavy nuclei should have broken up into "*many large pieces*" isotopes of well-known elements but that were far from uranium and thorium in the table of elements.

Nobody accepted this proposal and Amaldi describes what happened in Rome [24]: "*It seems to me to remember some discussions among the members of our group, including Fermi, in which the ideas of I. Noddack's were hastily set aside because they involved a completely new type of reaction: fission. Enrico Fermi, and all of us grown at his school followed him, were always very reluctant to invoke new phenomena as soon as something new was observed: New Phenomena had to be proved! As later developments showed, a much more fruitful attitude would have been to try to test Noddack's suggestion and eventually disproving it. But Fermi and all of us were, in this occasion, too conservative* [...] *Two reasons, or, maybe, two late excuses, why I. Noddack's suggestion was not taken more seriously neither in Rome nor in Berlin or Paris, are the following. Her suggestion of what has turned out to be the correct explanation* [and that is the "fission" of uranium and thorium], *appeared as a speculation, aiming more to point out a lack of rigor in the argument for the formation of element $Z = 93$, than a serious explanation of the observations. This remark seems to be supported by the fact that she never tried, alone or with her husband, to do experiments on irradiated uranium as certainly they could have done. Furthermore in those years the Noddacks had failed in some discredit because of their claim to have discovered element $Z=43$, that they had called 'masurium'.*"

As it is well known, in December 1938 Otto Hahn and Fritz Strassmann announced that uranium, when irradiated with neutrons, breaks up into two nuclei of intermediate mass; by applying refined methods of radiochemistry the authors had managed to show radioactive isotopes which had the chemical properties of barium and lanthanum.

A few days after the announcement of the discovery of nuclear fission, Otto Frisch and Lise Meitner interpreted the phenomenon with the "drop model" of the nucleus. Like a drop of liquid that vibrates can divide itself in two, so a heavy and unstable nucleus, capturing a neutron can start to vibrate in a way that the repulsive forces which act among the protons (which are all positive), end up by overcoming the forces of superficial tension, which tend to keep a spherical shaped nucleus. This was the first contribution that the discovery of fission gave to the nuclear theory.

The discovery of fission was the beginning of that very well-known chain of events which led Einstein to write to President Roosevelt on possible military uses of this phenomenon and saw Fermi make the first nuclear reactor in 1942.

2. – The sixties and the separation of particle physics

2·1. *The treatise "Nuclei and Particles" by Emilio Segrè.* – To describe the state of nuclear physics in the sixties let us be guided [25] by the famous book by Emilio Segrè "Nuclei and Particles", of which the first edition (754 pages) was published thirty-four years after Bothe and Becker's paper. The Italian version of "Segrè" was published several years later [26].

After the collaboration with Enrico Fermi and the discovery of technetium, made when he was a professor in Palermo, Segrè emigrated to the United States because of the racial laws and during the war participated in the Manhattan project. In 1946 he became a professor at Berkeley and in 1959 he shared the Nobel Prize with Owen Chamberlain for the discovery of the antiproton. Having taken part himself in both the research into nuclear physics in the thirties and in the new activities of particle physics carried out on the first large accelerators after the war, Segrè was certainly the experimental physicist who at the beginning of the sixties could best give the whole picture of the "two physics".

The index of "Nuclei and Particles" (which has a subtitle "An introduction to nuclear and subnuclear physics") allows us to realize the state of knowledge and importance that was given to the different subjects at that time.

1. *History and introduction.*

Part I - Tools

2. *The passage of radiations through matter.*
3. *Detection methods for nuclear radiations.*
4. *Particle accelerators.*
5. *Radioactive decays.*

Part II - The nucleus

6. *Elements of the nuclear structure and systematics.*
7. *Alpha emission.*
8. *Gamma emission.*
9. *Beta decay.*
10. *The two-body system and nuclear forces.*
11. *Nuclear reactions.*
12. *Neutrons.*

Part III - Subnuclear particles

13. *Muons.*
14. *Pions.*
15. *Strange particles.*

In chapter 3 the particle detectors in use at the beginning of the sixties are reviewed. Thirty years earlier most of them —semiconductors detectors, scintillators Čerenkov counters, bubble chambers, and spark chambers had not yet been invented.

The development of accelerators is even more impressive and chapter 4 is dedicated to this. Ernest Lawrence invented the cyclotron in 1929, but the first prototype worked only in 1932. Segrè describes the working of the cyclotron, the synchrocyclotron, the betatron and the synchrotron and stresses the fact that in thirty years the proton beams extracted from these accelerating machines have passed from 1-2 MeV to 6200 MeV —that is to 6.2 GeV— of the Berkeley Bevatron with which the first antiprotons were produced.

The rapid development of nuclear physics and in particular subatomic particle physics has only been possible through the invention of many new acceleration and particle detection techniques which we cannot even hint at here.

2·2. *Shell model.* – The second part of the "Segrè", 400 pages long, is the longest of the three; this demonstrates the central role that the author attributed to this chapter on subatomic physics. Two of the subjects discussed in chapter 6 must be recalled: the *shell model* and the *collective models.*

The *liquid-drop model,* which was introduced in the thirties to interpret the phenomenon of nuclear fission, explains the average properties of the binding energies of nuclei, but does not give any information about many observed regularities. In fact information that nuclei containing a given number of neutrons and protons (for example 8 and 20) are particularly stable, has existed since the thirties. On the other hand the *compound nucleus model* introduced by Niels Bohr in 1936 is based on the hypothesis that all nucleons take part in nuclear reactions in equal measure —and is in agreement with many experimental observations. For this reason Bohr, who had an enormous influence on all the physicists of his time, was resolutely against every model that considered important the behaviour of a single nucleon for the description of the nucleus *(single-particle models).*

Nevertheless, in 1949 Maria Goppert Mayer and Hans Jensen and collaborators independently showed that the particular stability of some nuclei can be explained by introducing into the binding energy of a single nucleon a term which depends on its spin orientation with respect to its angular momentum. Fermi is not a stranger to this development which happened when he was no longer working in the field of nuclear physics: at the end of her first paper in fact, Mayer writes "*I thank Enrico Fermi for his observation 'is there any evidence of spin-orbit coupling?' which was the origin of this communication* [27]."

Because of spin-orbit coupling for some special numbers of neutrons or protons — called "magic numbers"— the nuclei are particularly stable, as happens with atoms of noble gases. In this case the completion of the atomic layers which contain 2, 10, 18... electrons assures that these atoms are chemically inert. The numbers of nucleons that have "magic" characteristics are: 2, 8, 20, 28, 50, 82 e 126. The idea was fascinating and before Mayer and Jensen's papers it had been considered by various authors but they had always met with —and for many years continues to meet with— a lot of resistence because, as Segrè stated [28], as soon as it is formulated "*...we come up against a serious difficulty of principle: a central point does not exist in the nucleus which —like the nucleus for atomic electrons— produces a potential in which the single nucleons move. One can nevertheless suppose that all nucleons globally produce a potential well in which the single nucleons move. [...] Even without going into this very complicated subject, we must throw light on a serious difficulty concerning the shell model. How can a nucleon* *move in nuclear matter on an orbit* [of the shell model] *? The mean free path of a nucleon seems to be short with respect to the distance it has to follow in order to be able to speak of its 'orbit'. A partial answer to this difficulty is given by the fact that inside a nucleus the collisions are inhibited by Pauli's principle: the final states that colliding nucleons should reach are already occupied.*"

2·3. *The collective model.* – Both historically and in Segrè's description, the shell model of the atomic nucleus was the starting point of the development of all the collective models invented in the fifties and completed in the sixties. To introduce them Segrè writes: "*The shell model meets with a lot of success in the description of nuclei made up of a closed layer (that is complete with all its nucleons) to which only one or a few other nucleons are added. In the configuration of the completely occupied layer the nucleus is spherical; the addition of one or more nucleons produces only slight deformations. Nevertheless the situation is different half-way between the two layers: the nuclei deviate significantly from a spherical shape and collective motion involving many nucleons becomes important.*"

James Rainwater made an essential step forward at the Columbia University in 1950 when he explicitly introduced the assumption that the nucleon added to a complete layer could deform the rest of the nucleus (supposed to be perfectly spherical in the shell model) making it ellipsoidal [29]. At that time Aage Bohr, the son of Niels Bohr, who was at Columbia, was also present. Following many profound discussions, young Bohr published a paper in 1950 which began with an interesting analogy: the motion of the nucleons in a deformed nucleus can be compared with the motion of electrons in a non spherical molecule which rotates and vibrates [30]. Since the frequencies of rotation and vibration are much smaller than the frequencies of motion of the electrons, the movement of the molecule only influences the motion of electrons adiabatically. Similarly the collective motions which deform the nuclear core involves a very large mass and is therefore slow compared to the motion of every single nucleon.

Aage Bohr and Mottelson managed in this way to calculate both the energy of rotational and vibrational motions in many nuclei, of which the energy levels were up to that moment unexplained, and the deformation determined in even more sophisticated experiments in which the magnetic and electric momenta of a large number of nuclei were measured [31]. Bohr, Mottelson and Rainwater received the Nobel prize in 1975.

In 1964 Segrè ended the chapter dedicated to nuclear models commenting fig. 3, which is nothing but a chart of nuclides in which the horizontal and vertical axes represent the magic numbers [32]. The areas in black include the nuclides the energetic levels of which can be explained with the rotational model of nuclei: the other regions in which these nuclei can be found are enclosed by dotted lines.

Segrè ends the chapter dedicated to nuclear models with these sentences [33]: "*The study of the nuclear shell model introduces into nuclear physics many ideas familiar in atomic physics. Similarly the collective model introduces into nuclear physics ideas familiar in molecular physics; and ideas borrowed from the solid-state theory of super conductivity have found their application in nuclear models* [...] *Even stereochemistry*

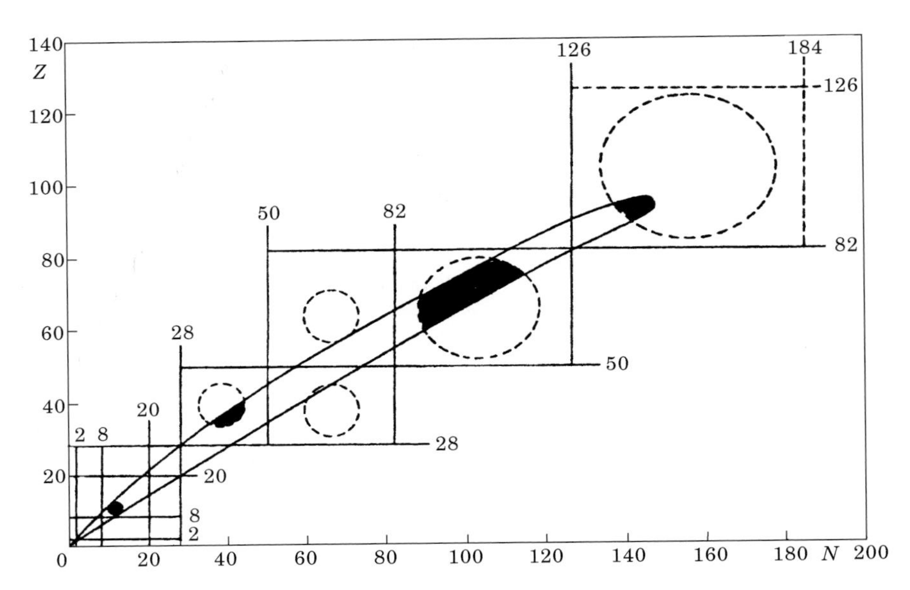

Fig. 3. – In the graph the number of neutrons N and the number of protons Z are shown on the two axes (so that $N+Z=A$) and the horizontal and vertical lines represent the magic numbers. The nuclei that have mean life greater than a minute are on the inside of a cigar shaped curve. In the sixties the collective model explained the properties of the nuclei that fall in the black areas.

has inspired an interesting and simple model, although of limited application. This is the alpha-particle model applicable to light nuclei having the same number of neutrons and protons, this number being a multiple of 4. Evidently one can think of these nuclei (^{8}Be, ^{12}C, ^{16}O, etc.) as being composed of ^{4}He nuclei. The interesting feature here is that such a simple idea can be used to predict various properties of these nuclei. We cannot further enlarge upon these rapidly developing subjects. However,it is noteworthy that they demonstrate once again the formal relations existing between apparently remote branches of physics."

2·4. *Nuclear reactions.* – Nuclear reactions are discussed by Segrè in chapter 11 which deals with the compound nucleus model, the optical model, Fermi's gas model and the statistical theory. We will only hint at some of these developments to give an idea of what was known of nuclear physics in the sixties.

In the optical model the ideas that are at the root of the shell model are applied to nuclear reactions. The incident particle penetrates a potential well similar to that used in the shell model and is absorbed by it; this absorption is formally described by adding an imaginary component to the potential well of the shell model. By solving Schrödinger's equation —that is the fundamental equation of quantum theory also used in atomic physics— starting from this potential one can calculate the scattering and absorption cross-sections as functions of the parameters of the optical potential, and in particular of the radius of the nucleus and the thickness of its surface. Naturally in this way we can only describe the average properties of nuclear reactions valid when the resonances are numerous and overlapping and the quantity which is really important is the density of the energy levels of the excited nucleus.

When one wants to calculate this average density directly, for a first approximation Fermi's gas model is very useful. Here, protons and neutrons occupy two overlapping potential wells in space but independent, given that the protons are subject to reciprocal Coulomb repulsion. In the fundamental state of the nucleus, the Z protons and the N neutrons occupy the lower levels in their respective potential wells; since they obey the exclusion principle, they behave like a Fermi gas degenerate at zero temperature. When the nucleus is excited, the gas heats and it is possible to quantitatively relate the excitation energy with a "temperature" of the nucleon gas. It is also possible to define the entropy of the nucleus, which is determined by the density of the levels. Segrè writes [34]: "*The thermodynamic approach can be extended by considering the emission of neutrons or of other particles by the excited compound nucleus as an evaporation process. The evaporated neutrons have an energy distribution corresponding to a Maxwellian distribution at the temperature of the residual nucleus. This has been demonstrated, at least qualitatively, by experiment.* [...] *If the initial excitation energy of the compound nucleus is sufficient, the evaporation of one particle leaves enough energy in the residual nucleus to permit the evaporation of a second particle, and so on* [...]." And this is a further example of a model which is applied in similar ways to different fields of physics.

2·5. *Subnuclear physics*. – The last three chapters of "Segrè" are dedicated to subnuclear physics and in about a hundred pages deal with muons, pions and strange particles. At that time these particles, discovered in cosmic rays, had been the object of experimental research for little more than ten years in that the particle accelerators had reached sufficient energy and intensity to produce them copiously and the bubble chambers were ideal detectors to study the reactions where these particles were produced or decayed.

As Segrè writes [35], half-way through the sixties "*one may consider four families of particles in order of increasing rest mass: the first contains only one member, the 'photon', a boson of spin* 0. *The second family, 'leptons' contains fermions of spin* 1/2, *lighter than the proton* [that is electrons and muons]. *Leptons are subject to electromagnetic and Fermi interactions only, not to the strong interactions. The third family, 'mesons', comprises bosons of spin* 0. *These are heavier than the leptons, lighter than the protons,* [in particular the Yukawa pion] [...] *The fourth family, 'baryons', comprises the proton and heavier fermions.* [...] *The discovery and classification of these particles is one of the major achievements of the decade 1950-1960.*"

In describing these particles and their interactions, Segrè uses the symmetries introduced by Murray Gell-Mann and others to frame mesons and baryons into a single scheme, symmetries that were brilliantly confirmed with the discovery of the *omega baryon* made exactly in 1964, the year in which the "Segrè" was printed, at the Brookhaven National Laboratory (Long Island).

These subjects lead us astray; it is enough to stress here that with the proliferation of the known number of subnuclear particles, towards the end of the fifties, particle physics broke off from the sturdy trunk of nuclear physics to make up a separate field of research, the success of which was strictly tied to the increased energy of the accelerators and growing sophistication of the detectors.

3. – Focal points of present-day research

3.1. *Nuclear physics at the Nuclear Physics Congress in Paris 1998.* – With another jump of thirty-four years, from the date of the "Segrè" publication and the discovery of the omega baryon, let us go to the World Congress of Nuclear Physics held in Paris in 1998; it is part of a three-year series of "International Nuclear Physics Conferences" and is known by the acronym INPC98. The list of the parallel sessions gives a precise idea of the subjects which are important for nuclear physics at the beginning of the twenty-first century [36]:

1. *Quarks and hadrons.*
2. *Deep inelastic scattering.*
3. *QCD at finite temperature and the quark-gluon plasma.*
4. *Nuclear structure.*
5. *Nuclear reaction dynamics at intermediate and high energy.*
6. *Nuclear reaction dynamics at low energy and fission.*
7. *Astrophysics.*
8. *Fundamental interactions.*
9. *Neutrinos.*
10. *New facilities.*
11. *Medical applications.*

In the first three sessions subjects that were not even imaginable in the sixties were dealt with. The most important concerns the *quark structure* of all the "hadrons" that is of the hundreds of mesons (pions, kaons etc.) and baryons (neutrons, protons, Λ and Ω, etc.) known today. The first are bound states of a quark and an antiquark while the second are made up of three quarks.

For example, a proton is made from two u quarks which have a 2/3 electric charge and a d quark which has a $-1/3$ charge, while a neutron has (udd) composition. The hundreds of hadrons created in the large accelerators during the collision of particles are made up of six types of quark and antiquark, three of 2/3 charge (u, c, and t quarks) and three of $-1/3$ charge (d, s and b quarks).

3˙2. *Quantum chromo-dynamics and the structure of nuclei.* – Quantum chromodynamics (QCD) is the theory of the "*strong forces*" which act on matter particles called "*quarks*" and are interpreted as the consequence of the exchange of virtual force-particles called "*gluons*". It is a field theory completely defined in which the fundamental fields (quarkon and gluon) strongly interact, in such a way that in general it is not possible to carry out a perturbation calculation similar to that used in quantum electrodynamics (QED) and which makes this latter theory extremely predictive. For example the boson and baryon masses are in principle calculable, but they cannot be obtained as the sum of a series of terms each one smaller than the previous one. Instead it is necessary to use non-perturbative methods, in particular numerical methods on a discrete lattice, which today give rather satisfying results when one computes the masses of the lighter bosons and baryons.

Going back to nuclear physics, it would certainly be illuminating to be able to obtain the properties of nuclei starting from QCD, but this goal is distant for two reasons.

First of all it is still not possible to calculate, starting from QCD that completely describes the strong force which acts among quarks, the nuclear force that acts between two nucleons. One can understand the difficulty of the problem by observing that such a globally attractive force is due to the fact that gluons —always present in large numbers because they are continually emitted and absorbed by the three quarks in each nucleon— constitute a kind of "nuclear glue" which, so to speak, overflows from each nucleon and influences the quarks in the nucleon nearby. The nuclear force between two nucleons is therefore the secondary effect of the strong force mediated by the exchange of gluons among the quarks.

Furthermore, since each pair of nucleons which interacts within a nucleon is embedded in nuclear matter, the force just described (the "bare" nuclear force acting between two free nucleons) is modified by the presence of the other nucleons. As illustrated in fig. 4, due to W. Nazarewicz [37], one needs therefore to build *two* bridges in order to obtain from QCD the "effective" force acting between two nucleons belonging to a many body system bound together by the exchange of gluons.

These bridges have still not been built and this remains a vast research programme projected towards the future. In fact today, making use the most powerful computers available, it is only possible to calculate precisely the binding energy and other properties of light nuclei starting from empirical expressions of interaction among two or three nucleons [38].

3˙3. *Deep inelastic scattering.* – The QCD theory has a unique characteristic: the elementary process of emission or absorption of a force-particle (a gluon) by a matter-particle (a quark) becomes *less frequent* when the energy and/or the momentum of the gluon increase. The indetermination principle links with a law of inverse proportionality this energy/momentum to the size of the region of space where the emission takes place. Thus, for small distances a quark, even when bound to a nucleus, is not usually accompanied by one or more gluons but, being "bare", it almost behaves like a free particle. To observe a quark in this state it is necessary to use instruments which have an adequate

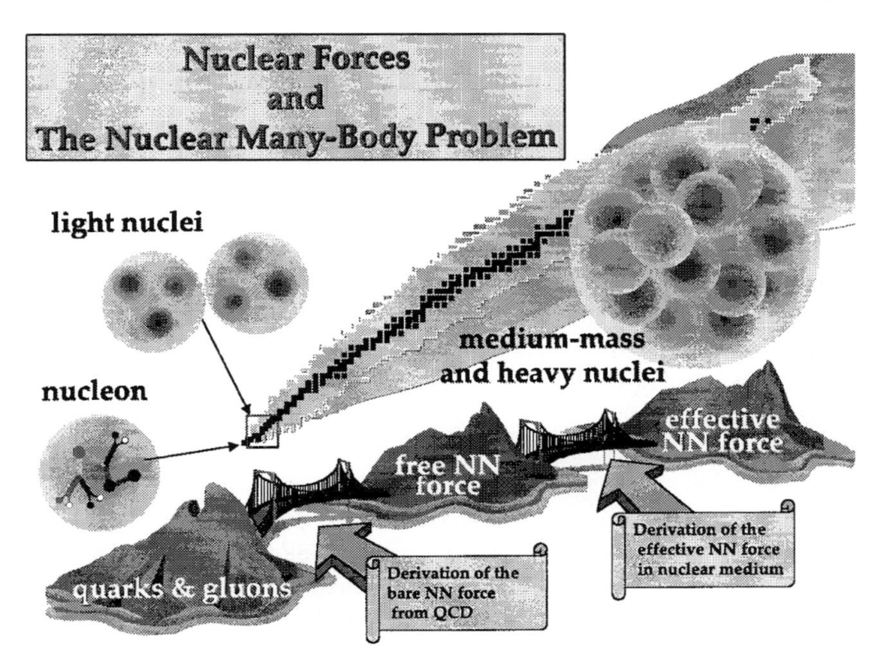

Fig. 4. – To understand the nuclear forces that determine the properties of nuclei starting from QCD, the theory of interacting quarks and gluons, it is necessary to build two bridges which allow one to derive the bare nucleon-nucleon force and subsequently, the nucleon-nucleon force inside a bound nucleon system [37].

spatial resolution, and that is a resolution inferior to 10^{-14} cm, corresponding to less than a tenth of the size of a nucleon.

In the second session of the Paris Conference entitled "Deep inelastic scattering" this topic was discussed: the high energy scattering of electrons and heavy electrons (muons) by nuclei.

When an electron hits a nucleus it changes direction creating other particles, the quarks that are inside the nucleus exchange a *gamma* quantum, which is the carrier of the electromagnetic force. This exchange takes place in a volume that is much smaller than the size of the nucleon when the energy is "high", that is when the energy of the electron is greater by about ten GeV. In this case the exchanged quantum can have an energy/momentum greater than several GeV and can catch the quark when it is not surrounded by gluons. In this phenomenon several complications disappear, and it is possible to make quantitative predictions using perturbative approaches; as we have seen, instead this is not possible when facing the problem of nuclear structure because the virtual gluons that are continuously exchanged in the nucleus usually have an energy lower than ten MeV.

With the *deep inelastic scattering* technique the motion of quarks inside nucleons and nuclei are studied today and the way in which their spins combine to give the global spin of the nucleon. In spite of the simplicity of the quark model and the possibility of describing these phenomena using the perturbation theory, there are many aspects of the quark structure of nucleons and nuclei that have still not been understood . For this reason extremely refined experiments on electron and muon scattering have been carried out at CERN in Geneva and at the SLAC laboratories in the United States.

3˙4. *Quark-gluon plasma.* – The third session at the Congress was entitled "QCD at finite temperatures and the *quark-gluon plasma"*. Experimentally this topic is studied by making two heavy nuclei collide at the maximum possible energy. In the central collisions the energy is distributed, in a short time, among hundreds of nucleons increasing the average kinetic energy. Since in a system composed of many corpuscles the average kinetic energy and temperature are proportional quantities, this nuclear physics is also called high temperature nuclear physics.

In a nucleus the average kinetic energy of nucleons is of the order of MeV and the temperatures involved are considered "low" in spite of the fact that 1 MeV corresponds to a temperature of as much as 10^{10} kelvin. In the central collisions of two lead nuclei that were studied at CERN, temperatures of the order of 200 MeV were reached for a very short time and without thermal equilibrium. These temperatures are more than a hundred times greater than the temperatures of ordinary nuclei and correspond to $2 \cdot 10^{12}$ kelvin.

At these temperatures QCD foresees that quarks, which at low temperatures are confined inside each nucleon of the nucleus, evaporate from their prison and mix together to form a new type of matter, a *nuclear plasma*, in the way that the atoms of a gas heated to a million degrees free themselves of their electrons and form an *atomic plasma*. Naturally all these quarks, that are not confined any more but are free to move, continue to exchange gluons so that the nuclear plasma is a homogeneous soup of quarks and gluons.

The transition phase that leads to quark and gluon plasma is often called "deconfinement" and represents the most interesting frontier of high temperature nuclear physics. This transition is usually analysed by describing the phenomena in a phase diagram such as that in fig. 5, in which the axes represent the density of the nuclear matter and its temperature.

When the density is that of an ordinary nucleus (value 1 on the abscissas axis in the figure) and the temperature less than ten MeV, we have a nuclear liquid to which the drop model can be applied. Outside this area there is a nucleon gas, which becomes hotter and hotter when moving upwards away from the abscissas axis. The half-moon area in fig. 5 is the area in which the QCD theory foresees deconfinement, whose critical temperature at low density is indicated with the symbol T_c. In the past ten years at the SPS accelerator at CERN very many experiments on the collision of lead particles have been performed at energies of several hundred GeV with fixed targets and the arrow labelled "SPS" shows qualitatively the phase space region which has been explored. In 2000 various phenomena observed have allowed us to come to the conclusion that a new form of matter (very probably the much sought after quark and gluon plasma) was produced at CERN.

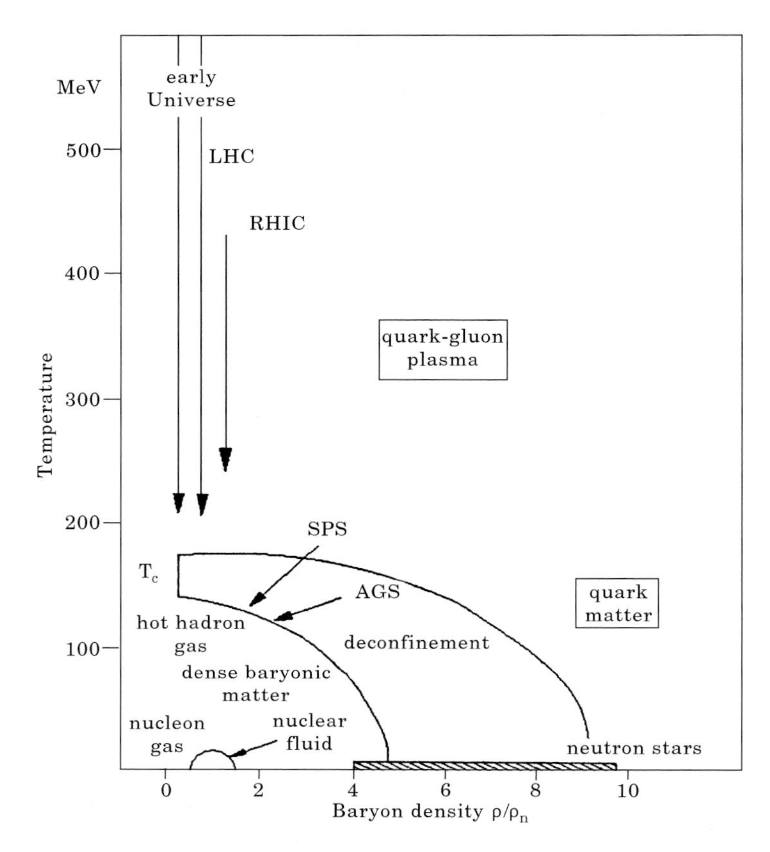

Fig. 5. – The phase transition ("deconfinement") is expected to take place in the half-moon area on this graph where the abscissas axis represents the density, normalized with respect to the density of nuclei, and in which the axis of ordinates represents the temperature of the system measured in MeV (100 MeV $= 10^{12}$ kelvin). In this context "baryon density" is synonymous with nuclear density. Neutron stars are cold and have a density of between 4 and 10 times the density of nuclei. The figure is taken from ref. [39].

At the end of 2000 at Brookhaven National Laboratory in the United States two interwined accumulation rings (called RHIC) were started up in which beams of gold nuclei circulate in opposite directions. In fig. 5 the arrow indicting "RHIC" shows the area of phase space which will be explored using this new accelerator for gold-gold collisions; in view of the much higher energy, it will be possible to confirm and study in detail phase transition expected from QCD and observed at the SPS at CERN.

This research is very interesting for astrophysics, because a billionth of a second after the Big Bang the Universe was a quark and gluon soup in that the average kinetic energy of the particles was about 1 GeV (1000 MeV). In expanding and cooling down our Universe followed the path indicated with "Early Universe" in the figure. It is therefore important to know if the passage from plasma to nucleon gas took place at a temperature of $T_c = 150$ MeV, which corresponds to about 20 milliseconds after the Big Bang. As the arrow indicated with "LGC" shows, the experiment ALICE —which from the year 2006 will study lead-lead collisions on the new Large Hadron Collider accelerator at CERN— and will reproduce the phase transition that happened at the beginning of the history of our Universe as accurately as possible.

3·5. *Nuclear structure and astrophysics.* – There is another theme in nuclear physics which is of great interest for astrophysics: the formation of the nuclei heavier than iron. In fact the nuclear fusion reactions occurring inside heavy stars which burn slowly lead to the production of only light and intermediate nuclei and they are interrupted when the star is made of concentric shells of hydrogen, helium, carbon, neon, oxygen, silicon and iron. The central iron core is the final stage of this evolutionary phase in stars because iron is the atom which has the maximum binding energy per nucleon. In fact the synthesis of heavier nuclei does not release energy but requires it.

If the mass of the star is more than one and a half times the mass of the Sun, the core collapses giving rise to a "supernova" which explodes ejecting into interstellar space not only the matter of which the star was made but also new nuclei produced during the explosion. With the powerful computers available today it is possible to carry out detailed simulations of the phenomena that take place in the few seconds such a cataclysm lasts for, but the phenomena involved and the nuclear data necessary to describe them are not sufficiently well known. This was discussed during the seventh session at the Paris Congress.

One of the greatest problems concerns the recognition of the chains of nuclear reactions produced, after the first collapse of the stellar core, from the enormous flow of neutrinos that it generates and which pass through the matter previously projected into the interstellar space. In the extremely hot layers that evaporate from the core, which collapses until it forms a neutron star, the neutrinos cause processes of synthesis in heavy nuclei through what has been called the *r-process*.

In this process, starting from nuclei with intermediate mass ($A < 100$), given the high density of neutrons, heavy nuclei with $A \sim 200$ are synthesized. The nuclei which have many neutrons then decay into stable, or almost stable, nuclei, through beta decay chains with emission of electrons. In the usual representation of fig. 6, these nuclei rich in neutrons are found between the grey area, wherethe well-known nuclei are found, and the broken line indicated as "neutronic dripline". In this area, shown as white in the figure, about 6000 nuclei are found of which for the moment we only know how to calculate a few average properties. To achieve a quantitative understanding of the abundance of heavy nuclei in the Universe it would be necessary to use experimental measurements of their properties and reactions.

In the last few years new techniques have been developed to produce nuclei rich in neutrons and make them collide with ordinary nuclei to study this phenomenology. As nuclei with many more neutrons than usual are unstable, today we talk about experiments with *radioactive nuclei.* Intense beams of radioactive nuclei are obtained in two very different ways. In the *in flight* technique the radioactive nuclei are produced in a thin target with the fragmentation of beams of heavy stable nuclei, which naturally contain many more

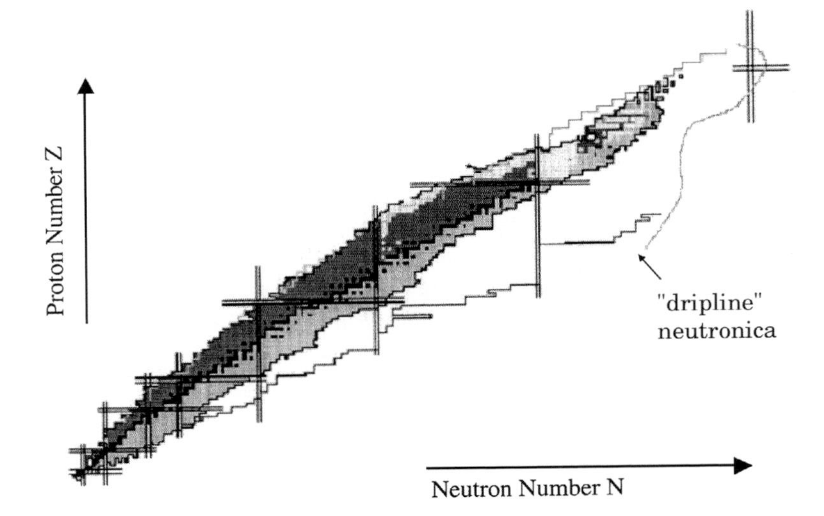

Fig. 6. – The small black squares represent the stable nuclei and the crosses indicate the magic numbers. In the grey area the well-known unstable nuclei are found. The nuclei rich in neutrons that intervene in the r-process and can be studied with beams of radioactive nuclei, can be found between the grey area and the broken line called neutronic dripline, below which the nuclei are not energetically bound.

neutrons than protons. Alternatively in the "ISOL" (Isotope Separation On Line) technique a beam of protons —for example— produces short mean life radioactive isotopes in a target. These exotic nuclei capture electrons and diffuse like a gas in the target; after extraction they are then brought to the desired energy with another accelerator.

At the beginning of the twenty-first century experimentation with beams of radioactive ions is now carried out in the most important laboratories in the world; in this way it will possible to explore a large part of the N-Z plane in fig. 6, which has up to now been inaccessible to experimentation in nuclear physics.

4. – Conclusion

Seventy years have elapsed since the first observation of penetrating radiation emitted by light nuclei was carried out in 1930 by W. Bothe and H. Becker. Enrico Fermi and his collaborators with their discoveries have given an important boost to the understanding of nuclear physics. Still, models are needed to explain the rich phenomenology because, in spite of the initial expectations, the problem of calculating the properties of nuclei from the knowledge of the nucleon-nucleon forces has still not been resolved because such forces are indirect consequences of the more basic strong force, which acts among quarks making up nucleons, and cannot be described in a perturbative way.

Though a general solution to the basic problem is still lacking, very refined models of nuclei and of the reactions that occurr among them have been developed. In this way the existence of magic numbers, the quark-like structure of nuclei, the transition from the state of nuclear plasma to nuclear gas, the collective motion of nuclei can be quantitatively explained and also many other phenomena which have not been described here. There are rather precise models for the synthesis of nuclei of heavy elements which occurred during the explosion of supernovas, but experimental data is missing; the new techniques of production and acceleration of radioactive nuclei open new prospects because they allow us to enter vast unexplored territories of the N-Z plane.

REFERENCES

[1] AMALDI E., *From the Discovery of the Neutron to the Discovery of Nuclear Fission*, *Phys. Rep.*, **111** (1984) 1-332.
[2] Ref. [1], p. 6.
[3] Ref. [1], p. 9.
[4] Ref. [1], p. 39.
[5] Ref. [1], p. 124.
[6] FERMI E., *Ric. Sci.*, **5** (1934) 330.
[7] AMALDI E., D'AGOSTINO O., FERMI E., RASETTI F. and SEGRÈ E., *Proc. R. Soc. London, Ser. A*, **146** (1934) 483, *Lord Rutherford's Communication*, 25 July 1934.
[8] Ref. [1], p. 135.
[9] SEGRÈ E., *A Mind Always in Motion* (University of California Press, Berkeley, Los Angeles) 1993, p. 93.
[10] Ref. [1], p. 152.
[11] CHANDRASEKHAR S., in Enrico Fermi, Note e Memorie (Collected Papers), Vol. **II** (Accademia Lincei and University of Chicago Press) 1965, p. 297.
[12] PONTECORVO B., Enrico Fermi (Edizioni Studio Tesi, Pordenone) 1993, p. 82.
[13] FERMI E., AMALDI E., PONTECORVO B., RASETTI F. and SEGRÈ E., *Ric. Sci.*, **5** (1934) 282.
[14] Ref. [1], p. 155.
[15] Ref. [1], p. 183.
[16] AMALDI E. and FERMI E., *Phys. Rev.*, **50** (1936) 899.
[17] BOHR N., *Nature*, **137** (1936) 344.
[18] BREIT G. and WIGNER E., *Phys. Rev.*, **49** (1936) 519.
[19] Ref. [1], p. 190.
[20] BREIT G., CONDON E. V. and PRESENT R. D., *Phys. Rev.*, **50** (1936) 825.
[21] YUKAWA H., *Proc. Phys. Math. Soc. Jpn.*, **17** (1935) 48.
[22] Ref. [1], p. 262.
[23] NODDACK I., *Ang. Chem.*, **47** (1934) 653.
[24] Ref. [1], p. 277.
[25] SEGRÈ E., *Nuclei and Particles* (Benjamin W. A., New York, Amsterdam) 1964.
[26] SEGRÈ E., *Nuclei e Particelle* (Zanichelli, Bologna) 1966.
[27] GOEPPERT M. MAYER, *Phys. Rev.*, **75** (1949) 1964.
[28] Ref. [25], pp. 248 and 251.
[29] RAINWATER J., *Phys. Rev.*, **79** (1950) 432.
[30] BOHR A., *Phys. Rev.*, **81** (1951) 134.
[31] BOHR A. and MOTTELSON B., *Danske Mat.-Fys. Meddv.* **27** (1953) No. 16.
[32] Ref. [25], p. 264.
[33] Ref. [25], p. 265.

[34] Ref. [25], p. 476.
[35] Ref. [25], p. 588.
[36] FROIS B., GOUTTE D. and GUILLEMAUD-MUELLER D. (Editors), *INPC/98, International Nuclear Physics Conference, Paris, France, Nucl. Phys. A*, **654** (1999) 1.
[37] NAZAREWICZ W., *Nuclear Structure*, in ref. [36], p. 195.
[38] PANDHARIPANDE V. R., *Quantum Monte Carlo Calculations of Light Nuclei*, in ref. [36], p. 157.
[39] AMALDI U., *The importance of Particle Accelerators*, in *Proceedings of the European Particle Accelerator Conference EPAC 2000, Vienna, June 2000*, pp. 3-7.

About the Author

UGO AMALDI was initially interested in nuclear physics and radiation physics at the physics laboratory of the "Istituto Superiore di Sanità" and from the 70's he worked at CERN in Geneva as research director on many experiments on particle physics. Among other things he was responsible for fifteen years for the DELPHI collaboration, which built and ran the large detector of the same name that gathered data from 1989 to 2000 from the electron and positron collider called LEP.

Amaldi has over 300 scientific works to his credit and published first with Edoardo Amaldi and then alone, twenty volumes of physics for secondary schools. He is a *Doctor honoris causa* of the universities of Helsinki, Lyon, Uppsala and Valencia. At present, as president of the TERA Foundation, he is working for the introduction into Italy and Europe of radiotherapy techniques for deep radio-resistant tumours called "*hadrontherapy*".

The birth of nuclear energy: Fermi's pile

CARLO SALVETTI

1. – The "puzzle" of uranium and the discovery of "fission"

In the days just before Christmas 1938 two German chemists, Hahn and Strassmann of the Chemistry Institute of the Kaiser Wilhelm Gesellschaft (from now on KWI) at Berlin-Dahlem, discovered, to their immense surprise but without a shadow of a doubt, that uranium (atomic number $Z = 92$), when hit by thermal neutrons from a radium-beryllium source, gives rise to the production of barium ($Z = 56$). To them the phenomenon was inexplicable because over nearly a decade of previous research on irradiation with neutrons the reactions (n, γ), and the subsequent β emission, gave rise to contiguous elements with an atomic number $Z + 1$, but never with atomic numbers so far apart, on the Mendelejev scale, as Ba and U with a value of Z 36 positions apart.

The results obtained by the two German chemists by means of refined techniques of fractional crystallisation and precipitation, were nevertheless beyond question. A further check proved that in the same U salt solution lanthanum atoms ($Z = 57$) were also present, produced by the β decay of Ba.

Hahn and Strassmann's discovery opened a new chapter in the extraordinary scientific event of the irradiation of U, which began with the publication in "Nature" in June 1934 of the famous article by Fermi and his group in Rome "Possible Production of Elements of Atomic Number Higher than 92" [1] in which, albeit with considerable caution, the hypothesis of the creation of transuranic elements was suggested. It was a real scien-

---

[1] The paper is reprinted at number 99 in the publication (2 vols.): "Enrico Fermi - Note e Memorie" (Collected Papers) Vol. I, published by the Accademia dei Lincei and the University of Chicago Press, 1962. Following a notation introduced by Segrè in his biography of Fermi these articles will be quoted from now on as FNM followed by the number.

tific puzzle with which the greatest European physicists and radiochemists engaged for almost five years [2].

At that time unfortunately Otto Hahn (future Nobel prize winner for this very discovery) was not able to count on the advice of his close collaborator and friend Lise Meitner, an Austrian physicist who had worked with him since 1935, but who had had to leave the KWI in July 1938 because she was Jewish and seek refuge in Stockholm where she took up her research activity again.

Hahn, who did not want to do without her advice, decided to write her a letter asking for enlightenment ("Perhaps you can suggest some fantastic explanation") and later sent her the carbon copy of the typescript of the article that he sent on the 22nd December to the German journal "Naturwissenschaften" where it appeared on the 6th of January 1939.

The much awaited reply arrived a few days late because at that very moment Meitner was on holiday in a ski resort in southern Sweden with her nephew Otto Frisch, who was also a physicist and in exile from Vienna at the Niels Bohr Institute in Copenhagen.

That skiing holiday turned out to be particularly fortunate and productive. Aunt and nephew, struck by the strangeness of the Berlin results, immediately began an animated discussion and reached the conclusion that what was happening was the fission of the U nucleus into two fragments of intermediate mass, one of which was barium. According to them the phenomenon was quite compatible with Bohr's drop model for the atomic nucleus. The binding energy of the captured neutron had provided the compound nucleus U + n with the necessary energy to set off a series of oscillations violent enough to produce a lengthening of the "drop" to the point where it assumed an unstable "dumbbell" shaped configuration (fig. 1). The two quasi-spheres at the ends had then moved apart due to electrostatic repulsion in the form of fragments, with the subsequent release of a considerable amount of energy. This energy had been provided as a consequence of the difference in mass ΔM between the U nucleus and the sum of the masses of the two fragments: on the basis of Aston's diagram, ΔM should be positive. Simply by using Einstein's famous equation $E = \Delta M c^2$, aunt and nephew estimated that the energy released by the fission of U would be of the order of about 200 MeV, a very high value since the loss of mass of ΔM was equal to about 1/5 of the mass of a nucleon. Thus Frisch and Meitner were able to explain the results obtained by the chemists at KWI in Berlin, anticipating the results that would be obtained many months later, with much more refined theoretical methods, by Bohr and Wheeler in a famous article published in "Physical Review" on 1st September 1939 [3].

([2]) In particular the papers of Ida Noddack, a German chemist known for discovering the new element rhenium in 1925 along with her husband and the French group of Irène Curie and Frédéric Joliot (both Nobel prize winners) in cooperation with the Yugoslav Pavel Savitch. For further information see U. Amaldi's chapter in this volume, p. 151.

([3]) The day the second world war broke out, as Wheeler himself pointed out to me personally many years later during the 1989 celebrations of the 50th anniversary of the discovery of fission. Incidentally I would like to point out that this was the last time that Edoardo Amaldi took part in a scientific meeting. He died in the following December.

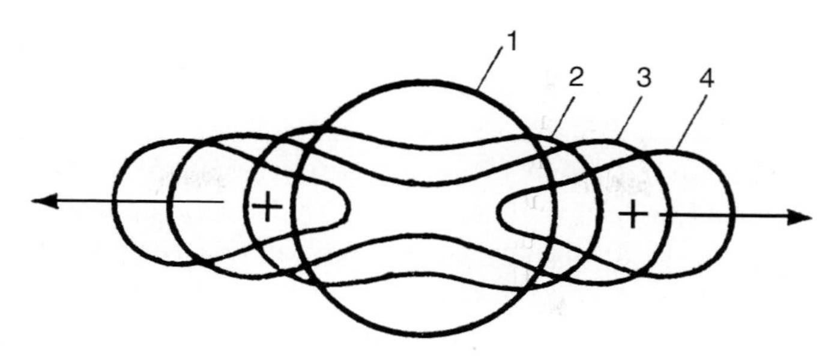

Fig. 1. – The mechanism of fission according to Bohr's drop model. (From Enrico Fermi, *Conferenza di Fisica Atomica*, Fondazione Donegani, Acc. Naz. dei Lincei, Rome, 1950.)

After ending their holiday and informing Hahn of their interpretation they agreed to publish it in the British journal "Nature", but not before listening to the opinion of Niels Bohr, the "grand old man" of physics. Frisch did this immediately on his return to Copenhagen, informing his aunt Lise, and setting to work writing a paper under their joint names for "Nature". There was very little time available to communicate a matter of such importance because Bohr was due to leave on 7th January to go to the United States for a theoretical physics convention in Washington. We shall see later how this trip was to have curious and unforeseeable consequences. So it was that Frisch took him a first draft to see on the evening of the 6th, the day before Bohr's departure, a first draft that Frisch would then hurriedly type in time to give it to Bohr on the morning of the 7th at the departure of the train for Göteborg where Bohr was to embark. During this brief meeting Frisch told Bohr that he also wanted to carry out an experimental confirmation of fission with "physical methods" since the purely "chemical methods" such as those used by the German chemists were traditionally less credible to physicists' eyes.

On his departure Bohr, without being asked, promised not to mention the "Nature" article to his American colleagues until publication had been confirmed. We shall see how this promise was not kept.

In the meantime, the previous day, the famous article by Hahn and Strassmann appeared on "Naturwissenschaften". After Bohr's departure Frisch immediately began his experiment. This consisted of detecting the fission fragments by exploiting the intense ionisation produced in an air ionisation chamber connected by means of a linear amplifier to an oscilloscope.

As we shall see, this equipment was to become the standard model for later experiments by other researchers, particularly the Americans. The equipment was ready on the evening of the 13th January and, as expected, as soon as the irradiation of a sheet of U with thermal neutrons began, Frisch saw vertical impulses appear on the cathodic screen, caused by the fission fragments, which in height towered over the numerous impulses at the "bottom" produced by the α particles of the U.

Working all night long, by the morning of the 14th Frisch was sure he had reached the "physical" confirmation of the two German chemists' discovery, and he immediately informed his aunt in Stockholm.

Frisch and Meitner decided at this point to send "Nature" not one article but two. The first, under both names, with the title "Disintegration of uranium by neutrons: a new type of nuclear reaction" and the other, under Frisch's name, called "Physical evidence for the division of heavy nuclei under neutron bombardment".

The articles, sent by Frisch to the journal on 17th January, appeared in the issues of "Nature" of the 11th and 16th February respectively: in these articles the term fission is used for the first time and it would then always replace the German "spaltung" and the English "splitting" ([4]).

2. – Bohr and Fermi in the USA: the strange events of January 1939

Enrico Fermi left Rome for good with his whole family (wife Laura and children Nella and Giulio) on the 6th December 1938, going to Stockholm to receive the Nobel Prize from the hands of King Gustav V on the 10th of December. After the ceremony, and a brief stop in Copenhagen for a couple of days as Bohr's guests, the Fermis boarded the "Franconia" at Southampton to go to New York. When they disembarked there on the 2nd of January they found waiting for them George Pegram, director of the department of physics at Columbia University and Gabriello Giannini, an Italian who had been resident in New York since 1929, who had taken care of the US registration of the patent for slow neutrons on behalf of Fermi and his team. It is worth taking a little more time to discuss Bohr's Atlantic crossing. Bohr boarded the "Drottingholm" at Göteborg with his nineteen year old son Erik and a young Belgian theoretician Leon Rosenfeld, who was also going to the Washington conference: as soon as they were aboard Bohr could not restrain himself from telling Rosenfeld the news about fission and the interpretation given by Meitner and Frisch, something he would later regret for reasons we will soon see.

Bohr and Rosenfeld continued to discuss the theoretical interpretation of fission throughout the crossing, intensely and exclusively as was Bohr's way (he even had a blackboard installed in his cabin's living room) despite the heavy seas and consequent problems.

On their arrival at New York, at one in the afternoon of the 16th of January, they found Enrico and Laura Fermi waiting for them, along with John Archibald Wheeler, a young theoretician from Princeton, who had worked with Bohr. After disembarking Bohr and his son, together with the Fermis, made their own way into town, while Wheeler returned to Princeton with Rosenfeld. Bohr, remembering his promise to Frisch, was careful not to mention Hahn and Strassmann's discovery and the consequent interpretation to Fermi or to Wheeler. However he had forgotten to warn Rosenfeld who, given the importance and novelty of events, told Wheeler as soon as they were aboard the train to Princeton. It so happened that that very evening (Monday 16th January) there was

---

([4]) According to Meitner the term fission was suggested to Frisch by a young American biologist William A. Arnold, his contemporary, at Copenhagen on a Rockefeller Foundation scholarship. When Frisch asked him what biologists called the subdivision of the cell into two parts he replied "binary fission". Thus only the term fission remained.

to be a meeting of the so-called Journal Club (a sort of informal and informative weekly seminar about new developments in physics) for which Wheeler was responsible. So he decided immediately to call on Rosenfeld to talk on the subject. What better occasion could there be to inform their colleagues? "The effect of my words on the American physicists was spectacular" Rosenfeld would recall later. No less spectacular were Bohr's surprise and disappointment when, the next day, he also arrived at Princeton and discovered what had happened. Despite his promise not to divulge the news before Meitner and Frisch's papers were published he now found himself the involuntary herald of fission in the USA! He tried to remedy the situation by hurriedly (in three days, so to speak...) ([5]) scribbling a brief note for "Nature" (in all 700 words) on the conclusions he had reached with Rosenfeld during the crossing: but the main aim of the article was to emphasise the priority of Frisch and Meitner's interpretation, before the news from Princeton could spread.

At Columbia Fermi remained in the dark about the news on fission until two Columbia physicists, I. I. Rabi and W. E. Lamb jr., both future Nobel prize winners, returned from Princeton at the weekend. Within a few hours Fermi learned from one (or both) of them ([6]) about the fission (still called splitting) of U. It is odd that Fermi, who had worked (and suffered) so much on the problem of U should have been the last of the European physicists who had emigrated to the USA to hear about fission. In any case, as his wife Laura said, he was very struck. *"It didn't occur to us"* he said to her, thinking maybe of the Nobel prize speech he had made at Stockholm just over a month earlier ([7]).

It is a fact that from that moment the question of the fission of uranium became an obsession for Fermi and the main theme of his early years of activity in America. It seems appropriate therefore, given the later consequences, to describe in some detail, as we have done for fission, the steps that characterized Fermi's research in the first half of 1939.

3. – A lot of physics and much excitement at Columbia

Less than a month after landing in New York, the last few days of January 1939 were very intense ones for him. On the 25th of January, on the eve of the theoretical physics conference in Washington, still unaware of Frisch's experiment, Fermi, together with John R. Dunning, in charge of nuclear physics at the Pupin Laboratories at Columbia and a young graduating student Herbert Anderson, planned a "physical" confirmation of fission.

The standard Frisch type of equipment (ionising chamber and linear amplifier) were already available because Anderson had built it for his thesis. The group had to abandon

([5]) Bohr's slowness both in speech and in writing was proverbial among physicists.
([6]) It is not clear from whom... there are different versions.
([7]) It should be emphasised that Fermi was awarded the prize for his work on artificial radioactivity and for slow neutrons and not for transuranic elements. These (called ausonium and hesperium) are only quoted briefly in Fermi's Nobel prize speech, given before Hahn and Strassmann's discovery. Fermi made the indispensable correction to the text of his speech later by means of a concise footnote (see Segrè's biography, p. 101).

their original intention to use the Columbia cyclotron and fall back on a much less intense neutron source, the traditional Ra + Be.

After Fermi's departure for Washington, Anderson immediately set to work and that same evening at nine o'clock he too saw on the oscilloscope Frisch's "peaks". Highly excited he showed them to Dunning who, it seems, at once informed Fermi in Washington. The Washington conference (26-28 January), supported jointly by the George Washington University (GWU) and by the Carnegie Institution (presided over by Vannevar Bush, whom we shall return to later), organised every year by George Gamow (but in practice by Edward Teller), was in its fifth edition and should have been devoted to the problems of low-temperatures physics. Instead, a few minutes before the inaugural session, Gamow learned about the European fission discoveries from Bohr and decided on the spot to ask him to give the opening talk. Although, notoriously, Bohr was no great orator it seems that his speech literally galvanised the audience. It is not difficult to imagine. Bohr was followed by Fermi who illustrated the significance of fission with his extraordinary clarity and referred to the Columbia experiment.

In the meantime two nuclear physicists from Carnegie, Roberts and Hafstad, highly excited by what they had heard, literally rushed out of the room and ran to their laboratory to carry out a Frisch type experiment, using the Van de Graaf accelerator at Carnegie's Atomic Physics Laboratory (APO). Unfortunately the usual gremlin in accelerators had got into the works: the filament in the source had burned out and the vacuum tube of the accelerator leaked. This caused an inevitable delay so the Carnegie physicists (Hafstad had gone skiing for the weekend and had been replaced by Meyer) were only able to show off the fission "peaks" to the various participants, with a touch of pride, on the evening of Saturday the 28th after the conference's traditional farewell dinner.

From the few lines of a letter Bohr wrote to his wife Margarethe it is easy to glimpse what was going through his mind as he was still unaware of the publication of Frisch's results.

It was in any case the third physical confirmation of fission. These things come in fours: nobody in Washington knew that Joliot, on the 26th January in Paris, had read the Naturwissenschaften article of the 6th of January and had observed the fission fragments using the same method as Frisch.

The Washington conference with its startling news about fission could not pass unnoticed by the press and so on the 28th of January articles appeared in Washington's Evening Star, in the New York Times and in an Associated Press agency flash. The same day the group of Californian physicists at Berkeley (Alvarez, Abelson, Oppenheimer), with understandable excitement, heard the news from the San Francisco Chronicle: Alvarez immediately asked Gamow for information and was told of the APO experiment. Alvarez began an intense discussion with Oppenheimer who saw the consequences and outlined disturbing prospects. Various nuclear physicists in the meantime began to be convinced that the fission fragments were not the only products of the neutron bombardment of the U because the excess number of neutrons compared to stable nuclei with the same Z should lead to neutrons being emitted both instantaneously and with a delay. If the neutrons were in number greater than one then, in a suitable structure and with a sufficient quantity of U, a chain reaction would be possible with production of energy and in certain conditions perhaps even with an explosive form ([8]).

Anyway, after the Washington conference Fermi returned at once to New York on the morning of Sunday the 29th, immediately summoned Anderson to Columbia, and proposed an experiment to show the presence of secondary neutrons produced by fission. It was conceptually an extremely simple experiment. The equipment consisted of a large container full of water (a throwback to the goldfish pond in Via Panisperna ?) with a large bulb, with an Rn + Be neutron source, suspended in a central position. The cavity layer between the source and the wall of the bulb could contain U oxide. The experiment involved measuring any increase in activity induced by thermal neutrons in a neutron detector (a thin sheet of rhodium, another throwback to experiments in Rome) placed in the container at varying distances from the bulb source.

An increase in the Rh activity with the bulb filled with U oxide would prove the existence of secondary neutrons, as indeed happened.

Strangely enough, while Fermi's experiment was being assembled in a basement at the Pupin laboratory, seven floors higher up in the same building two other physicists, Zinn and Szilard, were setting up an experiment, using different techniques, but with the same aim of showing secondary neutrons. Walter Zinn, a Canadian associate research physicist and Leo Szilard, a physicist from Hungary, were both destined to occupy roles of considerable significance in later nuclear developments in the USA.

Leo Szilard was a typical product of Mitteleuropa, a remarkable personality, cantankerous, disorganised in both his scientific and private life, brilliant, full of imagination, impatient of the routine of academic life, a typical free-lance. He arrived in New York after long wanderings from Budapest to Berlin (where he obtained brilliant results in the field of thermodynamics) and then, after Hitler's rise to power, to Britain where he achieved important results in the field of artificial radioactivity: the Szilard and Chalmers method of detecting radionucleides has remained famous. While in Britain he registered various patents; one of them, on (n, 2n) type nuclear reactions which was registered secretly at the Admiralty, stands out: in his imaginative intuition it would make chain reactions for nuclear explosives possible.

In New York Szilard stayed at the small Crown Hotel (where the Fermi family also stayed on their arrival), close to Columbia University which he visited occasionally, getting to know the Pupin physicists including Fermi. They soon came to appreciate his observations and brilliant intuition.

While Zinn was bombarding the U with 2.5 MeV neutrons produced by a small accelerator looking for secondary neutrons, Szilard suggested using thermal neutrons. Two essential ingredients were missing however for the neutron source: radium and beryllium; Szilard provided the beryllium himself with a hollow cylinder of metallic Be

([8]) As Fermi claimed some ten years later at a conference in Milan: "it was then that I realized that a phenomenon of this type could take nuclear physics out of the narrow field of pure research and transform it into something big." (Conferenze Donegani, Ed. Acc. dei Lincei, 1950, p. 97).

which belonged to him and which he had just had shipped from Britain (it arrived on the 18th February). The matter was more complicated as far as the radium was concerned because it was actually available on the market but the New York branch of Belgium's Union Minière demanded a rental of 125 dollars per month. Szilard was not discouraged and appealed for financial support from a well-off friend, Lewis Strauss (who after the war was to become the chairman of the American Atomic Energy Commission), who had already financed some of Szilard's inventions in the past, including a pulsed accelerator. Strauss however was tired of his friend's continuous requests for money and refused.

Szilard was not discouraged and, with his usual knack for tracking down funds, found them from a certain Benjamin Liebowitz who, without hesitation, wrote him a cheque for 2,000 dollars. Even so negotiations with the offices of the Union Minière dragged on, in part because Szilard had no official position. So he urged Columbia to give him an affiliation as an associate researcher with a short-term three month contract. In early March he finally obtained the source in the form of 2 grams of radium and his experiments with Zinn could begin. The delay was even more nerve-racking because of the news, told him by another Hungarian colleague, Edward Teller, a physicist at GWU, that two physicists at APO, Roberts and Meyer, were about to publish the detection of delayed neutrons as a consequence of fission in "Physical Review".

In the meantime at Princeton, in February 1939, by means of subtle theoretical arguments, after a series of discussions with Placzek and Rosenfeld, Bohr reached the conclusion that only the rare isotope U-235, found in natural U in the 1/139 ratio, undergoes fission with thermal neutrons, while the commonplace isotope U-238, and indeed Th-232 for that matter, can only be split by fast neutrons.

He also explained the nature of the so-called resonance absorption of U-238 of neutrons slowing down from their emission as fast neutrons up to thermal energies (the so-called epithermal neutrons).

In the meantime at Columbia the two teams, Fermi-Anderson and Szilard-Zinn, proved the existence of at least 2 secondary neutrons per fission. But in Paris, at the glorious Radium Institute, Frédéric Joliot, working together with the Austrian von Halban and the Russian Kowarsky, using a mud like mixture of U oxides and water, was ahead of the Columbia groups and published results in "Comptes Rendus" on the 3rd of March which matched the results of the Columbia physicists ("more than one neutron" for every neutron absorbed).

These important results sparked off in Szilard his old idea on reactions (n, 2n) which would therefore mean keeping the U research secret in order to avoid alerting the Germans. He had already been active on this matter, suggesting that the secret should be strictly kept even outside the United States. A passionate appeal of his to Joliot pleading for secrecy became famous ([9]).

The question of secrecy was becoming urgent because in the meantime two separate

([9]) According to Bertrand Goldschmidt the Radium Institute in Paris never received such a long telegram.

articles from the groups at Columbia were ready for publication in "Physical Review".

So on the morning of the 16th of March George Pegram, the head of the Physics Department at Columbia, called a meeting with Fermi and Szilard. At Szilard's insistence Eugene P. Wigner, a Hungarian refugee and professor of theoretical physics at the Institute of Advanced Studies at Princeton, also took part. The four physicists discussed secrecy for a long time, with Fermi firmly against the idea. In the end they decided to talk to Bohr, whom they knew to be firmly against for reasons of tradition and scientific ethics. In the meantime it was agreed that the two articles should be sent to "Physical Review" to have them "dated" but the editor should be asked to delay publication ([10]).

The four physicists also wondered about the future of the research as well as its secrecy. Wigner in particular expressed his considerable disquiet about the lack of certainty hanging over their work and asked that the government authorities be informed of their activity as well as the secrecy issue. The aim was to obtain government involvement and financial assistance. But who to contact? Pegram, who had contacts at the Navy Department, proposed that channel and since Fermi was going to be in Washington that very evening for a scientific seminar it was decided that he would act as go-between with the Navy in the capital. Without delay Pegram wrote a brief letter to Admiral Hooper, technical assistant to the chief of navy operations, in which, after a very flattering introduction of Fermi (professor at Columbia, Nobel prize winner, etc.) the possibility of using U as an extremely powerful explosive was hinted at in very prudent terms. This was the first, and for some time the only contact the physicists had with the American administration. Unfortunately Fermi's visit did not have the hoped for results, both because the admiral was not present at the lecture Fermi gave on the morning of the 18th at the Navy Department and also because of the excessive caution Fermi used when referring to the project's real possibilities of success. Fermi's experience was not completely wasted however because, thanks to the enthusiasm of a young physicist from the Naval Research Office who was present at the lecture, the Department contributed 1,500 dollars towards the research at Columbia. It certainly was not much, but the various groups could continue their research for the time being without depending exclusively on modest university funding.

4. – The circle spreads...

In the meantime two different and opposing lines of thought were developing on the whole U affair. Bohr, in accordance with his recent deductions about "235", claimed that a chain reaction, even if explosive, would only become possible by separating considerable quantities of the rare U isotope. He was very sceptical however about the chances of con-

---

([10]) The delay was not to last long. On the 18th of March issue of Nature an article on secondary neutrons by Joliot's group was published, in effect overtaking all of Szilard's arguments. It should be noted that in the same journal, Nature, a second article by the three Frenchmen appeared on the 22nd of April in which the number of neutrons released for every neutron absorbed was estimated to be 3.5.

structing a bomb because of the immense difficulties involved in the isotopic separation of U-235, given its rarity and the very small percentage difference between its mass and the mass of U-238. Fermi instead was obstinately convinced that a chain reaction could be obtained using natural U, a sufficiently pure moderator (graphite) and thermal neutrons.

It should be said in passing that Dunning also agreed with Bohr's idea, that is to say the use of U-235, and he was urging some specialists in isotopic separation, including Alfred O. C. Nier, a physicist at the University of Minneapolis ([11]), to tackle the problem. The conflict between these two approaches came to light in an unexpectedly public way at the annual Spring conference of the American Physical Society, so much so that the science correspondent of the New York Times made it the subject of an extensive article from Washington. The problem of the strategic choice between the two approaches had been raised and it was destined to influence and severely affect future choices, making Fermi's work more difficult. It is not that Fermi did not see the advantages of the possibility of using "235", in fact it seems that he too urged Nier to work on it, but he was aware not only of the technical difficulties of enriching U but also of the enormous disproportion between the means that would be required and the meagre resources then available. In any case, Fermi returned to Columbia full of confidence in the natural U approach. According to Anderson Fermi told him "Stay and work with me. You'll see. We shall be the first to bring about a chain reaction". The words would not seem very credible coming from such a self-controlled man as Fermi if it were not for the authoritativeness of the source.

In the meantime, in the month of April, Szilard, who was extremely active and skilled at this kind of thing, had managed to obtain more than 200 kg of uranium oxide from a Canadian company (the Eldorado Radium Corporation). So in June it was possible to plan with Fermi and Anderson a new experiment on secondary neutrons on a more significant scale. There was still the problem though of the passive resonance capture which threatened to alter unacceptably the results of the experiment, which was also based on the activation of a neutron detector, in the presence and the absence of uranium. This time, unlike on previous occasions, detecting the neutrons involved activating manganese dissolved in the form of sulphate in a large 540 litre water container (Szilard-Chalmers method).

The problem of capturing the epithermal neutrons seemed insurmountable. Anderson said (FNM 132) that it only took Fermi about twenty minutes to find a solution. It was to concentrate the U oxide in separate blocks (heterogeneous structure) instead of distributing it homogeneously in the water (as the French in Joliot's group had done). This would drastically reduce the capture of the neutrons as they slowed down. This precaution, which derived from Fermi's great mastery of the behaviour of neutrons, was a fundamental turning point in the story of nuclear energy. Without it it would never have been possible to reach a self-sustaining chain reaction with natural uranium. The heterogeneous

---

([11]) Nier had been the first to measure the 1/139 ratio of relative abundance of U's two principal isotopes: a third isotope, U-234, exists in nature in irrelevant quantities of around 1/17000.

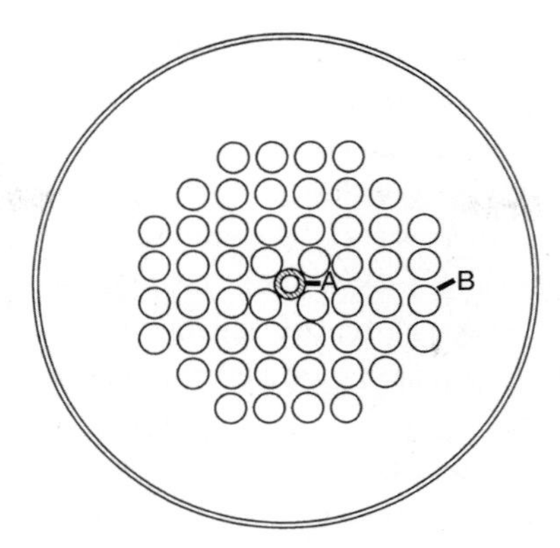

Fig. 2. – Fermi's first heterogeneous multiplying structure (Columbia University, 1939). A) Source of photo-neutrons of 2.3 g of Ra and 250 g of beryllium; B) one of the 52 pipes, 5 cm in diameter and 60 cm high, filled with U_3O_8. (From FNM 132.)

structure had the configuration shown in fig. 2, the U was contained in 52 metal pipes.

The results, obtained by comparing the activity induced in the Mn by the thermal neutrons in the presence and in the absence of U, were published, without any regard for secrecy, in "Physical Review" on the 3rd of July with the three signatures A.F.S. The effect of the presence of U was positive and was estimated at 1.2 secondary neutrons per neutron absorbed. The authors noted that with a further theoretical correction of the resonance capture "this number should be increased to perhaps 1.5", and they added "from this result we can conclude that a nuclear chain reaction could be maintained in a system in which neutrons are slowed down without much absorption until they reach thermal energies and are then mostly absorbed by uranium rather than by another element".

Another conclusion in the article was that, as a result of the absorption of the hydrogen in the water, it was doubtful that water could be used as a moderator for a chain reaction with natural U. In any case further measurements would be required.

The originality of the method is that it is an "integral experiment" in which, instead of measuring separately the 4 factors ϵ, η, p, f (which constitute the multiplying factor k_∞) because affected by too many measuring errors, the global result of neutron multiplication is sought. Integral measurements would be the strong point of Fermi's later research up to the so-called exponential experiments with U-graphite which would be the most significant and productive example.

It should be noted that this was the first and last article to be signed by both Fermi and Szilard. Their different concepts of experimentation and above all the distribution of work while the experiment was being carried out made it difficult for them to work together. Despite their very different characters their reciprocal esteem prevailed and they would continue to work together fruitfully for years. A proof of this is the intense correspondence after Fermi had left for what Segrè defined *"his beloved Ann Arbor (Michigan)"* for a summer course on meson physics. It was a fruitful exchange of letters during which the two men arrived at two independent estimates of the quantities necessary in a structure for a chain reaction. Szilard's was higher and more realistic (50 tons of carbon and 5 of U). As for the availability of the materials required he informed Fermi that he had obtained an offer for the supply of considerable amounts of graphite from the National Carbon Company *"at a reasonable price"* (the same old Szilard !...).

In their exchange of letters they both thought of heterogeneous U-graphite structures: Fermi thought of thick layers of graphite alternating with layers of U, Szilard thought of what was to prove the winning solution, a structure with small blocks of U immersed in graphite bricks. This structure, by analogy with crystals, was called a U-graphite lattice.

During the hot New York summer Szilard tried to persuade Pegram to undertake new experiments but, especially because of Fermi's absence, Pegram decided to postpone everything until the autumn.

5. – The story of a letter

Left alone in New York Szilard started meditating again about the best way to involve the American government in the project, since he had by now lost all faith both in the Navy and in possible contributions from private industry. He was supported in this by Teller, who was in New York for a summer course at Columbia, and by Wigner who often visited Columbia. One of the main worries of the Hungarian "triad" (or the "Hungarian conspiracy" as Tuve loved to define it) was that the Germans might lay their hands on large quantities of U produced in the Belgian Congo by the Union Minière. They had heard of the embargo placed by the Berlin government on the U from the Joachimstal mines in former Czechoslovakia, a clear sign that the Germans were also following the U path.

After several debates and contacts with some senior New Yorkers the three men reached the conclusion that the most efficient way to involve the US administration would be to send a signal to the highest level, *i.e.* to President Roosevelt. Yes, but how to reach him? The three knew that in the high spheres of Washington they were "nobodies". It was essential to get the backing of an unquestionable scientific authority. So Einstein's name emerged. He was on holiday though at Peconic, a small place in the north of Long Island and it was not easy to track him down. So it was that on the 16th of July Szilard and Wigner (who acted as chauffeur because Szilard did not know how to drive) managed to meet him in his small house in the country and talk to him about the Columbia experiments. Einstein, who had never heard of the possibility of a chain reaction, was completely taken aback: *"daran habe ich gar nicht gedacht!"* (I never thought of that!) At this point it seems that his scientific open-mindedness and his enthusiasm for new things took over. Not only did Einstein declare himself ready to cooperate but, by the end of the day, he had already agreed a first draft of the letter in German to Roosevelt with his young colleagues.

On his return to New York, while Wigner set off on holiday to California, Szilard wondered how to have the letter delivered to the president. Alexander Sachs, Vice-President of the Lehman Corporation, was a great help to him and suggested some substantial changes to the text.

Sachs, who was on very friendly terms with Roosevelt as an active supporter during the election campaign, offered to deliver the letter personally to the President. So it was that Szilard had to return to Einstein (this time with Teller as driver) for the necessary changes. The final text in English, approved by Einstein, was signed and dated on the 2nd of August 1939 and sent back to Szilard who received it the next day.

Alexander Sachs only received the final text on the 15th of August. Aware of the enormous amount of paper that arrived on the President's desk, it was Sachs's intention to deliver the letter personally to Roosevelt, or at least to read it to him out loud. With this in view he immediately set about arranging an appointment. Unfortunately, as history teaches us, the devil will interfere with the best laid plans, in this case in the form of Adolf Hitler. At 4.45 a.m. on the 1st of September 1939 the German armed forces invaded Poland. France and Britain declared war on Germany on the 3rd. There was therefore plenty to keep the President of the United States busy, with absolute priority, and to make him postpone the meeting with his friend "Alex", the purpose of which he had not been told.

From then on events in the United States were determined by, and even the pace was set by the timing of politics and the events of the war. The same happened to the "uranium" project and, as a result, to Fermi's work.

At this point I must give a warning. Even a summary description of the events that led up to the first chain reaction set off by man would require a text of a length which would not be compatible with the space available for this chapter. I therefore apologize to the reader if we are obliged to sum up the following events, condensing them into purely chronological order.

6. – In Washington things start to move...

As a first consequence Roosevelt only received Sachs on the 11th of October. The conversation, at which general E. M. Watson, a senior member of the presidential staff, was also present, had a successful outcome and it was decided that a consultative committee for uranium (Uranium Committee UC) should be set up under the chairmanship of Lyman J. Briggs. He was the Director of the National Bureau of Standards and as such was considered a sort of representative of American physicists in the administration ([12]). Sachs himself was also a member of the committee along with two military experts and the physicists that Briggs would nominate.

Briggs called a first meeting of the UC on the following 21st of October. Roberts from the APO also took part (representing Tuve), as did the three Hungarians Szilard, Wigner and Teller ([13]).

([12]) Differing views have been expressed about Briggs's role in following developments.
([13]) It is not clear why Fermi was not at the meeting. Maybe he was still resentful of the lack of success of his meeting with the Navy. In any case, at later meetings of the UC, as in other government committees, he seems to have had a cold detached attitude, very different from that

In an atmosphere of pronounced scepticism from the representatives of the armed forces, Szilard set out the prospects for a chain reaction and in particular the military aspects, emphasising the need to have considerable quantities of U and graphite available, a problem that he claimed was a first priority in order to obtain a better understanding of the nuclear properties of carbon. He did not actually make a specific request for funding however. At this point one of the two military representatives broke in, asking "Come on, how much money do you need?" Teller, taken by surprise, answered 6,000 dollars to which the other military man replied "All right, all right, you'll get your money". The consequences of the meeting were however much as could have been expected. Roosevelt received the report, ordered "File it" and the report remained filed until the following year.

Finally, in February 1940, Fermi received the 6,000 dollars for the purchase of 4 tons of the graphite Szilard had acquired and which was reserved for what at that time was called the "crucial undertaking", *i.e.* the measurement of the absorption of the neutrons by the graphite. On the seventh floor of the Pupin Laboratory a vertical "pile" was assembled (3 ft $\times$ 3 ft $\times$ 8 ft) made up of blocks of graphite, with the neutron source at the base and, at different heights, rhodium sheets as thermal neutron detectors. He was returning to the tried and tested methods of his experiments in Rome and the experimenters' famous sprints, rhodium foils in hand, in order to reach the detectors in time to measure the induced radioactivity, given the extremely short half life of rhodium, 44 seconds.

With this type of measurement, which he was so familiar with, his good humour also returned, encouraged by the result that the graphite would be compatible with a chain reaction with natural U. It should be noted that this was the first "exponential experiment" in which the cross-section of the capture of the neutrons was measured, with the appropriate corrections for lateral escapes, by means of the attenuation of the axial neutron flow, in accordance with the already tested philosophy of integral measurements.

There was an unexpected development at the end of February when Nier, at Minneapolis, after the pressure he had received, succeeded in separating modest but measurable quantities of U-235 and U-238, deposited on a sheet of nickel. Nier sent them to Dunning at Columbia, who immediately rushed to irradiate them with neutrons. It was thus proved that it really was only U-235 that underwent thermal fission ([14]). At the same time however the discussion was re-opened between the supporters of the two different paths, natural U and enriched U. And with it the perplexities about the success of Fermi-Szilard's U-graphite method. These doubts were destined to interfere heavily with Fermi's future work.

---

of the three Hungarians. It is also possible that as an Italian he felt a certain distrust towards him in some American circles. One should not forget after all that in May of that same year Mussolini had signed the famous "Pact of steel" with Hitler.

([14]) On the 14th of March 1940 Fermi wrote to Amaldi (in Rome) as follows: "The latest news is that the thermal neutrons produce fission only in the 235 isotope of uranium. Nier has separated about 0.3 micrograms of uranium with his mass spectrograph and only the 235 fraction gave fission." Note the date: only two months before Italy entered the war...

A first sign of this conflict emerged at the next UC meeting on the 27th of April 1940, at which Fermi was present and which ended inconclusively apart from a heartfelt exhortation from Sachs to carry on with the U-graphite experiments on an adequate scale.

It was clear that things were not going well. The real problem was more general though and concerned all scientific research in the USA. Some felt that the research was insufficient and inadequate given the problems that the nation would have to face because of the tragic events in the war in Europe. Particularly after the fall of France (June 1940) and the German threat to Britain, many Americans feared that their country would become involved in the conflict.

Vannevar Bush, the President of the Carnegie Institution, was particularly convinced of this. He was an open-minded man, an engineer with considerable experience in organising research (he had been Vice President of MIT). In discussion with a group of first class people, including Frank Jewett, president of the Bell Telephone laboratories and of the National Academy of Science (NAS), James Bryant Conant, a young and energetic chemist, president of the University of Harvard, Karl T. Compton, a physicist and president of MIT, Bush was convinced that sooner or later the United States would be dragged into the conflict and that this would require a reorganisation of the scientific and technological sector in order to have the right instruments for a war fought by technologically advanced means.

Thus persuaded Bush and the others thought of a new, mainly autonomous, organisation, working directly for the President, financed directly by the Government instead of by the military. Thus on the 12th of June 1940, with decisive help from Harry L. Hopkins, the Secretary for Trade and a man Roosevelt trusted, the National Defense Research Council (NDRC) was born. Its only aim was "to mobilise science for war purposes" (Segrè). Bush was named chairman at the same time and immediately called James B. Conant to join. Their first act was to absorb the already existing Uranium Committee. Briggs, while still keeping the chairmanship of the UC, would report directly to Conant.

The NRDC's mandate was thus research for war aims, but could Fermi's work really be of military interest? The prospects for a nuclear reactor seemed to be to produce energy, or at most to propel submarines, problems which were certainly not top priority, at least in Bush and Conant's view.

This was seen when Briggs, after an ad hoc meeting of the UC, made a written request to Bush for $140,000 ($100,000 for large scale experiments on U-graphite systems and 40,000 for cross-section measurements). He was only given $40,000 and with that Fermi and the Columbia physicists had to be content and wait for better times ([15]).

([15]) Things went better for the physicists working on enriching U. Through the US Navy's research laboratories they obtained $100,000 for research into isotopic separation. Separating U-235 in sufficient quantity for a bomb (the one which was later dropped on Hiroshima) was one of the most significant successes for American science and technology during the war. Unfortunately we cannot talk about it here.

7. – Times of war

An unexpected boost to the direction of Fermi's research came from the other coast of the United States, to be precise from Berkeley (CA). It was the discovery of a possible new nuclear explosive, plutonium.

It had been known since Hahn and Meitner's research in the early thirties that U-238, by capturing a neutron, formed an unstable isotope U-239 which transformed itself into a new element with atomic number $Z = 93$ by emitting an electron (with a period of 23 mins). Later the Americans would call the new element neptunium (Np-239). It was identified with certainty at Berkeley in the summer of 1940 by the physicists E. McMillan and P. A. Abelson.

Np-239 decays in turn, with a period of 2.3 days and emission of an electron, into a new element of $Z = 94$ called (by Sold in 1947) plutonium (Pu-239). This element made up of an even Z number of protons and uneven $A-Z$ of neutrons would presumably have characteristics similar to U-235, which is also even-uneven, and split under the action of thermal neutrons.

It was quite conceivable that, if it had a sufficiently long average life, it would be possible to separate it by chemical means and use it as a nuclear explosive instead of U-235 which was so difficult to separate. It was still unproven however that it really was a fissile element. If it were it could be produced in considerable quantity in a nuclear reactor. Fermi and Segrè discussed this possibility at length in December 1940 when Segrè visited Leonia (New Jersey) where the Fermi family had moved in the summer of 1939 ([16]). It is clear that this possibility, if confirmed, would give a considerable impetus to the line followed by Fermi and his friends at Columbia.

It was necessary however to prove that Pu-239 really was fissile. For this purpose it was necessary to prepare a quantity of it, even if only a minimum amount, in order to measure its nuclear properties. The new 60 inch cyclotron at Berkeley's Radiation Laboratory would be suitable for the purpose if it were available for irradiating the U. It so happened that Ernest O. Lawrence, the director of the Radiation Laboratory, was in New York and so on the 16th of December a favourable decision was made at a meeting at Columbia also attended by Pegram. Segrè returned to Berkeley and immediately set to work together with G. W. Kennedy and G. F. Seaborg, obtaining a mixed sample of Np and Pu. Later, on the 28th of March 1941 Segrè and Seaborg used a method perfected by Arthur C. Wahl and succeeded in separating the Pu from the Np. Thus it was possible, with a sample of only 0.28 micrograms, to measure the cross-section of the Pu-239 both with thermal neutrons and with fast neutrons.

Later, in May, Segrè and Seaborg showed that the thermal cross section for fission in the Pu-239 was 1.7 times greater than for U-235. Thus a new nuclear explosive was born, Pu-239. There was now the question of producing it in sufficient quantity, since it was

([16]) This discussion was reported personally by Segrè himself on page 121 of his book already repeatedly quoted here.

not possible to produce enough in the cyclotron. It became clear that Fermi's research into the uranium-graphite chain reaction might provide a solution to the problem. From this point of view the Columbia research became of military interest.

At the NDRC however considerable doubts remained about the possibility of obtaining results with military applications as suggested by the nuclear physicists who were working on the problem in ever greater numbers in various US laboratories. For that matter the British too, despite being under heavy German bombing, were organising U-related research with considerable determination, helped also by the emigration of European physicists fleeing from the Nazis. Amongst these were Otto Frisch, now working in Birmingham with Rudolf Peierls, a theoretician of German origin who had already calculated the possible configurations and sizes of an atom bomb, together with the French from Joliot's group, Goldschmidt, von Halban, Kowarsky who had brought safely to Britain the small (165 litres) but precious quantity of heavy water collected by Joliot before the German occupation. They were working mainly on multiplying U-D_2O systems.

The British activity was directed and coordinated by physicists of great prestige such as Chadwick, Oliphant, G. P. Thomson, Cockcroft and others. They were convinced that the bomb could be made with U-235 (since they were unaware of the existence of Pu), although they were equally convinced that the effort necessary for a U-235 enrichment programme would not be compatible with the resources available in the UK. Hence the British pressure on the Americans for a large scale programme for the bomb, to be built quickly since their main worry was that the Germans might get there first.

For his part Lawrence, who was ever more involved in these problems, when he heard of the British progress, impatient of Briggs's management of the UC (accused of excessive bureaucracy) attacked the doubtful Bush with a certain brutality. Bush then took the initiative of asking Jewett, president of the NAS, for an overall evaluation of the U project by the Academy. Jewett immediately created a Review Group which included Lawrence, William D. Coolidge, ex research director for General Electric and, as chairman, Arthur Compton, professor of physics at the University of Chicago and Nobel Prize winner for the quantum effect that bears his name.

On the 17th of May Compton sent the Review Group's report to Jewett who sent it on to Bush. The report put forward some options including mainly the chain reaction (not only in the U-graphite version but also U-Be or U-D_2O). As for the bomb, both the U-235 and the Pu (to be produced by chain reaction) options were considered and the report estimated that it would not be possible before 1945. It is important that, for the first time in a document, mention is made of the use of a pile for producing Pu and the necessary finance in order to continue Fermi's intermediate experiments is recommended.

These conclusions had been reached though through multiple and often conflicting steps (NDRC, NAS, and others). In any case the NDRC mandate only concerned scientific research and could only with difficulty be extended to cover the collaboration with industry essential for development for military goals. The military aims had become impelling as the situation in Europe evolved: the German invasion of Russia began on the 22nd of June 1941.

Bush thought of a new body, which would have more authority than the NRDC and would emanate directly from the President of the United States. Thus the Office of Scientific Research and Development (OSRD) was set up on the 28th of June 1941 on the orders of Roosevelt. Bush became its director, while Conant replaced him as chairman of the NDRC and the UC became a section S1 of the OSRD.

Over the following months there were a series of meetings and contacts with the Review Committee of the NAS, until the third and final report dated the 27th of November 1941 was sent to Roosevelt and approved by him.

The report formed the basis of the S1's discussions when it was called by Bush in Washington on the 6th of December.

The most important decision was a complete reorganisation and redistribution of responsibilities:

1. Harold Urey of Columbia was to develop the enrichment of Uranium using the gas diffusion method.
2. Lawrence at Berkeley would direct work on enrichment using the electromagnetic separation method.
3. Eger V. Murphree, the director of research for Standard Oil in New Jersey was to develop the enrichment process with ultracentrifuges.
4. Compton in Chicago was to be generally responsible for theoretical studies and for designing the bomb.

The influence of the British, who were now being carefully listened to in Washington, is obvious.

As you can see there is no mention of Pu or of Fermi's U-graphite method. Things were to go very differently for both projects, as later events will show, thanks above all to Compton's unstinting support.

The S1 meeting was called on the 6th of December and concluded with an agreement to meet again within a couple of weeks.

But on Sunday the 7th of December the Japanese launched their surprise attack on Pearl Harbour. The United States came into the war against Japan and its allies Germany and Italy. Fermi thus automatically acquired the status of enemy alien.

From that moment there was an enormous and tumultuous development of the "uranium project", forced by the development of the bomb. It would deserve an adequate description but unfortunately it would take too long. From now on therefore we will limit our attention solely to Fermi's research.

8. – The final experiments at Columbia: Fermi moves to Chicago

During 1941 Fermi had continued his measurements, while Szilard devoted himself even more, and with growing success, to obtaining the materials (U and graphite) in ever greater quantities and of ever higher purity. In the meantime Fermi planned large scale subcritical experiments to determine the critical dimensions of the U-graphite system, extending the neutron diffusion method already used successfully in the graphite column to the new structure. In August and September 1941 the necessary material began to arrive until it reached a quantity of 6 tons of U oxide (in the form of U_3O_8) and thirty tons of graphite blocks.

Given the size and weight of the material it was not possible to use the usual laboratories on the various floors of the Pupin Laboratory until the end of September when Pegram found a place of suitable size (the Schermerhorn room at Columbia). A large parallelepid shaped structure 11 ft by 8 ft was erected made up of graphite blocks and U oxide packed into cubic tin cans, 8 inches per side. A strong intensity neutron source (2 g of Ra + Be) was placed at the base. The neutrons were detected by the activation of sheets of In distributed appropriately in the structure.

So Fermi's first exponential pile was born. On the basis of his calculations with the theory he had worked out it was possible to measure the multiplication factor k_∞ which turned out to be 0.87.

The result is given in a secret report dated 26th of March 1942 which was signed not only by Fermi and Anderson but also by Bernard Feld, George L. Weil and Walter H. Zinn (see FNM 150).

The result was not exactly encouraging since the k_∞ was 13% lower than the acceptable minimum, but Fermi planned new experimental arrangements that would allow the results to be improved, starting with the elimination of the tin cans and the consequent elimination of the Fe which absorbed neutrons.

Fermi thus built a new exponential pile. This time the U was inserted into the graphite housing in the form of small cylinders (which approximated the ideal spherical shape) with a diameter and height equal to 3 inches and a weight of 1795 grams. To increase the efficiency of the system and reduce the passive absorption due to the hydrogen in water the U_3O_8 was pressed and dried by being heated to 250 °C.

Moreover the entire structure was completely encased, with some difficulty, in a covering of welded steel plates. This technique meant that the humidity of the pile and the absorption of nitrogen from the air could be reduced by means of a partial vacuum. In this way the result matched expectations with a $k_\infty = 0.918$ as recorded in a report in April 1942 (see FNM 151). This was Fermi's last exponential experiment at Columbia.

In the meantime it was necessary to follow up quickly on the decisions taken at the S1 meeting on the 6th of December. The committee met on the 18th of December under the chairmanship of Conant. Compton took part and fixed some concepts for the groups entrusted to him in a memorandum to Conant on the 20th of December and laid down a code. To start with he stated his belief that a Pu bomb could be made and that its critical mass should be about half that of U-235. The difficulty would be the need for a remote location for the industrial production plant given the high levels of radioactivity of the involved products. Compton also drew up an expected time table.

1. 1 June 1942: acquisition of the fundamental data for the chain reaction.
2. 1 October 1942: achieving a chain reaction
3. 1 October 1943: a pilot plant for Pu production
4. 31 December 1944: mass production of Pu.

It is clear that the timing estimates, which were fairly optimistic, were also designed to send a strong signal to Conant and Bush that the bomb would be available in time to influence the outcome of the war.

Compton drew up a cost estimate of 1.2 million dollars, a figure he put forward with a certain hesitation used as he was to the relatively modest university funding of a few thousand dollars a year. There was no hint however of the problem of coordinating, and preferably bringing together at a single site, the activities scattered between Columbia, Princeton, Chicago and Berkeley. It was a question that was a matter of animated discussion in January between those directly involved. Finally on the 24th of January '42, in a heated debate at a meeting with Szilard, Lawrence and Alvarez in Compton's house in Chicago, it was decided, above all because of Compton's stubborn obstinacy, to concentrate all efforts connected to the Pu-pile project at the University of Chicago, with the conventional name of Metallurgical Laboratory. Fermi was far from pleased when he was told the news, particularly since the experiments at Columbia had by now reached a considerable size. Anyway the decision had been taken. Fermi, who in the meantime had sent some of his younger colleagues to Chicago, was forced to "commute" between New York and Chicago for a few months, with all the difficulties inherent in his status as an "enemy alien", until he finally moved to Chicago at the end of April ([17]).

9. – Chicago becomes the centre for research

In the meantime the Met lab group had grown in number as Seaborg and his young colleagues from Berkeley moved to Chicago as a result of Compton's decision to put the activities of the Pu group and of Fermi's group alongside each other. Some samples of Pu were in fact already available in Chicago (obtained by irradiating 150 kg of uranyl nitrate with the cyclotron of the University of Saint Louis) in sufficient quantity for the measurements essential to develop the industrial production phase. As decided by Washington, a Boston engineering company, Stone and Webster, had been entrusted with the design for this plant. The arrival of the first engineers, completely unprepared for the task ahead of them, was not welcome in Chicago, not only because of their unpreparedness which made it difficult for them to work with the physicists, but also perhaps because the physicists were afraid that they would be overtaken by the engineers and end up playing second fiddle. By the end of the summer there was considerable malcontent, with Leo Szilard, increasingly concerned by the delays and worried that Germany's physicists might get to the bomb first, acting as spokesman. It took all of Compton's patience and

([17]) It cannot be said that the U-graphite process was looked on favourably in Washington. As late as the 23rd of May 1942, in a special meeting of the S1 programme chiefs called by Conant to identify the most promising method of building the bomb, to then move on to the pilot plant stage and then on to the following industrial development, the U-graphite pile was competing with no less than three other 235 enrichment methods (ultracentrifuges, gas diffusion, electromagnetic separation) and with the U-D_2O pile (always supported by Urey). Conant tried to estimate the funding commitment: in any case however Roosevelt's orders were to spare no expense.

authority to calm the peaceful revolt.

The whole U project, which was being worked on by now in various different universities and institutions scattered around the United States, required a more authoritative and efficient organisation since the state of war made communications more difficult, and hence management too. Vannevar Bush reached this conclusion and thought of involving the armed forces, with the designation of a high ranking officer to direct the project. He spoke to General Sommerville who was responsible for army supplies. He started looking amongst the senior officers of the Corps of Engineers, the most technical of the army corps, and found the person he was looking for in Colonel Leslie Richard Groves. On the 17th of September 1942 Sommerville had him declared responsible for the bomb project, without even informing Bush, who was furious when told after the event. It was an excellent choice however. Groves had all the qualifications (a degree in engineering, two years specialisation at MIT) and top class experience, most recently as the army's Deputy Chief of Construction he had completed the construction of the Pentagon in Washington.

Groves discovered that he had had a predecessor, a colonel in the New York detachment of the Corps of Engineers, responsible in August for coordinating the then very modest activities of the U project entrusted to the military. He had set up an office in New York for the task with a deliberately bland name, the "Manhattan Engineer District". Groves kept the name and the U project would be referred to from then on as the "Manhattan Project".

Groves was promoted to general and he named as his deputy Col. Kenneth D. Nichols, a competent representative in Washington of the old Manhattan District. Groves immediately showed how active he could be. On the 18th of September he sent Nichols to New York to buy a 1,250 ton shipment of pitchblende, packed in 2000 steel drums, with a content of 65% U oxide, which had been sent to the USA by the Union Minière in 1940 to prevent it falling into German hands and which was still lying in open air storage on Staten Island. On the 19th of September Groves had the activity of the Manhattan District declared top priority (AAA) by the person responsible at the War Production Department. On the same day he concluded the purchase of a vast area (about 62,000 acres) in western Tennessee, called site X, which, under the name of Oak Ridge, was set aside for the large U-235 enrichment plants using gas diffusion.

A few days later, on Bush's initiative, Groves was summoned to lay out his ideas and to receive overall instructions from a top level committee (the Military Policy Committee) which was made up of Bush himself, the Secretary for War Stimson, the Chief of General Staff Marshall (who was to give his name to the famous plan), Conant and some senior officers. The U project passed at that moment into the hands of the greatest organisation for the conduct of the USA's war strategy.

10. – The pile

In the meantime in Chicago Fermi had returned to building exponential piles on a large scale: he had been provided with a vast space previously used as squash courts under the western stands of the University's playing field (Stagg Field). The first results were already so encouraging that in May they achieved $k_\infty = 0.995$.

By improving the quality of the graphite and using denser metallic U which was more efficient than the oxide, it would be possible to hope for $k_\infty > 1$. It was then that Fermi, convinced of the possibility of a chain reaction, began to think of a pile that could go critical.

It would be of considerable size and spherical in shape, to reduce as much as possible the leakage of neutrons at the edge and it would therefore have an effective multiplication factor k_{eff} closer to k_∞. Given the unknown factors involved in the working of the first nuclear reactor it was *a priori* unthinkable to build it at Stagg Field in the middle of the city, or so they thought at the Met Lab. Compton found a suitable site about twenty miles southwest of Chicago, in the Argonne Forest. The site was bought thanks to Nichols' rapid intervention and the works contract was awarded to Stone and Webster.

In the middle of August Fermi's group obtained a k_∞ close to 1.04 with a graphite and U oxide structure. A further improvement of around 1% of the k_∞ value would be obtained by eliminating the absorption by the nitrogen in the air. Unlike the previous experiments at Columbia it was not possible to enclose this large structure in a box in order to operate in a rarefied atmosphere; so it was decided to wrap the whole thing in a covering similar to a cubic aerostatic balloon (it seems that the idea was Anderson's) with sides twenty-five feet long. The technicians at the Goodyear Rubber Co. were astonished at the unusual shape but carried out the order without asking embarrassing questions about its possible use.

Between the middle of September and the middle of November Anderson and Zinn (responsible for materials) assembled a further twenty exponential piles, testing the material as it arrived. The graphite was in the form of bars while the U oxide to insert in the holes cut deliberately in the graphite bars was compressed into "pseudo spheres", that is to say cylinders with the bases replaced by spherical semicaps.

Cutting the graphite bars into blocks weighing nearly 10 kg each was extremely heavy work. When the work was in full swing about ten tons a day were produced. On the 5th of October General Groves paid a quick visit to sound out the opinions of Fermi and the other physicists at the Met Lab about the cooling system for the future plutonium generating pile at Hanford for which he was negotiating with the chemical industrial colossus Du Pont De Nemours. Groves left Chicago very satisfied with the Met Lab's work and convinced, as he himself admitted, that *"the plutonium process seemed to offer us the greatest chances for success in producing bomb material"*. Everything seemed to be going for the best when there was an unexpected setback —a strike for an indefinite period of Stone and Webster's employees who were constructing the building for the Argonne pile which according to the contract should have been completed by the 20th of October.

In early November, a few days before beginning to build the pile the grave problem thus arose of where to build it. Fermi tackled Compton decisively about the situation and proposed building it in those same squash courts at Stagg Field which, unfortunately, was located in a densely populated area not far from the middle of Chicago.

It was a really very difficult decision. In the previous exponential experiments, even with $k_\infty \geq 1$, the critical point could never have been reached because of the small size.

But the pile was different. Despite all the numerous automatic control systems there were fears that, with $k_\infty > 1$ an uncontrolled chain reaction might be set off with all its disastrous consequences.

Fermi though was convinced he was right. The delayed neutrons (discovered by Roberts at the APO back in 1939) would, according to the theory, slow down the dynamics of the reaction as long as k_{eff} was only slightly greater than 1. Once again Fermi's mastery of neutron science and his familiarity with the engineering problems proved decisive ([18]). Compton was convinced and gave the OK to building the pile at Stagg Field. Afraid of a refusal, and to be on the safe side, he refrained from informing the President of the University of Chicago.

Assembling the pile began on the 6th of November 1942 (FNM 181). Fermi organised the work in two teams: one on the day shift run by Walter Zinn (who kept his responsibility for materials procurement), the other on the night shift directed by Herbert Anderson. Fermi named Volney Wilson, a dynamic young student of Compton's, as responsible for instrumentation and for the pile control system.

The pile would be put together under the stands, in front of a balcony that had been used by spectators watching the squash games. From the very beginning the balcony proved useful, for example during the assembly of the Goodyear "cubic balloon" which had a front face left free to allow the pile to be built. Initially a first circular layer of blocks was laid, made up of only graphite, which constituted the reflector at the base. The rest of the structure was put up alternating graphite only layers (inert graphite) with two layers of (active) graphite containing blocks of U oxide. In this way a U-graphite multiplying device with a lattice made up of cubic cells was formed. The outermost layer of the pile, which had to act as a neutron reflector was made up only of inert graphite. The whole structure was supported by strong wooden scaffolding erected a bit at a time as the assembly proceeded.

Assembling the blocks proved to be not only tiring (they reached a rate of two layers a day) but also very delicate because the alignment had to be done extremely accurately in order to leave the (horizontal) housings for the control bars free. The bars were made of sheets of cadmium (a strong absorber of neutrons) fixed to wooden planks 12 ft long. They were manoeuvred by hand and only extracted when measurements were taken. Normally they stayed completely inserted and locked in place with padlocks to which only Anderson and Zinn had keys.

The neutron flux measurements began once the 15th layer was reached and were carried out by Zinn and Anderson at the end of each shift, always in the same central spot in the pile. They were done with boron trifluoride counters, built by Leona Woods, a brilliant graduate student of Compton's, and calibrated every day by checking against

---

([18]) That Fermi possessed remarkable engineering-technological talents is proved clearly by his contributions to the Manhattan Project after the Chicago pile, for which there is an ample bibliography, starting with Segrè's book. What is even more remarkable is his ability and readiness to use those talents.

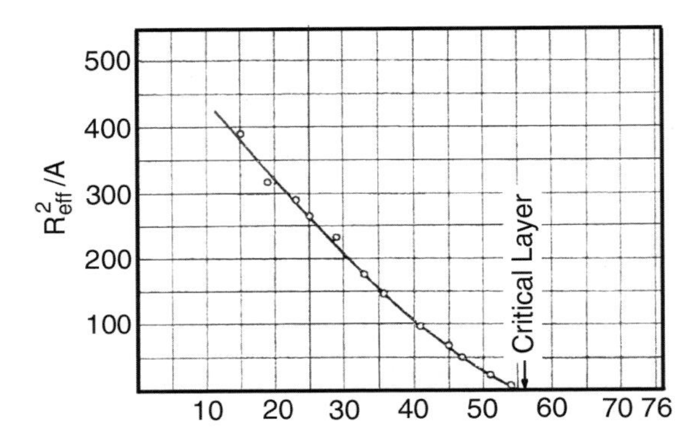

Fig. 3. – The pile's approach to critical level during construction (Chicago, Nov.-Dec. 1942). (From FNM 181)

sheets of indium irradiated with the neutrons from the pile. Of course the measurements were taken with all the control bars extracted, at least in the initial phase. It should be noted that as well as the control bars there was also a safety device consisting of a few bars, controlled electromagnetically, called ZIPs, fitted with a counterweight at one end and held in position by an electromagnet.

When the pile was working all that had to be done to carry out a fast shutdown (scram) of the reaction was to cut off the solenoid's electric current and the bars would be inserted by gravity. The scram would be set off by ionisation chambers calibrated for high neutron flows, while the BF_3 counters were used for low flows.

For further safety a ZIP bar was held outside of the pile by a rope tied to the balcony rail. The rope just had to be cut to achieve a scram in an emergency.

Finally, as an extreme safety measure on the day the critical level was reached, three of the younger physicists were to stand on the platform of the elevator used for the assembly and stay close to the top of the structure, each with a bucket of Cd sulphate solution, ready to pour it onto the pile in an extreme emergency.

As the number of layers grew the intensity increased. This was caused by the neutrons in the counters and in the indium. With a simple formula Fermi calculated the radius ($R_{\rm eq}$) of a spherical shape equivalent to the structure being put up. To forecast when the critical point would be reached Fermi chose the size $R^2_{\rm eq}/A$. As the critical point approached it should tend towards zero. According to his calculations the spherical structure should reach the critical point at the 76th layer.

After the assembly had already begun new and much better material began to arrive however. It was 250 tons of very pure graphite which had been produced in the meantime by National Carbon Co and over 6 tons (from Ames-Iowa) of extremely pure metallic U in the shape of cylinders suitable for insertion into the holes that had been made in the graphite. The metallic U was in any case preferable to the oxide because of its greater density. The material was extremely useful and would allow the $k_{\rm eff}$ multiplication factor to be improved. It was therefore placed in the centre of the structure. Fermi calculated

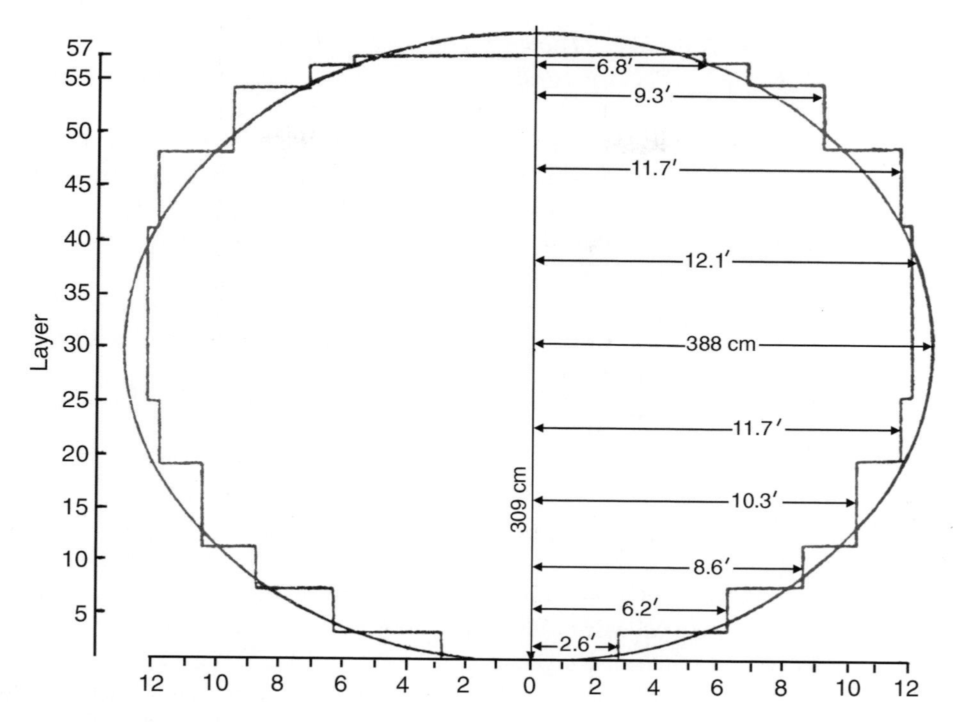

Fig. 4. – Vertical section of the Chicago pile. (From FNM 181).

that it would no longer be necessary to use the Goodyear balloon (and its vacuum) but that, above all, it would be possible to save a score of layers. The extrapolation of his measurements gave an R^2_{eq}/A close to zero, and thus a $k_{\text{eff}} = 1$, in an area between the 56th and the 57th layer (see fig. 3). In consequence the pile would not be spherical but would have a shape which was approximately an ellipsoid with a squashed rotation, with semi axes of 388 and 309 cm respectively (see. fig. 4).

The planned final configuration was achieved by Anderson's team the night of the 1st of December. Anderson himself said that he was tempted to extract the last Cd bar and make the pile go critical. But he had promised Fermi that he would not do it and so, after finishing the measurements and fixing the usual padlocks, he went home that night to sleep.

So the fated 2nd of December dawned. There are countless versions of that day and the various events, nearly all based on the personal recollections of people present at the experiment. They contain details often described but not always in agreement between themselves. It is understandable given the emotional state of those present and the lapse of time between the events and the description of them. In many cases the descriptions could only be made after the relaxation of the rigid discipline imposed by security classification. We therefore quote the simple words Anderson devoted to the event (introduction to FNM 181): *"The next morning, December 2, I was on hand, bright and early, to tell Fermi that all was ready. He took charge then. Fermi had prepared a routine for the approach to criticality. The last cadmium rod was pulled out step by step. At each step a measurement was made of the increase in the neutron activity, and Fermi checked the result with his prediction based on the previous step. That day his little six-inch pocket slide rule was busy for this purpose. At each step he was able to improve his prediction for the following. The process converged rapidly, and he could make predictions with increased confidence of being accurate. So it was that when he arrived at the last step, Fermi was quite sure that criticality would be attained then. In fact, once the cadmium rod was pulled out entirely, the pile went critical, and the first self-sustaining chain reaction took place."*

The pile run for twenty-eight minutes, with a $k_{\text{eff}} = 1.0006$ at a maximum power of 1/2 watt.

Abbreviations and acronyms.

APO	(Atomic Physics Observatory of the Carnegie Institution)
FNM	(Fermi, Note e Memorie - quoted in footnote 1, p. 177)
GWU	(George Washington University)
KWI	(Kaiser Wilhelm Institut)
Met Lab	(Metallurgical Laboratory, cover name for the Chicago section of the Manhattan Project, directed by A. H. Compton)
NAS	(National Academy of Sciences)
NDRC	(National Defense Research Council)
OSRD	(Office of Scientific Research and Development)
S1	(ex UC, section of the OSRD)
UC	(Uranium Committee)

Amongst the various books on the subject the following are recommended:

– DE WOLF SMYTH H., *Atomic energy for military purposes* (Princeton University Press) 1945. This is the American government's first official report, published in August 1945, just after the bombing of Hiroshima and Nagasaki.

– SEGRÈ E., *Enrico Fermi Physicist* (The University of Chicago Press, Chicago, London) 1970. A complete scientific biography written by Segrè, Fermi's student, colleague and friend, who also won the Nobel prize for physics. It contains an extremely extensive bibliography.

– FERMI L., *Atoms in the family: my life with Enrico Fermi* (The University of Chicago Press) 1954. Enrico Fermi seen through his wife's eyes in an affectionate and shrewd biography.

– RHODES R., *The making of the atomic bomb* (Simon Schuster Publ.) 1988. The most exhaustive and well documented history of the American atomic bomb, which constituted a constant and valuable reference for the Author.

– ALLARDICE C. and TRAPNELL E. R., *The first pile*, U.S. Energy Commission, Report TID 292. The construction and starting up of the 1st pile, written in 1946 on the basis of the recollections of a dozen people present at Stagg Field on the 2nd December 1942. (The report was translated and printed by the Forum Italiano dell'Energia Nucleare in 1982, on the 40th anniversary of Fermi's pile.)

– *Symposium dedicated to Enrico Fermi* (Rome, Accademia Nazionale dei Lincei) 1993. Papers by Fermi's colleagues and by other international experts on the 50th anniversary of the 1st reactor.

– DE MARIA M., *Fermi, un fisico da Via Panisperna all'America.* In the collection *I grandi della Scienza* (Le Scienze Editore) April 1999. An extremely recent biography rich in documents and archive photographs.

About the Author

CARLO SALVETTI is a university professor, an author and promoter of studies, research and projects in the field of Nuclear Energy. He has held prestigious posts in Italy, at the CNEN (National Council for Nuclear Energy), and abroad, as Chairman of the Board of Governors of the United Nations' International Atomic Energy Agency (1963-64).

From the Chicago Pile 1 to next-generation reactors

AUGUSTO GANDINI

The aim of this contribution is that of presenting a simple, elementary description of the nuclear reactor physics, a science which had its beginning more than half a century ago with the Enrico Fermi and his collaborators' pioneering work, culminated with the construction of the Chicago Pile 1 reactor. Starting from those first experiments, the developments that followed, and those foreseeable in the next future, are shortly discussed.

1. – Exponential experiments

The first investigations on the propagation of neutrons concerned experiments which are called, for reasons we shall see shortly, exponential. In these experiments a source of neutrons (for instance, polonium-berillium) is inserted into a diffusing medium (made up of graphite, natural uranium or other materials) for studying their propagation properties. In their diffusion process, neutrons are subject to collisions with the nuclei present in the medium. In these interactions they are absorbed, in case of a capturing collision, or are deviated from their path, in case of an elastic or inelastic scattering. In the following, we shall give an elementary description of the neutron diffusion in media. The discussion will consider average values, not taking into account space, or time microfluctuations of the quantities involved, at their turn subject to specific analysis ("noise analysis").

Let us consider a plane source of S monoenergetic neutrons emitted per second and per square centimeter from one side (of area A) of a parallelepiped of length H, and let us try to determine the fraction of them arriving without undergoing collisions at the (generic) distance x (see fig. 1). The neutrons emitted from a source generally diffuse in all directions. For the purpose of this description, we shall assume that they are emitted following parallel routes in the direction of the opposite side of the parallelepiped. We introduce then the concept of "cross-section" associated to each nucleus of the medium (supposed to be made up of one element). We shall interpret this quantity as the area

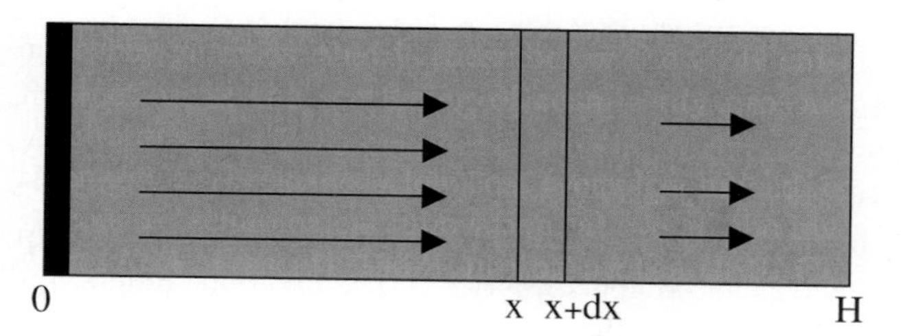

Fig. 1.

of such a nucleus projected on a plane perpendicular to the x-axis. We shall denote this elementary area as σ ("microscopic cross-section") ([1]). Let us define the neutron current density $j(x)$ as the number of neutrons emitted by the source S and crossing the transverse area at point x per second and per square centimeter without having undergone capture, or scattering collisions. The number of all these uncollided neutrons arriving at point x may be then written as $Aj(x)$. The number of those that reach point $x + \mathrm{d}x$ (where $\mathrm{d}x$ is a very small quantity) will be given by this same value diminished of the number of collisions which will have taken place in the interval $\mathrm{d}x$, always within a unit time. These collisions are assumed to occur every time the trajectory of a neutron crosses a nucleus, *i.e.*, intersects its "cross-section". Assuming a density of N nuclei per unit volume, the collision probability will then be given by the ratio between the sum of all cross-sections (elementary areas) of the nuclei within volume $A\mathrm{d}x$ and the transverse area A, *i.e.*, by $N\sigma\mathrm{d}x$. Consequently, the number of collisions in the $\mathrm{d}x$ interval in a unit time is $Aj(x)N\sigma\mathrm{d}x$. A number of neutrons per second and per square centimeter of the transversal surface will then reach point $x + \mathrm{d}x$, corresponding to the expression

$$j(x + \mathrm{d}x) = j(x) - j(x)N\sigma\mathrm{d}x.$$

This can be written, indicating with $\mathrm{d}j(x)$ the difference $j(x + \mathrm{d}x) - j(x)$,

$$\frac{\mathrm{d}j(x)}{\mathrm{d}x} = -j(x)N\sigma,$$

which is the expression of a simple homogeneous linear differential equation if $\mathrm{d}x$ is assumed as a differential, *i.e.*, as a quantity tending to zero. The solution is

$$j(x) = S\mathrm{e}^{-\Sigma x},$$

where S is the intensity of the neutron plane source at point $x = 0$, while Σ denotes the "total macroscopic cross-section" (*i.e.*, relative to collisions both of capture and

([1]) In reality, a correct cross-section definition would require the knowledge of quantum mechanics. The analogy with the geometric definition above is given for illustration.

scattering), given by the product $N\sigma$. Therefore, the greater the density of the element, its cross-section and the length H, the smaller the number of neutrons that will reach the opposite side of the parallelepiped without undergoing collisions ([2]). The expression obtained, of an exponential type, has suggested the name of these types of experiments.

In reality, the first experiments of neutron propagation were made up of various materials in homogeneous and heterogeneous structures, aimed at understanding the diffusion properties of neutrons, in particular in multiplying media. In these cases, natural uranium was used (in which the fissile element uranium 235 is present in the fraction of 0.7%). A typical exponential column could be made up of elements (pellets) of uranium oxide inserted within graphite bricks.

In the simplified scheme shown above we have considered only neutrons coming from a source at point $x = 0$. In reality, in studying the neutron propagation in a medium, in the generic point x it is necessary to consider, in addition to the neutrons coming from the source, those coming from (elastic and inelastic) scattering collisions and those originated from fission capture, if a multiplying medium is concerned. These events will take place both on the left and on the right of a given point and the neutrons produced will be emitted in all directions. The expression of the neutron current $j(x)$ would then result from the sum of different contributions. Furthermore, in the above description, all the neutrons have been assumed to be monoenergetic, that is, having the same kinetic energy. In reality they are characterized by speeds, and therefore energies, that may largely differ from each other. Consider in fact that neutrons, when subject to fission capture, may have speeds up to several thousand times smaller than those they had at birth. This is due to the multiple scattering collisions that gradually slow them down.

Rather than the current density, the neutron flux $\varphi(x)$ is commonly used, defined as the product of the neutron density (number of neutrons/cm^3) by their speed. It is related to the neutron current by the equation

$$j(x) = -D\frac{\mathrm{d}\varphi}{\mathrm{d}x},$$

where D, commonly defined "diffusion coefficient", is a quantity (given in cm) also depending on the characteristic of the medium, that is, on the cross-sections and densities of the elements that compose it.

The flux φ is obtained from the solution of the balance equation relevant to the events (in the unit time) which occur in the volume element $A\mathrm{d}x$. At stationary conditions (far from the source), we may write:

(incoming neutrons − outgoing neutrons) + neutrons born from fission = neutrons captured by fission and parasitic absorption (that is without generation of new neutrons),

---

([2]) This attenuation property is exploited in shielding against radiation, using "absorbent" materials characterized by large capture cross-sections.

i.e., introducing the (macroscopic) parasitic and fission capture cross-sections Σ_{a} and Σ_{f}, respectively, and the number (ν) of neutrons born on average from every fission,

$$A\left(D\frac{\mathrm{d}\varphi(x+\mathrm{d}x)}{\mathrm{d}x}-D\frac{\mathrm{d}\varphi(x)}{\mathrm{d}x}+\nu\Sigma_{\mathrm{f}}\varphi\,\mathrm{d}x\right)=A\mathrm{d}x(\Sigma_{\mathrm{f}}+\Sigma_{\mathrm{a}})\varphi\,,$$

which, dividing by $A\mathrm{d}x$, making $\mathrm{d}x$ tend to zero and reordering, becomes the diffusion equation:

$$D\frac{\mathrm{d}^2\varphi}{\mathrm{d}x^2}+\nu\Sigma_{\mathrm{f}}\varphi-(\Sigma_{\mathrm{a}}+\Sigma_{\mathrm{f}})\varphi=0\,. \tag{1}$$

If in this equation $\nu\Sigma_{\mathrm{f}}$ is smaller than $\Sigma_{\mathrm{a}}+\Sigma_{\mathrm{f}}$, *i.e.*, if the ratio

$$\frac{\nu\Sigma_{\mathrm{f}}}{\Sigma_{\mathrm{a}}+\Sigma_{\mathrm{f}}}\,, \tag{2}$$

which corresponds to the so-called infinite multiplication factor ([3]) (k_∞), is lower than unit, the number of neutrons produced per unit time by fission would not exceed the number of those captured, always per unit time, by fission and parasitic absorption. In this case the solution, far from point $x=0$ where the source is placed , and from the end point $x=H$, will still be of an exponential form, *i.e.*,

$$\varphi=a\exp[-\kappa x]\,,$$

a being a constant coefficient and κ a positive quantity given by the expression

$$\kappa=\sqrt{\frac{\Sigma_{\mathrm{a}}-\Sigma_{\mathrm{f}}(\nu-1)}{D}}\,.$$

We may see that the exponential behavior of the solution corresponds to those cases in which the radicand at the second member has a positive value (and, therefore, the root κ has a real value), that is, as seen above, the number of neutrons born from the fission multiplication is smaller than that of capture events. A system under these conditions may be defined "undercritical", irrespective of its size. The constancy of the flux level is assured by the presence of an "extraneous" source (so called to distinguish it from the fission one) which compensates the deficit of the neutron balance.

It may occur that a system is undercritical even in the case in which the value of the root κ is not real, that is, in the case in which the events of multiplication prevail over those of capture, provided that the disappearance of neutrons escaping from the system (process called "leakage") prevails over their surplus. In these cases the neutron flux, in

---

([3]) This quantity is so defined since it refers to the multiplication factor (see section 2) relevant to an infinite medium.

regions far from the source and from boundaries, with geometries like that of the case considered, assumes a convex cosinusoidal shape.

In a structure composed of natural uranium, with a "moderator" ([4]) such as graphite most of the fission events occur through neutron capture by the U-235 isotope at relatively low speeds (about 10^{-3} m/s, *i.e.*, at values almost at thermal equlibrium with the medium ([5]). A small fraction (ϵ) of neutrons, also called "fast fission factor", are born from the more abundant U-238 isotope. This occurs at high energy values, close to those at which neutrons are generated.

It must also be added that, from quantum mechanics considerations, the value of the capture cross-sections of many isotopes may vary remarkably as a function of energy. This is due to the presence of the so-called "resonances" in the epithermal energy range where they may in fact reach extremely high values within relatively narrow energy intervals. This particularly concerns the U-238 isotope. In these cases the possibility for the neutrons of not being (parassitically) captured in their slowing-down process from fission to "thermal" energies, at which most fissions take place, depends on the probability that an elastic scattering collision with the moderator allows them to pass through the resonance energy band ([6]).

To increase this effect, and therefore improve the value of κ_∞, preserving the average composition of the system, a heterogeneous reactor core is considered, made up of fuel elements surrounded by a moderator, the whole structure forming a lattice of regular cells. The probability for the neutrons of being parassitically captured at resonance energies is then reduced (such captures, in this case, occurring only at the outer layers of the fuel elements). The neutrons that remain in the surrounding moderator may then "jump" the resonance band and enter then into the fuel elements at a greater depth, producing a relatively higher number of fissions. It is exactly the intuition of this mechanism, called "lumping", which made criticality possible in systems fuelled only with natural uranium ([7]).

([4]) So called, since it slows down the neutrons born at high energies from fission.
([5]) For this we speak of "thermal" neutrons, in that the speeds of these neutrons are (almost) at thermal balance with those of the atoms (or molecules) of the medium with which they collide in the process of scattering, speeds which follow the Maxwell-Boltzmann distribution law (derived from the kinetic theory of gases).
([6]) It is noteworthy to remember that the neutron slowing-down process was thoroughly studied by Fermi, who gave it an elegant theoretical interpretation. According to it, the neutrons, on their slowing-down process, are associated with a quantity (called "Fermi age") related with the kinetic energy lost, starting from the fission event.
([7]) Apart from the first reactors moderated by heavy water (element characterized by a very low capture cross-section) and the first ones moderated by graphite, the other types of "thermal" reactors, moderated by "light" water, and the more recent graphite-moderated ones, are fuelled with uranium enriched with the U-235 isotope. The fraction of this isotope then passes from the 0.7% abundance in natural uranium, to a few per cent.

2. – The four-factor formula

Instead of expression (2) we may more generally write the expression of the infinite multiplication factor κ_∞ (relative to an ideal infinite medium) as the product of four factors, that is

$$k_\infty = \eta f \epsilon p\,, \tag{3}$$

where, assuming the presence of natural uranium (U-238+U-235), graphite (C) and structural elements (Fe):

$\eta \left(= \frac{\nu \Sigma_f^{\text{U-235}}}{\Sigma_f^{\text{U-235}} \Sigma_f^{\text{U-235+U-238}}}\right)$ is the "reproduction factor", *i.e.*, the number of neutrons produced on average for each "thermal" neutron captured by the fuel (U-235 and U-238). For natural uranium its value is around 1.3;

$f \left(= \frac{\Sigma_f^{\text{U-235}} + \Sigma_a^{\text{U-235+U-238}}}{\Sigma_f^{\text{U-235}} + \Sigma_a^{\text{U-235+U-238}} + \Sigma_a^{\text{C}} + \Sigma_a^{\text{Fe}}}\right)$ is the "thermal utilization factor". It expresses the probability that the capture event at thermal energies takes place inside the fuel, rather than in the moderator (graphite), or in the structural elements (iron);

ϵ is the "fast fission factor" previously encountered. It takes into account the probability that fission occurs with fast neutrons through capture by U-238. Its value is about 1.03;

p is the "resonance escape factor", which expresses the probability that neutrons escape resonance capture.

Expression (3), commonly defined as "four-factor formula", represented a fundamental step towards the understanding and characterization of multiplying media. It allows to obtain the number of neutrons generated (by fission) for every neutron that disappears (by fission and parasitic captures) in the medium under study.

Up to now we have limited ourselves to considering undercritical systems. It is evident that, at particular conditions, such as a better arrangement ("lumping") of the fuel, or an increase of the "enrichment" in the U-235 isotope, the multiplication factor κ_∞ increases correspondingly. For $\kappa_\infty = 1$ we say that the fission reaction in a medium of supposed infinite size sustains itself without the help of an "extraneous" neutron source. If this is present, a steady, linear increase of the flux with time would take place. In fact, in this case, for each neutron that has vanished through fission or parasitic capture, besides the birth of a new neutron, there would be, on average, a fraction born from the extraneous source. This process clearly diverges.

In reality, all systems are of finite size. As we have already mentioned, the leakage of neutrons from a multiplying medium is associated with this circumstance. These neutrons are definitively lost and, therefore, do not contribute to maintaining the fission chain. If we indicate by P_{NL} the probability for a neutron of not being subject to leakage, we may define the "multiplication factor" for finite media

$$k_{\text{eff}} = P_{\text{NL}} k_\infty = P_{\text{NL}} \eta f \epsilon p\,. \tag{4}$$

The P_{NL} value decreases as long as the size of the system considered is reduced, as we may infer from the expression from which it is obtained, in relation to a monodimensional plane geometry (infinite slab):

$$P_{\mathrm{NL}} = \frac{1}{1 + L^2 \frac{\pi^2}{H^2}},$$

where L, commonly defined as "diffusion length", is a quantity (given in cm) depending on the cross-sections and on the density of the elements that make up the medium, while H is the slab thickness.

3. – Approach to criticality

As said above, for a system of finite size it may happen that it is still undercritical in the presence of a κ_∞ value that exceeds unity. At such conditions, as the system gradually increases in size, for example by adding new elements, κ_∞ approaches unity, that is, the criticality conditions. This is what happened with the Chicago Pile 1 experiment. The approach to criticality was evaluated on the basis of the increasing levels reached by the flux (in the presence of the extraneous source) with the increase of the core size, and therefore of κ_∞.

To analyze this process more closely, let us look at eq. (1). Instead of the number ν of neutrons that are born from fission, let us consider the product νP_{NL} (so taking into account the losses that take place at thermal and higher energies through leakage). Likewise, instead of the source S of extraneous neutrons, supposed to be emitted with the same energy distribution of fission neutrons, let us consider the product SP_{NL}, so taking into account the probability also for these neutrons not to escape from the system before and after reaching thermal energies. We may therefore replace eq. (1) with the following one, averaged over the whole system and supposing for simplicity that the extraneous source (expressed now as neutrons per second per cubic centimeter) is uniformly distributed in the medium,

$$\nu\epsilon p P_{\mathrm{NL}} \Sigma_{\mathrm{f}} \varphi - (\Sigma_{\mathrm{a}} + \Sigma_{\mathrm{f}})\,\varphi + SP_{\mathrm{NL}} = 0.$$

We may easily infer from it the proportionality expression

$$\varphi \propto \frac{S}{1 - k_{\mathrm{eff}}}, \tag{5}$$

with κ_{eff} given by (4). This expression indicates that, as long as the system approaches criticality conditions, that is, as long as κ_{eff} tends towards unity, in the presence of an extraneous source the flux tends asymptotically to infinity.

It is this law of flux increase for increasing P_{NL} values in the presence of an extraneous neutron source which has been (and is still now) used in the process of bringing the reactor pile to criticality. In fact, the start-up operations begin always from reactor

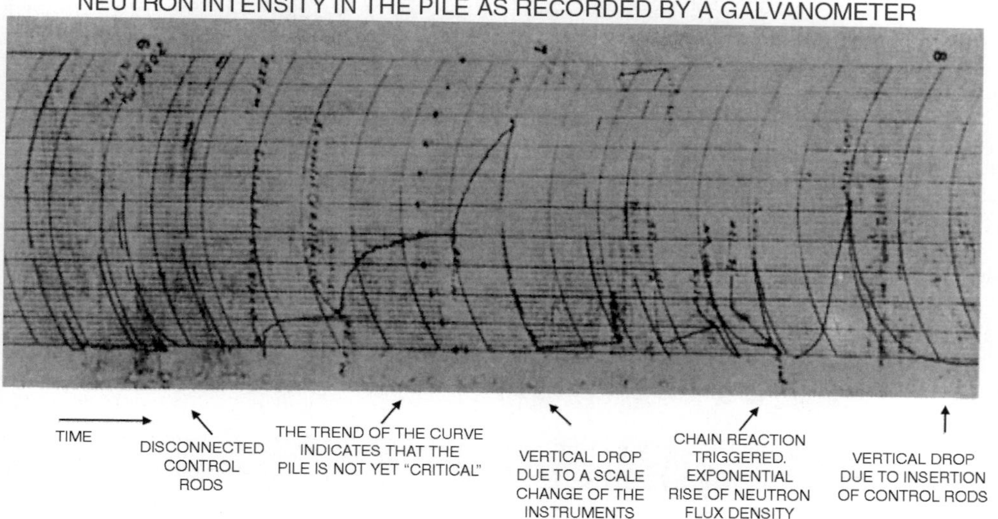

Fig. 2.

configurations unconditionally undercritical, that is, characterized by κ_{eff} values which are much smaller than unity. It is then waited until the flux level reaches an equilibrium condition, which corresponds to eq. (5). In the subsequent step, further fuel elements are added, to which a higher κ_{eff} value, and then, after a while, a new flux level will be obtained (see fig. 2, relevant to the first critical experience, indicating the behavior of the neutron flux density *vs.* time). This operation proceeds until criticality is reached. It is realized that this has occurred when the flux, always in the presence of the extraneous source, continues to increase without arresting at a maximum level.

If the criticality level is exceeded, *i.e.*, if the multiplication factor κ_{eff} is greater than unity, the growth in time of the flux would follow an exponential law of the type

$$\varphi = c \exp\left[\frac{\rho}{\ell}t\right], \tag{6}$$

where c is a constant coefficient depending on initial conditions and ℓ is the average lifetime of neutrons, *i.e.* the average time that elapses between their birth and their disappearance (by capture, or leakage from the system), while the quantity ρ, denominated "reactivity", is given by the difference $\kappa_{\text{eff}} - 1$. Since the value of ℓ for systems of the type considered is of the order of a thousandth of a second, for κ_{eff} values even slightly greater than unity the exponential growth of the flux, and, therefore, of the reactor power, would become uncontrollable. Luckily, while most of the neutrons are emitted

Fig. 3. – A sketch of the CP-1 pile based on the description of those present, in the absence of photographers. State secrecy prevented the equipment of the Metallurgical Laboratory from being photographed. The man in the centre is shown operating one of the control rods by hand.

immediately after fission, with times of the order of 10^{-14} seconds ([8]) a fraction of them (about 0.75%), also called β fraction, are born with a time delay, on average, of the order of ten seconds. These neutrons are called delayed neutrons and it is exactly their delayed birth that allows nuclear reactors to be controllable. In fact, until the excess of κ_{eff} with respect to unity is kept below the fraction of the delayed neutrons (*i.e.*, for uranium-235, as long as κ_{eff} is less than 1.0075) the flux increases exponentially but gradually. Instead of (6), in this case, its growth is given by the expression

$$\varphi = c \exp\left[\frac{\rho}{\ell_{\text{rit}}(\beta - \rho)}\right] t ,$$

where ℓ_{rit} (of the order of ten seconds) represents the average delay time of delayed neutrons.

Therefore, there is enough time for the intervention of safety mechanisms and then reducing the neutron multiplication. This is carried out through the prompt introduction of materials containing elements with large neutron absorption cross-section (like the B-10 boron isotope), which drastically reduce the κ_{eff} value. In last-generation reactors

([8]) For this reason these neutrons are called "prompt" neutrons, that is, with instantaneous emission.

these safety mechanisms consist of rather sophisticated devices with a high degree of reliability. At the time when the first pile became critical, there was full awareness of the consequences of a reactivity accident, however remote, and a system of regulation and intervention essentially based on three independent devices was arranged. The first one of these, of regulation, was made up of a rod (operated by George Weil) for keeping the reaction under control (see sketch, fig. 3), the other two were safety devices. The first one was made up of automatic rods, the second one of an emergency rod the operation of which was assigned to Norman Hilberry. A rope was attached to one end of this rod, running along the whole length of the pile, and carrying a large weight at the opposite end. The rod (called "scram") was pulled out from the pile and tied to a balcony with another rope. Hilberry was ready to cut this cable with an ax if something unexpected happened...

The description of the first critical experiment on the reactor pile in Chicago, and the facts that preceded this historical event are amply covered in the preceding monograph by Carlo Salvetti. The author, who is a renowned lecturer and scientist, had the privilege of meeting and knowing personally some of the protagonists, in particular Enrico Fermi. His intervention is therefore particularly important, as a testimony from direct sources of that exceptional event.

4. – Nuclear power reactors

After that first experience, carried out with greatest secrecy, the events that followed and led to the construction of the atomic bomb are well known. It was also in that period that the foundations were laid for the industrial applications of atomic energy. The first conference on its pacific uses, held in Geneva in 1955, finally made known this scientific information, up to then "top secret", to the academic and industrial world ([9]). The amount of studies and experiments that followed in laboratories and universities all over the world was impressive, certainly for the high value of what was at stake, future sources of energy, but also for the variety of problems relevant to the physics and engineering of nuclear reactors. The enormous quantity of physical data required, the campaigns of measurements for their definition, the complexity of their treatment in the calculations of physical and engineering projects justified large programs of research and development during the following decades in various fields and disciplines: from nuclear physics to calculation algorithms and tools, thermo-hydraulics, metallurgy, chemical processes, instrumentation, safety and control analysis. With the large mainframe processors, in rapid expansion at that time, there was a great synergy. The development of machines with ever increasing performances allowed in fact to cope with analysis and design problems more and more sophisticated; in particular, it allowed passing from calculations relevant

([9]) Let us recall the first two basic "official" treatises on reactor physics which were published during those years: "The Elements of Nuclear Reactor Theory", by S. Glasstone and M.C. Edlund (Van Nostran) 1952; "The Physical Theory of Neutron Chain Reactors", by A.M. Wienberg and E. Wigner (The University of Chicago Press) 1958.

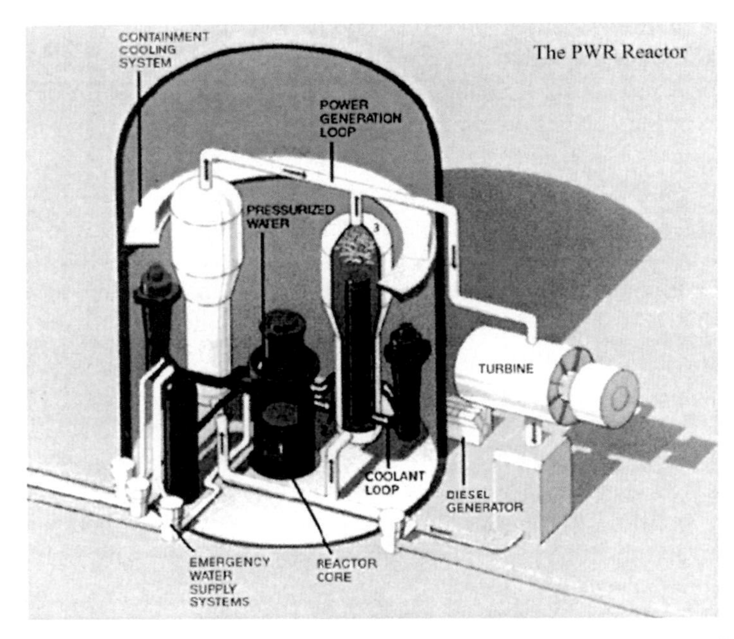
The PWR Reactor
CONTAINMENT COOLING SYSTEM
POWER GENERATION LOOP
PRESSURIZED WATER
TURBINE
COOLANT LOOP
DIESEL GENERATOR
EMERGENCY WATER SUPPLY SYSTEMS
REACTOR CORE

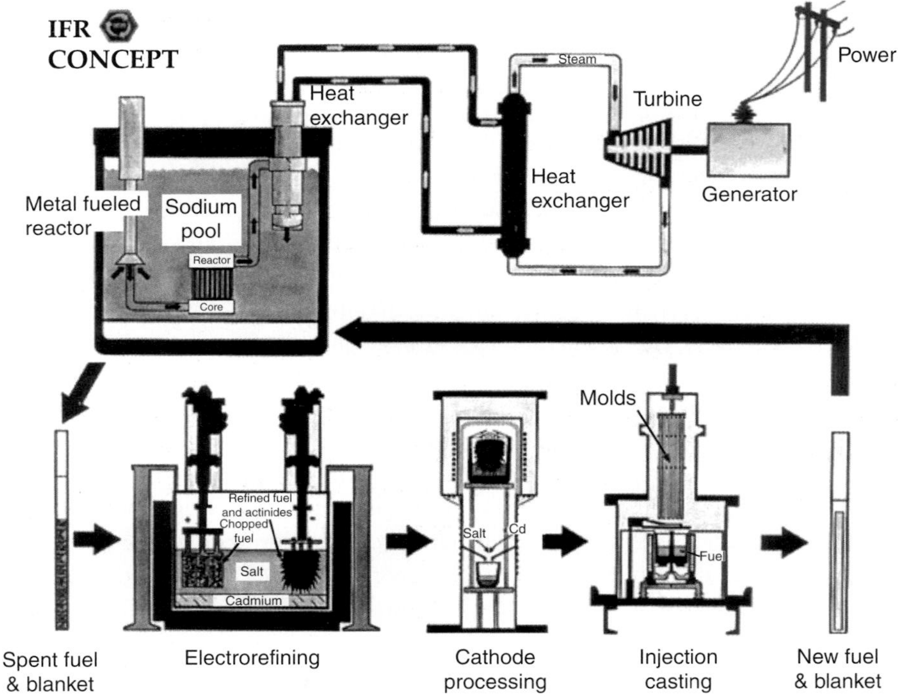
IFR CONCEPT
Heat exchanger
Steam
Turbine
Power
Heat exchanger
Generator
Metal fueled reactor
Sodium pool
Reactor
Core
Refined fuel and actinides
Chopped fuel
Salt
Cadmium
Salt
Cd
Molds
Fuel
Spent fuel & blanket
Electrorefining
Cathode processing
Injection casting
New fuel & blanket

Fig. 4.

to systems described by simplified geometries to others in which their characteristics were reproduced in greater detail. This is the period in which the large American laboratories (Los Alamos, Argonne, Hanford, Idaho, Oak Ridge, Brookhaven, to quote only the most important ones) greatly expanded, together with those in Canada, in Europe (France, England, Russia) and in Asia (Japan, India, China).

The history of nuclear energy, like the history of many important technological innovations, is bestrewn with successes and failures, achievements and accidents. We shall omit comments on its non-pacific use, since we would go out of context. We just remind that scientific achievements are ineludible, they are part of life and, then, of man's history. It is our responsibility to direct its use towards useful applications and, therefore, towards progress.

In the second half of the last century, through continuous research, conspicuous investments in men and resources, and extensive experimentation, gradually a kind of natural selection of the most promising systems occurred. Among the "thermal" nuclear plants we may recall the water-cooled BWR and PWR, *i.e.* the boiling water and pressurized water reactors (see fig. 4, upper part), the Canadian heavy water reactors (CANDU) and the high-temperature gas-cooled graphite-moderated reactors (HTGR). Among the "fast" systems cooled with liquid metal we may recall, in particular, the EBR II reactor of the Argonne National Laboratory. This reactor was aimed at the demonstration of the Integrated Fast Reactor (IFR) concept, a modular system at the same time energy producer and "burner" of transuranic elements (plutonium, neptunium, americium, curium), with the fuel cycles processes integrated at the same site (see fig. 4, lower part). Each one of these systems has been studied in different versions, depending on the type and shape of the fuel, the geometry, the cooling systems, the safety devices and the fuel cycle management. Every accident, large or small, has nevertheless served to increase the understanding of the mechanisms involved, allowing to avoid in the new plants the inconveniences noticed in the previous ones, and consequently raise the efficiency and safety levels. This is what generally occurs in the history of almost all important technical-scientific innovations which mark the path of progress.

In the following monograph by Maurizio Cumo, eminent lecturer and expert in reactor engineering, existing nuclear power plants are amply dealt with and discussed under the design, operation and safety point of view. Here they have been mentioned to illustrate the evolution between the first experiments in the 1940s and the future scenarios envisaged with the so-called innovative (or fourth-generation) reactors described in the following section.

5. – Innovative reactors

There are three major issues we are confronted with at present in relation to nuclear reactors: the problem of plant safety, the problem of (fresh or spent) fuel fabrication and transport and the problem of radioactive waste disposal, besides, of course, that of the economy of the energy produced.

To these issues we should add that of eliminating the large stocks of plutonium, both of civil and military origin (produced through the neutronic capture in the fertile isotope U-238). Phasing out drastically from nuclear energy would leave this serious problem unresolved.

From what said before, the improvement of the last-generation plants has made them extremely safe for their intrinsic physical characteristics. In particular, only negative temperature coefficients ([10]) are allowed, so to drastically reduce the consequences of a hypothetical accident. Moreover, the criteria by which the reactor buildings are planned make them earthquake, missile, air crash and flood proof, with protection barriers that can contain all the radioactive material released in the case of a serious improbable accident. The problem of safety associated with the nuclear technology remains then mainly linked to the transformation process and transportation of the fuel.

The above issues are now under extensive investigation at different laboratories throughout the world. An international cooperation is underway for the research and development of the so-called fourth-generation reactors. It is believed that these new systems will arrive at an industrial stage within a few decades, *i.e.*, at times when serious difficulties in finding new sources of energy are expected, following the exhaustion of traditional ones (oil, coal and gas), and the increasing need of reducing the pollution emissions will become a most serious concern (it is well known that nuclear energy from this point of view, is "clean"). Research in areas of alternative energies (in our opinion, not decisive in the long term but which may cover an important complementary role), associated with a persisting political and economical resistance, could postpone in a few countries the time at which nuclear energy is taken up again on an industrial scale. In any case, it remains a precious reserve available in the future in a developing world, and an essential means for eliminating, by incineration, the existing stocks of plutonium.

We conclude by giving a quick look at some of the perspective new-generation reactors (which will cohabit for a certain period with the present-generation ones). Some of these concepts have old roots, they were studied at the beginning of the nuclear era, but were then discarded, being considered too expensive, or too difficult to be realized with the existing technology. We may mention:

1) *Molten salt thermal reactors (MSR).* These are reactors in which the fuel is dissolved in salt (chloride of fluoride) and flows continuously through the core. It passes from this to an external heat exchange circuit and, via a by-pass, basing on a technology still under development, it should be subject to a process of separation from the fission products before re-entering into the core. These systems were studied in the past at the Oak Ridge Na-

([10]) Temperature coefficients are defined as changes of the multiplication factor per one degree change of the temperature. Temperature changes affect, in fact, the macroscopic cross-section values through density changes of the component material, or through the capture resonance broadening by the so-called "Doppler effect". The reactor feedbacks associated with a temeprature change following a system alteration (such as during an accident) depend on these coefficients.

tional Laboratory and, at a certain point, were dismissed because of vessel corrosion problems. At present, it seems that these problems may be solved (in the fluoride salts version). These systems present several important advantages: they are characterized by an almost minimal mass configuration (height-to-diameter ratio of the cylindrical core close to unity) and, because of the continuous removal of fission products, by a limited reactivity loss during burn-up. This means that an accident of maximum gravity (vessel failure) would certainly lead to subcritical configurations, and then to a power reduction, while, on the other hand, a serious reactivity accident would be practically impossible, both for the fact that the temperature reactivity coefficients in these systems are negative, and for the relatively small reactivity margin required for the reactor operation during burn-up. Molten salt reactors could operate exploiting the so-called uranium/thorium cycle. This cycle, apart from allowing the utilization of the existing large deposits of thorium (a fertile element which through neutron capture produces the fissile isotope U-233), has the merit, with respect to the uranium/plutonium cycle, of reducing the production of transuranium elements, emitters of alpha particles (of concern for the long term radiological risk). These systems are being extensively studied in France, Japan and the Unites States.

2) *Pebble bed reactors (PBR).* These are helium-gas–cooled systems which were studied for a long period in the past (in Germany). The fuel consists of "pebbles" with a diameter of about 6 cm, made up of a low-density graphite matrix. In each pebble about 11,000 microspheres with a diameter of about one millimeter (called "TRISO fuel particles") are inserted. They are formed by a uranium oxide core surrounded by a low-density "buffer" layer (in which the fission products are retained), a second layer of pyrolitic carbon, a third layer of silicon carbide and finally a fourth layer again of pyrolitic carbon. The peculiarity of this fuel is that of being capable, on the one hand, of easily retaining the fission products up to very high "burn-up" values, and, on the other hand, to be able of withstanding, without serious damage, temperatures up to 1600 °C. These pebbles are recycled continuously extracting them from the bottom of the core, putting aside (storing for final stockage) those that have reached the assigned burn-up limit, and then reinserting the remaining pebbles on the top, together with fresh ones, so to reintegrate those removed (see fig. 5). Because of the continuous fuel recycling, these reactors have the advantage of requiring, as the MSR systems, a low reactivity margin for their operation during burn-up, which greatly reduces the probability of a reactivity accident. It has been demonstrated that small modular reactors of this type (of the order of 100 MWe) can easily withstand the worst forseeable risk: pump arrest without power shutdown. In this case the temperature of the core would reach values up to 1500 °C and the heat produced would be dispersed by radiation and conduction through the plant structures to the surrounding environment. In short, there would not be a serious nuclear accident. The reason for which these systems have been again brought to our attention depends also on the great improvement of the gas turbine technology in the last decades. This type of reactor is being studied extensively in England, South Africa, Japan and China.

3) *Undercritical reactors (ADS).* These are systems which, being undercritcal (*i.e.*, with $\kappa_{\rm eff}$ smaller than unity), need an external neutron source for maintaining the flux at an assigned power. This source is supplied through a beam of protons which, having

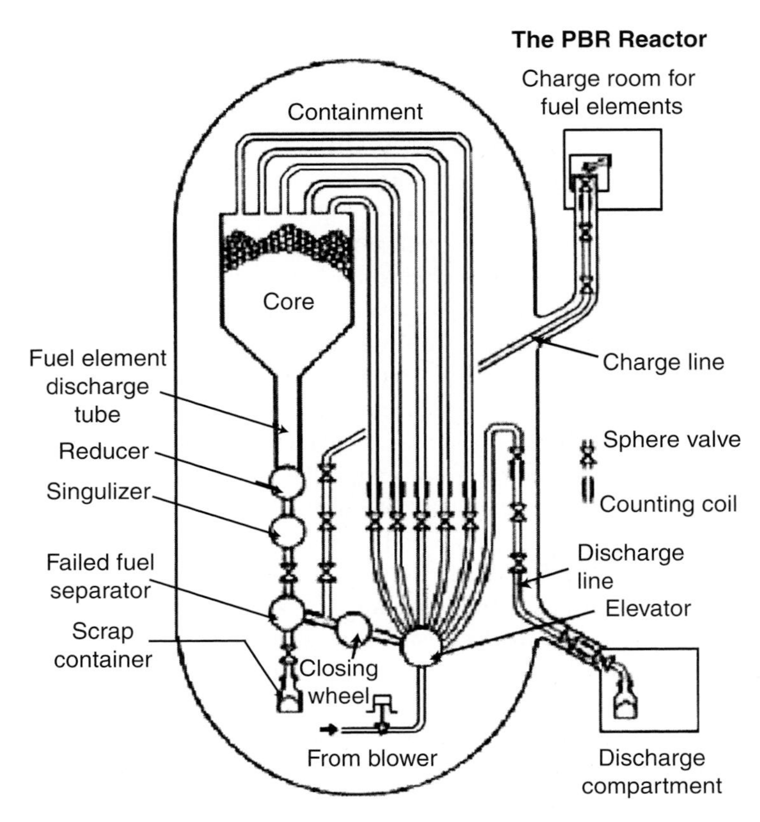

Fig. 5.

reached very high speeds inside an accelerator, collide with a metallic target in a central area of the core, producing the so-called "spallation" neutrons. The undercriticality considered, that is, the difference $(1-\kappa_{\text{eff}})$, is usually around a value of 0.05. The distance from critical conditions makes these systems relatively safe. One disadvantage is that a significant fraction of the electric energy produced must be used for the generation of the proton current. Besides the lead-cooled fast reactors presently under consideration ([11]), other systems, in particular the MSR and PBR ones mentioned above, may be also conceived as ADS systems.

([11]) See, in particular: C. Rubbia *et al.*, *Conceptual design of a fast neutron operated Energy Amplifier*, CEN/AT/95-44 (ET) (1995). These systems adopt many important features of the metal-cooled Integrated Fast Reactor (IFR) concept, developed at the Argonne National Laboratory (ANL) from the 1960s (see fig. 4), in particular: the closed, on-site fuel cycle strategy reprocessing and the repeated recycling of actinides, up to their elimination by fission. ANL built in those years the fast experimental reactor EBR II which demonstrated the potentiality of this concept.

In the above-mentioned MSR and PBR reactors the fuel would not be subject to a separation process from heavy elements, minor actinides, that are formed during the burn-up, differently from what occurs in current reactors, in which these elements are periodically separated and processed as radioactive waste to be geologically confined. This would also alleviate the risk connected with the fuel transportation to/from the reprocessing plants. Since these elements undergo fission, or transmutation into other fissile elements through neutronic absorption, after a certain period, at constant feed, a condition of (quasi) equilibrium is established in which the isotopic composition of the fuel does not vary significantly with time. In a world under continuous transformation, these are fascinating challenges full of surprises. Perhaps the future will reserve us different solutions. The routes of science and technology, in fact, are seldom linear. However, the future has its roots in the past. And modern reactors, and those to come, are the continuation in time of the experiment made more than half a century ago by Fermi and his team on the "Squash court", under the stands of Stagg Field at Chicago University.

APPENDIX

A singular technical term

Sometimes names originate from particular situations, events that were important but are now forgotten. Their frequent use makes them "normal" and rarely an evident reference to their original meaning remains. This is the case of the term "scram", normally used in technical jargon to designate the release of the safety rods for activating the interruption of the reactor power. The frequent use of this word has made it lose its original meaning. It is worth quoting a few passages on this subject from an article by Raymond L. Murray of North Carolina State University, referring to a letter sent to him by Norman Hilberry, and published in August 1988 in "Nuclear News". The sentences mentioned are highly informative on the spirit that guided that exceptional team of scientists.

From Raymond L. Murray, "Nuclear News", page 105, August 1988. The word "scram" is commonly used in place of "trip" in reactor operation; in fact, *Webster's Ninth New Collegiate Dictionary* defines scram as "a rapid emergency shutdown of a nuclear reactor." No origin of the word is provided.

One might conjecture that it was an adaptation of the conventional usage of the slang word that means "to go away at once," as if operators suddenly left the scene of a reactor accident. But the dictionary also notes that "scram" is an abbreviation for "scramble," one definition of which is "a rapid emergency takeoff of fighter interceptor planes." This analogy is too farfetched, and thus there must be another explanation... The best source of information on the derivation of "scram" is Norman Hilberry. He was a member of Enrico Fermi's team on the first reactor, later was director of Argonne National Laboratory, and on retirement became professor of nuclear engineering at the University of Arizona. In a January 1981 letter, he explains that on the original pile, there was a shim rod going

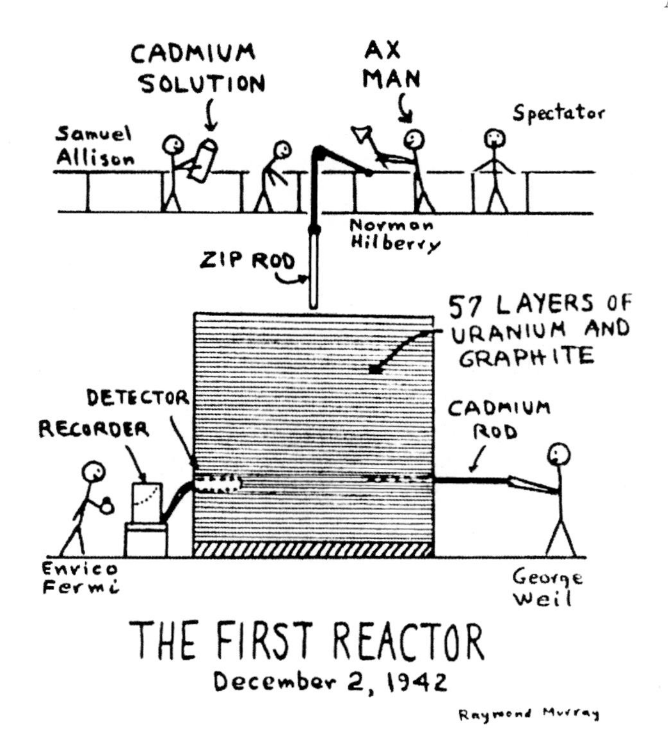

Fig. 6.

through the middle of the pile and two safety rods, plus carboys of cadmium solution to be dumped on the pile if a blue glow appeared. A third safety rod was attached to a rope over a pulley. In the design sketches of the system, there was the legend SCRAM, standing for "Safety Control Rod Ax Man." Hilberry had the assignment of standing with the ax, ready to cut the rope if necessary. He learned much later that his colleagues had called him "Mr. Scram." No photographs of the startup appear to be available, but from accounts by participants, one can draw the accompanying cartoon (see fig. 6, drawn by Murray).

Hilberry in his letter mused about the situation: *"I don't believe I have ever felt quite as foolish as I did then. Clearly if there were any real reason for my standing there with that ax, we should not be doing what we were doing—at least not where we were doing it. The fact that any misadventure there would almost certainly result in tipping off the Germans as to what we were doing and how far we had gotten was at least as worrisome as was any possible damage to our neighbor's health. Concerned as we all were with the absolute need for secrecy, we none of us would have been there had we had any doubt as to the outcome."*

"By the afternoon, there had been some forty exponential experiments with successively larger geometries done. The nuclear physics was well established, the microprocesses well

known... There was no scientific excuse for the ax man. The possibility of a power failure, of course, did exist but as I stood there it sure seemed remote. The argument then as now was if the scandalously remote should occur and you had not taken every conceivable safety precaution, whether reasonably justifiable or not, you would certainly be adjudged to have been remiss. The philosophical roots of some of our problems extend way back to our very early days."

Finally, he says, *"... the real safety measure that day was that the night before Fermi had computed what the activity readings should be for every position of the shim rod that he called for so that if any conceivable deviation from the accepted theoretical behavior should occur it would become obvious well before criticality would be reached."*

The story has two lessons for us. The first is that there is an ancient precedent for redundancy in safety equipment that can respond to apparently inconceivable events. The second is that we cannot expect our operators and supervisors to have the stature of Fermi, but we can expect them to be extremely knowledgeable of the fundamentals of the behavior of the reactor system.

About the Author

AUGUSTO GANDINI, physicist. After research experience on nuclear reactors at Argonne National Laboratory in the United States, he was director of CNEN and scientific advisor for ENEA. He is a lecturer on reactor physics for a degree course in Nuclear Engineering at "La Sapienza" Rome University. He is the author of 2 treatises, 10 monographs and over 100 scientific articles.

Reactors and nuclear technology: Development in the world

Maurizio Cumo

1. – Electronuclear energy in the world

There were 436 working nuclear power stations in the world at the end of the last century globally producing 350 electric GW and with a production of about 17% of the total electricity consumed.

83% of this electronuclear power is localized in developed countries; the highest regional percentage of nuclear electricity is 43% in Western Europe which has 151 reactors. In North America 118 reactors supply 19% of the electricity in the USA and 12% in Canada. In Eastern Europe and in the new independent states (NIS) 70 reactors are active with a production, in the Russian Federation, of 13% of the electricity. In the Far East and in Southern Asia 90 reactors work: this is probably the only region where the use of nuclear energy is destined to grow over a short period. Latin America and Africa together have less that 2% of electronuclear power installed at present. The operating experience in the world has exceeded 10,000 reactors per year. From this picture and from the evolution that electronuclear energy has had in the same macroregions over the last decade we can see substantial stagnation in its prospects of growth which cannot manage to keep up with the growth of global request for electricity. OCSE even foresees a reduction of the nuclear quota in the world from the present 17%, within the space of two decades, to 10%, maintaining the installed power substantially constant. The reasons for this stagnation are mostly economical. A balance between the cost of kilowatt-hour produced by electronuclear power stations and fossil fuel power stations, often indicated, is represented by that of petrol equal to 15 dollars per Brent barrel. Of the three components of cost per nuclear kilowatt-hour (capital cost, fuel cost and operation and maintenance of the power stations) the International Energy Agency (OCSE) now estimates 60% capital, 25% for personnel and 15% for fuel while, for a fossil fuel plant, it is the cost of fuel which prevails with an incidence of 45% for plants run on coal and 70% for those run on natural gas. A combined cycle gas plant, with Brent barrel costs around 10 dollars, as those which were verified up to a short time ago, shows undeniable convenience, also taking into account that the capital cost is a quarter of nuclear plants and construction time (passive interests) is a third. This is the reason why today, in western countries, electricity companies prefer gas plants with combined cycles with high thermodynamic efficiency [1].

We can add to this the less complicated authorization procedures (faster and less uncertain). The plants are easier to place and therefore more easily accepted by public opinion. Bearing also in mind that the initial rapid growth of nuclear power came about in a period in which, in many countries, electricity companies were nationalized, with electricity prices controlled by governments. Now many of the monopolies have been broken down into smaller private companies in competition among themselves and intent on making profits over a brief period. These market laws do not take into account the so called "external cost" of electricity, that is, the impact that the whole cycle of production has on the environment and on the health of the population. ISO (International Standard Organization) is completing a Life Cycle Impact Analysis which is using consistent evaluation requirements of the various energetic cycles and their damaging emissions, their waste products and the necessity to impose a "carbon tax" is being discussed. This takes into account the greenhouse effect and acid rain, based on the principle of "who pollutes pays" to repair the damage done. But this, in spite of the promises in Rio and Kyoto Conferences, is a long process of uncertain international settlement. The comparison between the quantities of fuel necessary to feed 1000 MWe power stations for a year is significant. For a coal run station 2 million tons and 57000 railway wagons are needed, for a fuel oil run station a million and a half tons and 48000 wagons, for a nuclear power station 30 tons and one wagon. 57000 wagons means a thousand trains with 57 wagons and if the trains are of that length, this means three trains a day including holidays. Under the same conditions of energy production, natural gas emits less CO_2 than coal or fuel oil but the inevitable losses of gas during its transport, up to 5% in some areas, taking into account the greater shielding power of methane molecules with respect to CO_2, greatly reduces the minor contribution to the greenhouse effect. For the same kilowatt-hours produced the damage provoked by the gas cycle and coal is substantially the same. At a world-wide level the "energetic" emissions of CO_2 have risen from 5 billion tons (Gt) per year in 1950 to 20 Gt/y in 1990 up to 24 Gt/y today. Without the contribution of nuclear energy this value would increase by 9%; the spontaneous increase rate, without strong preventive measures, would take us to 28/30 Gt/y in 2010. Concerning the waste produced, the large amounts from fossil fuel plants must be got rid of using a d-d strategy (dilute and disperse) in the place of the c-c strategy (concentrate and confine) used for the tiny volumes of nuclear waste.

As far as the resources of uranium are concerned, the known reserves of four million tons, at the present rate of consumption, are sufficient for another seventy years, while further probable reserves for 16 million tons would last for another three hundred years. In comparison, the duration time of fossil reserves (ratio between reserves and annual consumption) is estimated at 45 years for petrol, 60 years for natural gas and 220 years for coal.

The use of fast breeder reactors, advised by Fermi and still not perfected up to commercial standards and not on the market, allows a factor of 60 increase of the energy extractable from uranium, converting non fissile uranium into plutonium. This would take the nuclear energy duration up to 4000 years, but taking into account that also thorium, three times more abundant than uranium could be used, we can conclude that with fast breeder reactors, already significantly experimented and available, mankind could have all the energy necessary for thousands of years. Our generation and Fermi *in primis*, will be remembered as that which discovered this new fire. During fissions about 190 MeV of energy is produced and transmitted, with thermal effect, in the core of the reactor. This means that with the fission of 1 gram of uranium 235, 22000 kWh of energy is provided, that is 2800000 times the energy that is obtained by burning the same quantity of coal. Fission has been recognized in this way as a tremendously powerful source of energy, since the first experiment of the Chicago Pile 1 in 1942. This energy unfortunately has been used for evil purposes too, right from the beginning, like other discoveries, and this has given it a characteristic of danger and risk, but evil is in man and not in his discoveries. If one day a dangerous meteorite should come near the earth we should risk ending up like the dinosaurs: the only way we would have to deviate its trajectory would be to use even this military aspect of nuclear energy. And if we want to explore space and widen our frontiers, the use of nuclear energy is practically the only method we dispose of, thanks to Fermi's work.

The social acceptability of nuclear energy, substantially reduced by the terrible military application at the beginning, has often resulted difficult even in socially developed countries. Sweden and Germany are two examples near to us. To be confirmed as a source of energy in competition with fossil fuel sources in the production of electricity nuclear energy must satisfy several important requirements:

- it must have above all present economical advantages for the producers, advantages which, as we have seen, have not revealed themselves in recent years;
- it must win the trust of the population concerning two important aspects: the safety of the reactors and their operation (the ghost of Chernobyl) and the certainty of the perfect confinement of radioactive waste produced (we will go deeper into this aspect). The constant and meticulous diffusion of all the initiatives of nuclear activity and how they will be carried out, explained in simple terms to the public, has been shown to be of fundamental importance when trying to win substantial consent. We must take this to heart to be able to find a suitable site in Italy to safely store radioactive waste that we have accumulated from old power stations and fuel cycle plants let alone from all the medical, industrial and research activities. These last activities not only continue but are rapidly developing all over the world. To quote a few figures that illustrate the volume necessary for a national repository (at

present the waste, even though it is well kept for a short term, is spread over about ten different sites) we can talk of 100000 cubic metres for the old energy activities and 1000 cubic metres per year for medical-industrial activities in progress.

Perhaps a deeper consciousness of the risks we run with the greenhouse effect and acid rain will help us to appreciate in the future a source of energy that is completely exempt, like nuclear energy, but it is worth remembering that it is not a question discussed among experts that counts, but rather the large numbers that influence political orientation and choices. Sometimes, as the experts teach us, between a real risk and a perceived risk there can be a difference of a factor of one thousand!

2. – Planning criteria for nuclear reactors

Several aspects must always be kept in mind for the operation of nuclear reactors based on chain reactions:

- the ionizing radiations must be adequately shielded;
- the fission products must be well confined;
- the natural dynamic process must by adequately controlled;
- the heat generated cannot be reduced immediately and there must be adequate cooling for a long period.

These conditions deeply-rooted in the process can be satisfied in different ways, so that different types of reactors have been produced, following common criteria. The first aim is to avoid the release of radioactivity during normal functioning and during transition phases, both foreseeable and supposed.

Above all, accidents must be prevented, supposing that they could actually happen, and the measures to mitigate the consequences must be foreseen. This requires the highest guarantee of quality and control applied to all the components and systems, during building and trial, during normal and abnormal operation conditions. In particular, the control and cooling systems are critical in preventing overheating of the reactor core. The control system must guarantee the interruption of nuclear chain reaction and intervene with insertions of negative reactivity before the power of the reactor can reach dangerous levels, in case of an accident like that which on 26 April 1986 took place at Chernobyl in the Ukraine. This can be obtained using various devices during planning which, for example, foresee counter-reactions of negative reactivity even linked to the physics of the process, using the Doppler effect of U-238 isotope, which when the temperature rises absorbs more neutrons than those produced by fission.

In normal working conditions the thermal power produced in the core is considerable and the cooling capacity must be adequate. If there are leaks in the cooling system (for the breakage of tubes, pumps and their engines and of the secondary circuit of heat exchange) the fuel could rapidly overheat and melt, spreading radioactive products inside the pressure boundary of the reactor. This is prevented by a rapid shut down of the reactor through the insertion of the control rods (or by an alternative system that injects neutronic poison) and the activation of an emergency cooling system.

Even when the reactor has been shut down as we have seen, the heat from the radioactive decay must be removed. There are appropriate systems which intervene to guarantee the removal of the heat decay and prevent the core from melting.

In general we can say that the safety is guaranteed by the intrinsic physical characteristics of the process and by the appropriate systems mentioned above, of both passive and active nature, organized in various combinations. The intrinsic physical characteristics resort to the laws of nature: for example, the Doppler effect, as we have seen, arranges that the excess of temperature, consequence of an unwanted power increase, is followed by a power decrease at least equal to the initial increase itself, or else by the circulation of the coolant by natural convection, without pumps, which removes the residual decay heat from the core. Among the safety systems of passive nature we must enumerate the many barriers interposed to prevent the release of radioactivity into the external environment. Other examples of this type are the control rods which are immersed by gravity into the core and the accumulators containing emergency coolant (water) inside gas pressurized containers, which can inject it into the core without the use of a pump, simply through the energy previously accumulated in the compressed gas.

Finally, the active safety systems need a signal to trigger them off, which requires a source of energy. The use of auto-control characteristics in the various systems lessens the effect of errors or accidents, freeing the operators in the plant from taking quick decisions under conditions of stress. A widely diffused concept in various types of reactors is that of multiple barriers against the release of radioactiviuty . The inmost barrier is the fuel itself, of ceramic structure, which holds most of the fission products present in it. The second barrier consists of the metallic sheath around the fuel, resistant to corrosion. The third one consists of metal walls which surrounds the pressurized circuit which contains the core and primary coolant. The fourth barrier is the outer building which is sealed and capable of resisting the pressure and temperature that can be induced by an accidental release of primary coolant. It must be pointed out that the presence of this barrier made the fundamental difference between the external consequences of the Three Mile Island accident (practically zero) and the extremely serious ones at Chernobyl.

To hinder the passage of radioactive products there are also filtering and purifying systems situated among the different barriers; for this reason, the release of radioactivity to the outside, under normal working conditions, is extremely reduced.

The integrity of the different barriers, constantly checked, is put to the test by various forces both internal (mechanical and thermal stress, chemical attacks, radiation etc.) and external (earthquakes, floods, explosions, etc.). The fundamental aim is therefore to protect the barriers from all attacks which come about in accidental conditions. This is carried out using the so called "defence in depth", with three levels of safety measures: the first are *preventive*, the second *protective*, the third *mitigative*.

The first level preventive measures must ensure that all the causes that can lead to accidents (the so called "initiator events") are avoided and their effects cancelled. The second level protective measures must foresee the occurrence of faults in the first level and prepare a second defence level to correct and stop the evolution of eventual accidents which could be triggered off.

The third level mitigative measures are expected to limit the consequences of the accidents that in spite of the intervention of the second level, might happen. Some systems intervene both at the second and the third level.

The safety measures are carried out using planning principles that can schematically fall into three categories: *redundance, diversity and physical separation.* Redundance ensures the presence of extra components or subsystems, in general from two to four in a safety system; in this way safety does not depend on the functioning of a single unit. Diversity concerns two or more safety systems, based on two different physical principles, but which guarantee the same function (the shut down of a reactor, for example, can be simultaneously guaranteed through the falling of control rods or by injection of liquid poison). Physical separation imposes that the components and systems which have the same function are not exposed to the same common cause of breakdown (*e.g.* fire or flood), but must be kept in separate environments with adequately resistant physical barriers.

Finally, we remember the application of the principle by which, if a safety system breaks down, the reactor must drop in a safer condition. If the electric energy is cut off, the control rods fall by gravity because they are no longer held up by an electromagnetic field and the electric current is not needed to make them fall, and so on. Moreover the electric supply is guaranteed by several parallel diesel generators and a system of batteries for the essential functions.

There is increasing attention in considering the human factor as one of the most delicate and important with regard to safety; for this reason the limitations and capabilities of each operator are taken into account, to be able to avoid incorrect manoeuvres. In spite of the fact that the operators are extremely adaptable to unexpected circumstances, it is preferable for this reason to concentrate their efforts on functions that answer their specific capabilities and do not put their limits to the test (like making prompt decisions under stress, checking an excessive number of instruments, promptly discriminating between coherent and incoherent signals, etc.) In modern nuclear power stations in the West the operator's role is to be in charge of gathering information, planning and decision making, and only occasionally to be involved in prompt inspection if the plant should not be in normal working conditions (transitory operational, abnormal events etc.). The plant is protected by automatic control systems which are highly reliable and which the operator must not substitute. The inspection of the many interacting variables of the plant is more easily carried out by computers programmed for this purpose, which show simplified situations on the terminals, indicating the presence of eventual inconveniences and the best way to correct them. The operator can in any case shut down the reactor almost immediately and use his experience as a further guarantee to cope with unforeseen events during planning.

3. – Evolution in the planning of nuclear plants

In the last few years, two different directions have been followed in the world for the development of new reactors: an "evolutionary" direction, based on minor modifications that improve reactors already working, with an approach that takes a step at a time, and an "innovatory" direction which introduces greater modifications. The "for" and "against" of these two directions can be summarized in this way:

- in the first case we fully take into account past experience by improving the pre-existing safety level, but gradual improvements are counterbalanced by persistent support for old plants;

- in the second case the plant can benefit from new, original ideas but it is not borne out by test experience and it is generally necessary to build a prototype.

Two examples of plants of these two types are supplied by the EPR [2] reactor and the MARS [3] reactor.

The EPR (European Pressurized Water Reactor) plant was carried out by a cooperation between Framatome and Siemens through their subsidiary NPI (Nuclear Power International) with considerable commitment by the French and German electricity boards, to represent the interests of the future EPR operators. The evolutionary approach was chosen for three main reasons:

- NPI could be based on the experience acquired from one hundred reactors already built or being built in different countries in the world;

- with gradual planning evolution, the risks of making wrong moves could be avoided or minimized, keeping up and continuously improving developments confirmed by experience;

- the evolutionary approach would minimize the risks of not getting the necessary approval from safety authorities, risks that could arise with a completely new design.

In the meantime the main European electricity boards have brought out some technical manuals which highlight, in particular, safety requirements. These manuals called EUR (European Utilities Requirements), represent the direction taken by the EPR plant where safety is concerned. In fact the prevention of possible accidents, even though extremely improbable, has been pointed out as well as the mitigative action of their possible consequences.

In the extreme case of core melting and perforation of the primary pressure vessel the melted core would be recuperated, confined and cooled at the bottom of the reactor building preventing it from perforating the base and from contaminating the surrounding environment in a lasting way. For this reason the bottom of the cavity of the reactor has been planned to resist melted material up 2000 °C by means of a refractory lining and with a suitable slope to let the melted core slide down and disperse, confining it, in a

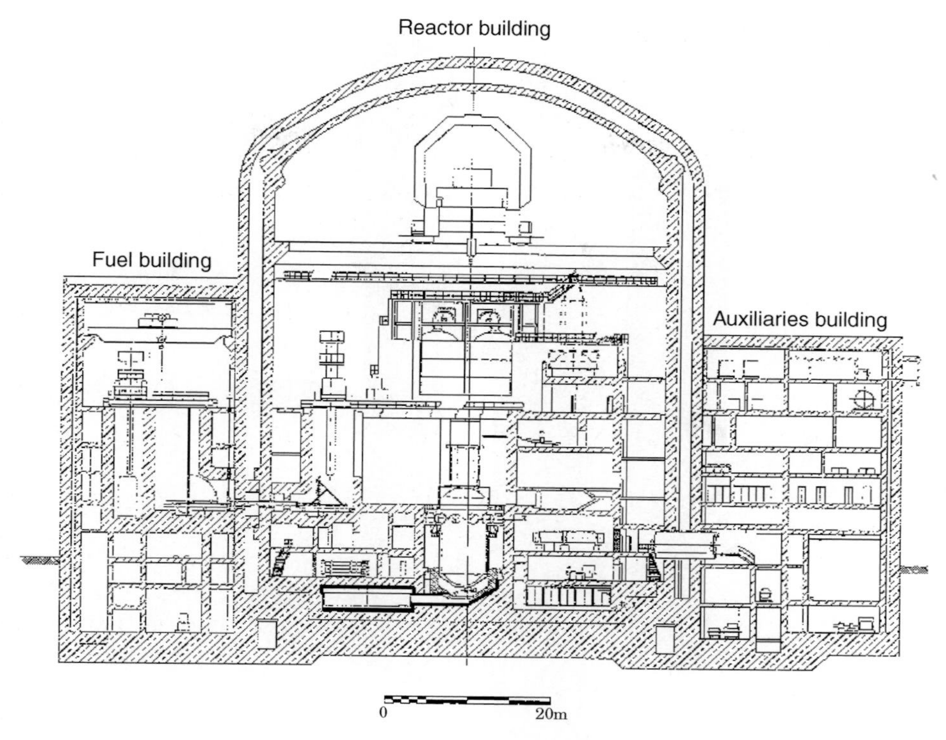

Fig. 1. – Sketch of the European Pressurized Reactor with the reactor cavity for the recover and cooling of the melted core to halt and mitigate the maximum conceivable accident.

space provided, where water is automatically let in and it is possible to cool it avoiding explosions of steam (see fig. 1).

The reactor building has a double wall: an internal one in precompressed armoured concrete, planned to resist explosions of hydrogen and steam pressure in the case of breakage in the primary circuit, and an external one in armoured concrete, planned to resist every possible type of external aggression (plane crash, explosion of inflammable cloud, etc.). The EUR manuals impose the adoption, in the analysis of safety, of deterministic methods, reinforced by probabilistic ones, to identify possible accidents. These are subdivided into two categories:

- accidents included in the basic conditions of the design (DBC);
- accidents included in the "extended" conditions of the design (DEC).

The latter conditions are extended to consider very serious accidents (*e.g.* melting of the core) and to mitigate their consequences in such a way that, in the worst possible case, the external radioactive contamination would have a limited time extension without necessitating the evacuation of the population living outside the plant.

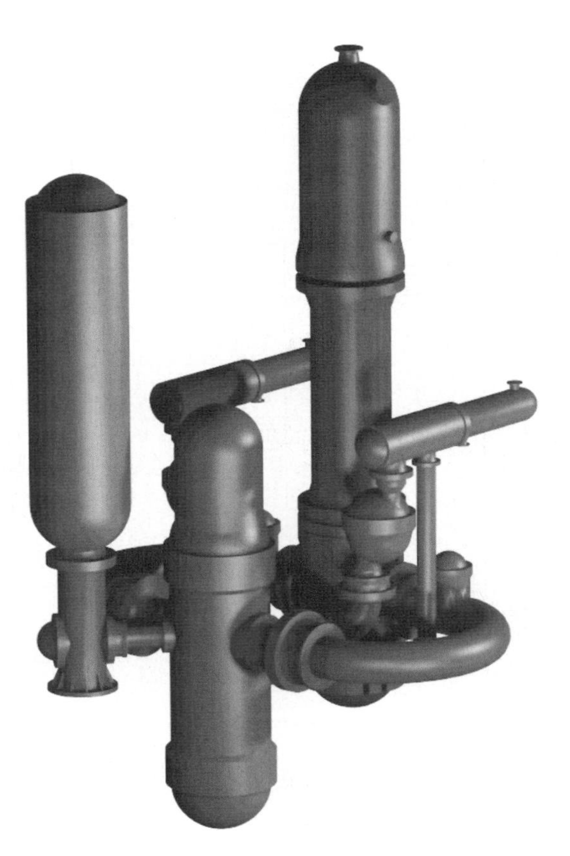

Fig. 2. – MARS reactor primary loop.

Considering the second design line, the innovative one, there are examples concerning pressurized water reactors (MARS and PIUS), run on helium (HTR 100), sodium (PRISM and SAFR), and others. These reactors, also said to be "intrinsically safe", have essential safety functions based on indisputable laws of nature. That is to say the balance of active and passive systems which characterize the first generation reactors is substituted by intrinsic characteristics which, based on natural laws, guarantee the quenching of the chain reaction (first function) and the removal of residual decay heat (second function) independent from the activation of the system, of the availability of energy sources, and above all from any intervention by man. From the shut down of the reactor, which must happen as soon as the neutron flux , temperature or pressure go above pre-established safety levels, the removal of decay heat must be carried out over a sufficiently long period, several days at least. This condition is usually the one that limits the unit power to around a few hundred MWe. Reduced size is generally unfavourable for the cost of energy production although an extended simplification can even invert this effect [4].

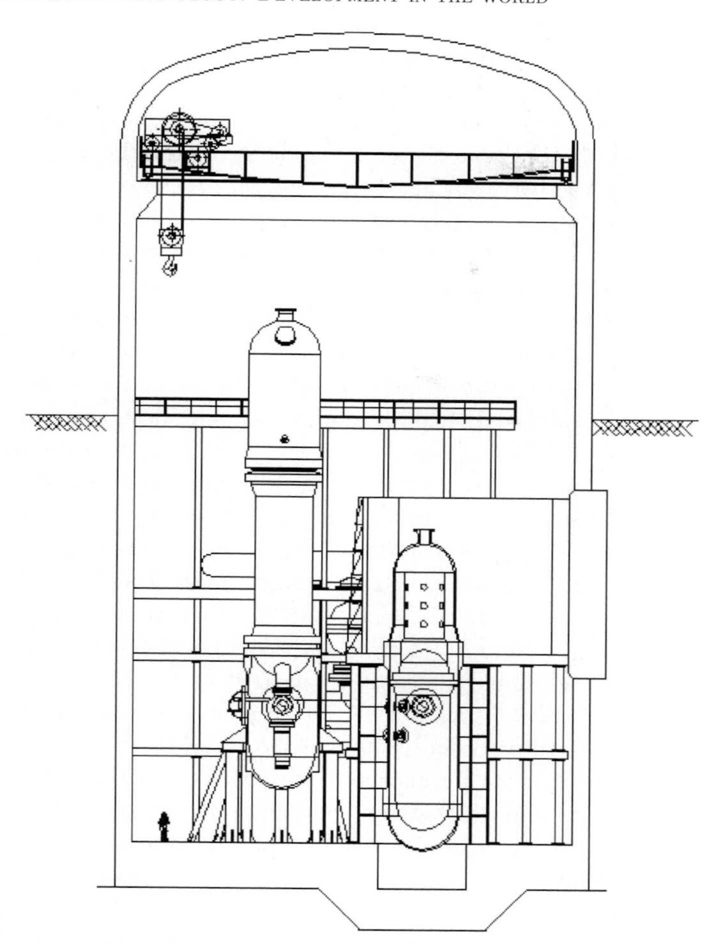

Fig. 3. – Section of MARS reactor building (a man is sketched in the bottom left).

The Italian design MARS (University of Rome "La Sapienza" and ENEA), the power of which is 200 MWe, is a pressurized water reactor of Westinghouse type, with a single primary circuit equipped with an emergency cooling circuit for removal of decay heat. It therefore makes use of the thirty years development experience of the series of pressurized water reactors in a modified and simplified version. It has been modularly devised for use in generating electricity and heat together, above all for desalting sea water (about a third of the world population suffers from insufficient availability of drinking water and this happens above all in countries with inadequate electricity supplies which therefore need small size modular plants).

The plant is all made of steel (see fig. 2 and fig. 3) and can be disassembled completely (the principal link-work is flange-mounted instead of soldered on the site) and this allows its transport in blocks, pre-assembled in the factory, from the factory to the site of the plant, together with the auxiliary systems pre-assembled in suitable groups.

The substitution of the on site construction with the construction in the assembly shops is the main characteristic of MARS. It greatly reduces the construction time and consequent interests that come between, and permits better quality control in clear conditions. Complete prefabrication (apart from the civil engineering) and the simplification of the design permit very competitive costs in the production of energy. With reference to a power station with three modules of 150 MWe each the cost of kWh produced results, from a detailed analysis [5], 0.032 Euro/kWh during the first twenty years in operation and 0.012 Euro/kWh for all the successive years, when the capital investment has been redeemed. In the MARS ("Multipurpose Advanced Reactor inherently Safe") besides the normal control system for pressurized water reactors there is also a special automatic scram system for the chain reaction which works by differential dilation of bi-metallic pairs inserted into the (traditional) fuel assemblies. To eliminate the main cause of accidents, the loss of coolant through ruptures of various size in the primary circuit, the outside of the primary circuit is also pressurized at the same pressure as the inside (about 70 atmospheres), using water at low enthalpy. Even the dangerous accident of expulsion of the control rods is avoided in this way. Furthermore the cancellation of the difference in pressure permits all the components to be flange-mounted and therefore facilitates their maintenance and substitution, prolonging the life of the reactor at extension of 100 years and more. This is very important from the economical point of view. When the "capital cost" has been redeemed, the production cost of nuclear energy strongly decreases. It is enough to think that in the USA, up to now, the owners of 26 of the oldest reactors (out of a total of 101 working reactors) have asked the Control Authorities the authorization to continue their use for another twenty years over the initial project limit. In England gas reactors of MAGNOX type, contemporaries of our old plant in Latina, are still working. The life expectancy of a MARS type reactor can in this way be greatly prolonged (even its pressure vessel can be substituted) also because the instrumentation itself can easily be dismantled and substituted with more up to date components. In this way the final dismantling is rapid and easier, it is a question of disconnecting and removing steel pieces and components with the same lifting and handling machines that were used to assemble them, and therefore very cheap. As a result this modular and co-generating reactor shows interesting characteristics for a market area that includes small electricity companies, with modest networks and the necessity to gradually build up power in their plants, and countries that particularly need to provide themselves with drinking water. On the other hand its electricity production costs are competitive, on the current market, even in countries with large power stations and strong interconnections.

4. – Nuclear energy for naval and space uses

Naval nuclear propulsion has had numerous uses in the military area (compact pressurized water reactors of 50 MWt for submarines, 150 MWt for battle cruisers, 450 MWt for aircraft carriers) but few uses in the civil area, essentially for economical reasons linked to the complexity of running them. In the case of submarines, a nuclear reactor allows months of autonomy with a speed of 30 knots, at a considerable depth. The energy produced also permits the production of oxygen from the sea water, which is introduced into the ventilation system, while the carbon dioxide produced by the crew is absorbed by batteries of purifying elements, allowing artificial renewal of the air inside. Even drinking water is distilled from the sea allowing the above mentioned autonomy.

The main nuclear uses in space are energy sources for fuelling satellites and probes for mission times that last for a few years and thrust engines for rockets. In nuclear propulsors unlike chemical ones (liquid or solid) the propulsor fluid (generally hydrogen) collects energy through thermal exchange instead of a process of combustion. Up to now it has been possible to increase the outflow speed with respect to chemical propulsors by an order of magnitude.

The specific impulse of the rocket is proportional to $\sqrt{T/M}$ where T is the absolute temperature of the gas discharged from the nozzle and M its molecular weight (here is the advantage of using hydrogen). Assuming the same weight between the liquid fuel and ^{235}U, one would draw energy from the nuclear fuel which is two million times that which would be drawn from the reacting oxygen-hydrogen pair.

Fifteen kg of ^{235}U are sufficient to guarantee 1000 kW of electric power in a small reactor, specially made for the purpose, for ten years.

Spatial nuclear generators can be divided into two categories: generators based on thermoelectric or thermoionic conversion of the heat produced from the decay of alpha-radioactive material (RTG, Radioisotope Thermoelectric Generators) and generators based on the conversion (static or dynamic) of heat produced by fission reactions. Many generators have already been launched into space.

RTG were made in the USA using ^{238}Pu decay heat for electric power up to about 1 kWe (Voyager mission). For greater power it is better to use nuclear reactors launched from the ground with a virgin core and activated in space using remote control, with orbits high enough (over 800 km) to guarantee re-entry only after several centuries, when the level of radioactivity is greatly reduced.

A critical parameter, given the launching problems, is the mass per unit of electric power which is less, even more than 4 times smaller, for reactors compared to RTG. In Russia, generators have been made with fast reactors (ROMANSKA) and ceramic type fuel (40 kW of thermal power, 1800 °C of maximum temperature in the fuel, direct conversion).

A French project (ERATO) foresees 200 kW of electric power using a fast neutron reactor with a diameter of 320 mm and the same height, made up of small U rods enriched up to 93% and sheathed in a molybdenum-rhenium alloy, cooled by lithium and working up to 1200 °C. The core is foreseen to remain undercritical even in the cases of faulty launches and falling back on to the Earth. The total mass of the reactor and the associated thermal cycle (Brayton) is of about 7 t and is compatible with the rocket vector Arianne.

An Italian design (MAUS [6]), foresees 25 kW of electric power using a fast cylindric reactor (diameter and height of about 30 cm), cooled by a sodium-potassium alloy with a solidification point of -11 °C and working by direct thermionic conversion (see fig. 4). The fuel is 95% enriched uranium oxide, sheathed in tungsten with a superficial

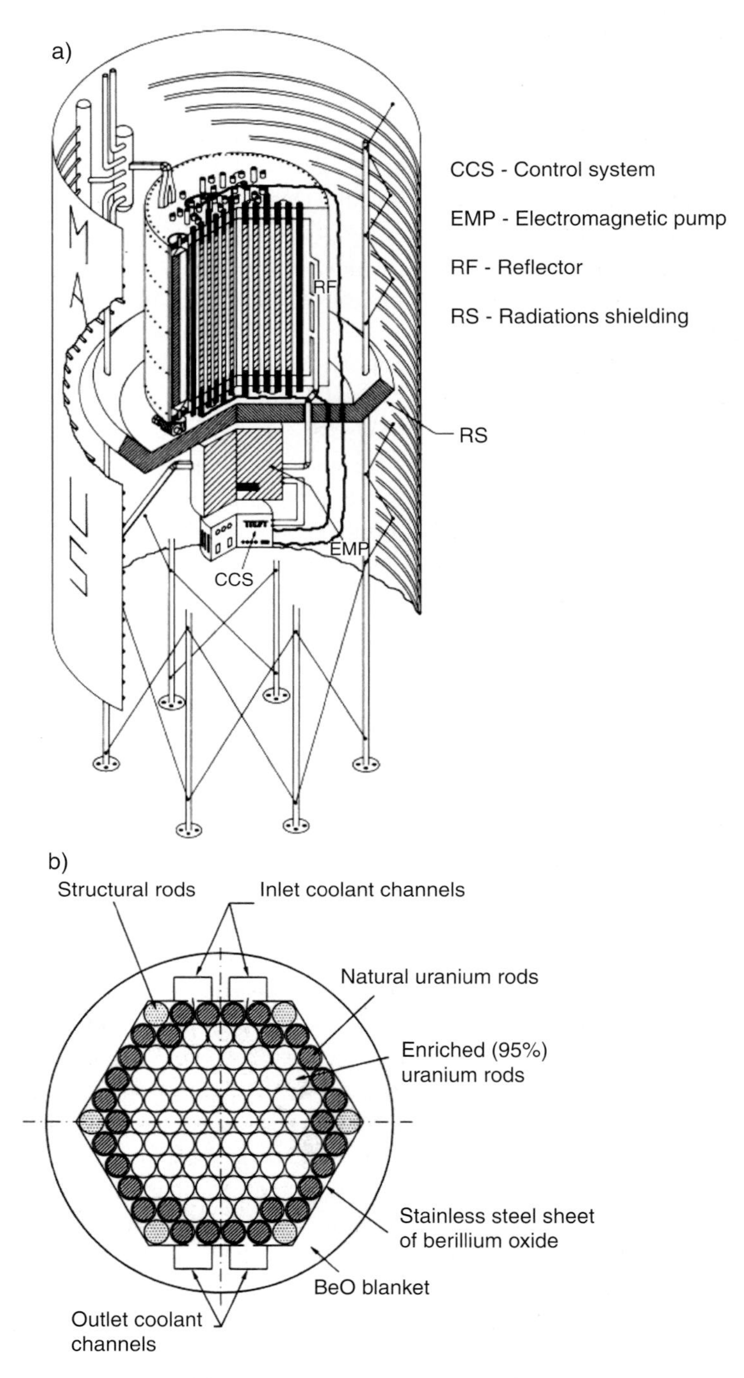

Fig. 4. – MAUS space reactor (a) with its core cross-section (b).

temperature of 1300 °C; it can run for seven years.

At the end of the mission period the energy still available is enough to allow it to reach an escape speed, with special propulsive systems to send it into space in order to eliminate once and for all the radioactive waste accumulated in the core. This solution is still being studied. Coming back to nuclear propulsion, as many as 19 experimental reactors have been built and tried out in rapid succession in the USA (NERVA program).

In one of the latest prototypes the fuel, formed by solid dispersion of uranium carbide and zirconium in graphite, is moulded into hexagonal elements containing longitudinal canals to allow the passage of hydrogen coolant; it reaches a maximum temperature of 2880 K. The hydrogen enters the core at a "cryogenic" temperature and goes out at 2700 K, then it expands in a nozzle where it provides a thrust of about 73000 N with a specific impulse of 875 s. At much higher temperatures the hydrogen can be heated by slowing down fission fragments emitted by very thin layers of fissile matter (a few microns thick) bombarded by thermalized neutrons. A recent project by Prof. Rubbia [7] uses ^{242}Am to take hydrogen up to 15000 K in a nuclear rocket to the planet Mars. The reactor is also covered with a graphite casing loaded with ^{242}Am.

Other studies plan to feed the neutrons necessary to fission the pellicular layer of ^{235}U with a high neutronic flux service reactor, uncoupling the neutronic generation function from that of fission fragment emission to heat the hydrogen flux which will then be expelled from a nozzle at extremely high temperatures.

5. – The storage of radioactive waste

The exploitation of nuclear energy produces radioactive waste which must be suitably stored until the decay of its radioactive level reaches values comparable to those in the natural geological formations from which it derives. A large 1000 MWe power plant running continuously produces about two hundred cubic metres of low level radioactive waste in a year, a hundred cubic metres of intermediate level waste and four cubic metres of high level waste, if immobilized in blocks of borosilicate glass (or thirty tons of depleted fuel elements, if not reprocessed in appropriate reprocessing plants). The latter is placed, after about fifty years, in suitable deep hollows (500 or more metres) in appropriate geological formations (granite, clay, salt layers that water has not touched for millions of years). Having extremely long radioactive decay time (hundreds of thousands of years) it obviously has to be placed in cavities in formations that have remained unaltered for millions of years and for which there is no reason to suppose there will be modifications in geological times.

Radioactive substances can be roughly grouped into two kinds: nuclei resulting from fission (fission products of which, among the most dangerous ^{90}Sr, ^{137}Cs and ^{85}Kr isotopes, with halving periods of not more than thirty years) and nuclei which form through neutronic absorption in fissile and fertile elements (actinides with much longer halving periods).

For "management of radioactive waste" we mean the measures as a whole concerning collection, transport, treatment or conditioning, storage and final disposal of the radioactive waste. We can identify three basic principles used in the management of radioactive waste.

1. dilution and dispersion in the environment in the form of effluents containing radionuclides in a number inferior to the established limits of radio-protection;

2. storage and decay of waste that contains only short life radionuclides;

3. concentration and confinement of waste that contains large quantities of long life radionuclides.

The extreme measures that can be adopted are evidently: maximum dispersion in the environment, maximum isolation. Between these two extremes there are some intermediate solutions (*e.g.* provisional storage followed by unloading outside). The choice of one or the other type of disposal depends above all on the level of contamination of the waste and the characteristics of the environment. For example, the remainder with very low activity can be emitted into the environment within the receiving limits of the latter, while highly contaminated waste like that coming from irradiated fuel that has been reprocessed, can only be subject to complete confinement.

About 2000 different radionuclides are known today, characterized by chemical properties, decay mechanisms, types of radiation and their energetic spectrum. Obviously, the longer the average life is, the weaker the specific radioactivity. About 70 radionuclides have an average life longer than a few months: only these are important for radioactive repositories. Highly active waste contains 95% of the total activity of all waste in concentrated form.

The main purpose of radioactive waste treatment is to concentrate part of the activity in a reduced volume, which can be stored more easily. The purpose of conditioning instead, is that of putting the waste that must be contained, in such a way that the reliability of the storage is increased. For waste disposal we distinguish between temporary (or *interim*) storage, with space equipped to recuperate it, and definitive disposal.

More often than not, the activity in the waste is too high to allow it to be unloaded, or it has too great a volume to be stored. In these cases it is necessary for it to undergo a treatment capable of concentrating as much of the radioactive substance as possible in volumes smaller than the initial ones (phase of increased activity) and therefore more economically stored or more easily isolated in remote localities, so as to be able to freely dispose of the rest in the environment (phase of diminished activity).

Every operation carried out on waste is both risky and expensive. Good management of waste evidently allows to reduce risks at a reasonable cost. For example, in the case of highly active waste produced from irradiated fuel which has been reprocessed, a period of "rest" in the reprocessing plant is generally considered appropriate, since it permits the decay of nuclides which have a short life and therefore reduces the production of heat in the waste. This greatly simplifies the following phases of definitive disposal of the waste itself.

Its radiotoxicity diminishes, after several tens of thousands of years, below the typical level of original ores. The protective mechanisms do not have to guarantee absolute isolation, but merely limit the transmigration of radionuclides from the repository to the biosphere. A dangerous means of transport is water, the absence of which must be guaranteed by choosing suitable geological formations. In the case of clay formations, even if these can contain up to 20-30% of water, their mobility and therefore the migration of eventual dissolved radionuclides is extremely low. The ionic exchange properties of clay contribute to the reduction of migration. In the case of crystalline formation the greatest danger comes from the possibility of fractures or fissures which could form preferential ways for the migration of radionuclides.

Artificial barriers created by man for such underground repositories have certain features which allow them to remain efficient for thousands of years, in the absence of serious geological disruptions. They are made up of special borosilicate glass, with very low leaching level, inside steel containers, with special materials that fill the spaces between the containers and the armoured cement that lines the tunnels in underground repositories.

Some very interesting research has been done on the diffusion and storage of radioactive products generated, within the space of a billion and 800 million years, following natural nuclear reactions which happened far back in the past in Oklo (Gabon). It concluded that most of the fission products formed remained in the place where they had originally been generated, without migrating or dispersing, for nearly two billion years; this allows us to optimistically study the techniques for the definitive elimination of radioactive waste by means of natural analogues.

To have an idea of the size of the problem of placing waste, with an example from near home, a few years ago the annual production of industrial waste in the European Community was about 1000 million cubic metres, of which about 10 million were classified as "toxic" (with infinite decay time). The total production of radioactive waste amounted to 80000 cubic metres (less than 1/100 of that value), of which with high radioactive level (and millennial decay times) about 150 cubic metres (1/100000 of that value). Only the latter need geological repositories while the remaining 80000 cubic metres can be placed in definitive repositories that are superficial or sub-superficial (ten of more metres below ground) with a supervision period of three hundred years, after which it has substantially decayed and can be left unattended (obviously inside concrete chests and the inert matrixes that contain it).

The quantity of storage space needed in geological formations is therefore so small that an interesting separation process becomes questionable (that of long life actinides) and their nuclear transformation into stable isotopes or much shorter decay times, a process which is in any case thoroughly evaluated. It is based on a type of nuclear reactor called "undercritical" because it is not able to maintain the chain reaction alone.

Right from the beginning of the development of nuclear plants the possibility of making undercritical reactors has been considered. These would be able to work thanks to an external neutronic source, with neutronic flux values and power density comparable to the fission reactors used in present power stations. This solution would have the advantage of increasing the possibility in the choice of geometrical parameters and materials used in the core and reduce the reliability requirements for the neutronic control system.

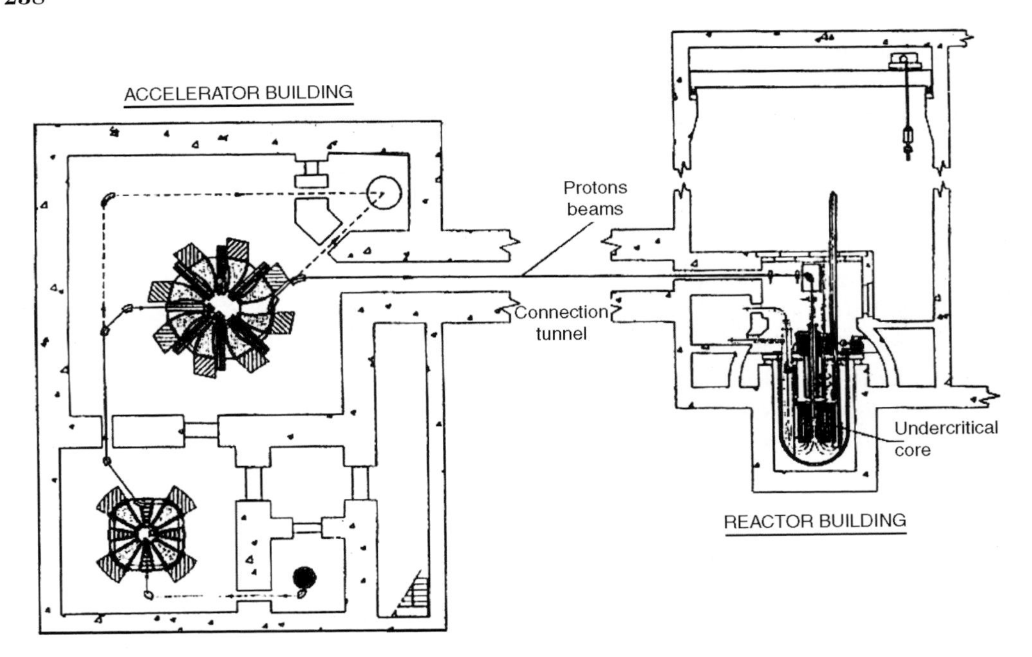

Fig. 5. – ATW type plant following Prof. C. Rubbia's proposal.

Concepts for undercritical reactors have been developed both for problems of production of fissile matter (in competition with fast breeder reactors), and for the destruction of long life high radioactive waste (nuclear transformation of waste).

The concept of these undercritical reactors has followed two streams:

- undercritical systems coupled to a proton source carried out through accelerators (electronuclear channels)
- undercritical systems coupled to fusion reactors (hybrid fission/fusion reactors) which supply the neutronic source.

The advantage of an independent neutronic source lies in the inherent safety system: it is enough to interrupt an accelerated proton beam to shut down the reactor, where shutdown comes about in milliseconds instead of in seconds as it does in present reactors. Furthermore reactivity trips are not physically possible since it runs in definitely undercritical conditions (maximum $\kappa_{\rm eff} = 0.97$). The neutronic sources now considered to be able to guarantee a long term operation with high intensity and constant value are those of the "spallation" type. Protons or deuterons accelerated at high energy ($\sim$ 1 GeV), with an associated wavelength of the same order of magnitude as the dimensions of the target nucleus, produce from 20 to 50 neutrons which escape from more external levels of the atom. A project by Los Alamos National Laboratory foresees proton accelerators from 1 GeV and 250 mA with beams that carry 250 MW of power to the target.

These machines are indicated by the initials ATW (Accelerator driven Transmutation technology for Waste). An Italian project by Prof. Rubbia [8], foresees a compact 3 MW accelerator and a Pb-Bi target which also works as coolant for an undercritical fast spectrum reactor (see fig. 5).

REFERENCES

[1] SALVINI G., *Il nostro futuro energetico, La Termotecnica*, maggio 2000, anno LIV, n. 4 (see also *Atti del Convegno "Energia e Ambiente"*, Accademia dei Lincei, Rome 8-9 March, 2000).

[2] *Nuclear reactors: here comes the EPR, Framatome Outlook News and Views*, n. 3, June 1999.

[3] Università di Roma "La Sapienza" and Dipartimento di Ingegneria Nucleare e Conversioni di Energia, *600 MWth Nuclear Power Plant Design Progress Report 1997*, Rome, July, 1997.

[4] CUMO M., *Inherent and passive safe nuclear reactors: steps towards a second nuclear generation*, Fermi lecture, the Italian Academy for Advanced Studies - Columbia University, New York, Jan. 26, 1995.

[5] CUMO M., NAVIGLIO A., SIMBOLOTTI G. and SORABELLA L., *Produzione di energia elettrica da fonte nucleare. Competitività economica e semplificazione impiantistica, La Termotecnica*, marzo, 1999.

[6] CAIRA M., CUMO F., GANDINI A. and NAVIGLIO A., *MAUS: a fast nuclear reactor for space electric generation, Energia Nucleare*, Anno 10, n. 2, maggio-settembre 1993.

[7] RUBBIA C. *et al.*, *Report of the Working Group on a preliminary assessment of a new fission fragment heated propulsion concept and its applicability to manned missions to the planet MARS* (ASI, 15 March, 1999).

[8] RUBBIA C. *et al. Conceptual design of a fast neutron operated high power energy amplifier* (CERN/AT/95-44) (ET) 29 September, 1995.

Recommended for further reading

– WEINBERG A. M. *"Immortal" Energy Systems and Intergenerational Justice, Energy Policy* **13** (1) (1985) 51-59.

– WEINBERG A. M., *The "Immortality" of Nuclear Systems*, Oak Ridge Associated Universities, *Atti del Convegno "Energia e Ambiente"*, Accademia dei Lincei, Rome 8-9 March, 2000.

– MARLAND G. and WEINBERG A. M., *Longevity of Infrastructure*, edited by ANSUBEL J. and HERMAN R., *Cities and Their Vital Systems* (National Academy Press, Washington, D.C.) 1988.

– VENDRYES G., *Nuclear energy resources and perspectives*, in *Atti del Convegno "Energia e Ambiente"*, Accademia dei Lincei, Rome 8-9 March, 2000.

– WEINBERG A. M., *The first nuclear era* (American Institute of Physics Press, New York) 1994.

– I.A.E.A. Board of Governors, *Nuclear Technology Review*, 2000 GOV/2000/28, 10 May, 2000.

– CUMO M., *Safety in nuclear plants and waste disposal*, in *Atti del Convegno "Energia e Ambiente"*, Accademia dei Lincei, Rome 8-9 March, 2000.

– Fornaciari P., *Nuclear power plants operating to-day*, in *Atti del Convegno "Energia e Ambiente"*, Accademia dei Lincei, Rome 8-9 March, 2000.

– *Low-level radioactive waste repositories: an analysis of costs*, in NEA Policy papers (1999) of the Nuclear Energy Agency/Agence pour l'Energie Nucleaire, www.nea.fr.

– *Progress towards geologic disposal of radioactive waste: where do we stand? An international assessment*, in NEA Policy papers (1999) of the Nuclear Energy Agency/Agence pour l'Energie Nucleaire, www.nea.fr.

– *Commission Nationale d'Evaluation relative aux recherches sur la génération des déchets radioactifs instituée par la loi 91-1381 du décembre 1991, Rapport CNE n. 4, Octobre 1998, La Gazette Nucleaire*, www.resosol.org/Gazette.

About the Author

Maurizio Cumo, Professor of nuclear plants at "La Sapienza" University, Rome, has worked on the study of heat transfer phenomena in extreme and transient conditions and on the design of inherently safe nuclear reactors.

The scientific legacy of Fermi in particle physics

MAURICE JACOB and LUCIANO MAIANI

1. – The legacy of Fermi

The year 2001 will mark the hundredth anniversary of the birth of Enrico Fermi. Fermi was a physicist of spectacular achievements with many lasting contributions despite his untimely death at the age of 53. As Pais said in his book "Inward bound": *"Physics thus lost one of its great figures, a man at home in the instrument shop, in the laboratory and in theoretical physics.... His teaching inspired two generations on two continents."* Particle physicists can hardly spend a day without referring to fermions, to distances measured in fermis or to something else bearing the name of Fermi. His legacy is overwhelming.

One of the last but important acts of Fermi as a physicist was his teaching at the Varenna Summer School in 1954, that famous school of physics which now bears his name. He also taught at Les Houches Summer School that very year, a school in the French Alps which played a very important role in the revival of theoretical physics in continental Europe after World War II. In both schools, his teaching covered his pioneering work on π-meson scattering, which he had been performing in Chicago and to which we shall return at the end of this essay.

Fermi's list of achievements is impressive. One can quote:

i) The introduction of Fermi statistics for half-integer spin particles now called fermions (1925). It led to the concept of Fermi surface in condensed matter and nuclear physics.

ii) A very important contribution to the development of quantum electrodynamics as it was being formulated in the late twenties and early thirties.

iii) The vector coupling theory for β-decay (1933), formulating the proper structure of the weak interaction. Whereas this structure, which he proposed in analogy with electrodynamics, was challenged for some time and validated experimentally only by the end of the fifties, it always remained the term of reference for researchers in that field. It is known as the Fermi interaction with its Fermi constant, measuring the strength of the coupling.

iv) The introduction and study of neutron-induced radioactivity, complementing and replacing that induced by alpha particles (1934).

v) The study of slow-neutron interactions, starting in 1934 and extending up to 1939. The Rome group with D'Agostino, Segrè, Amaldi, Rasetti and Fermi was much contributing to the coming age of nuclear physics.

A concept, which is often referred to in connection with nuclear structure, is that of Fermi motion, the quantum motion of the protons and neutrons bound in a nucleus. This motion allows for production processes below their classical thresholds.

Fermi was awarded the Nobel Prize in 1938 for *"his demonstration of the existence of new radioactive elements produced by neutron radiation, and for his related discovery of nuclear reactions brought about by slow neutrons."*

Fermi then left Rome for the United States, first to the Columbia University. As it was said at the time, he "took the wrong boat" on his way back from Stockholm. At Columbia Fermi immediately engaged in fission work and in the detailed study of the slowing-down of neutrons. The latter domain makes great use of another concept once again introduced by him. This is known as the "Fermi age" and traces down the kinematics history of neutrons slowing down in a moderator.

At this stage one may recall an amusing and instructive story. Fermi had many admirers and, among them, far away ones, who eagerly looked at his published papers for inspiration or guidance. This was the case of Flerov, a young Russian physicist who discovered spontaneous fission in 1940. He was drafted in 1941 as Germany attacked the Soviet Union. By 1942, during a furlough, he could rush back to a physics library and read the "Physical Review", looking for the latest papers by Fermi. It was a very great surprise for him not to find anything from Fermi and this all through the 1941 issues! His immediate conclusion was that fission research had become a secret research in America and that the United States were seriously engaged in a nuclear bomb programme. Since he had some special relations with the top authorities, he immediately wrote to Stalin to urge an assessment of Soviet possibilities along that line. This had some effect and, already by 1942, a sizeable effort on the chain reaction was started, under the leadership of Kurchatov. As this story shows, lack of appearance can sometimes be a precious source of information! At that time the key question was not so much whether it was possible to build a bomb, but rather whether it was worth engaging during the war in the very big effort which was clearly necessary.

Let us continue with the long list of Fermi's achievements. One now finds:

vi) The construction of the first nuclear reactor, in Chicago, achieved in 1942, with the famous (coded) message: "*The Italian navigator has reached the New World*".

After working at Los Alamos, Fermi returned to Chicago, to the newly founded Institute for Nuclear Studies. Under his guidance that Institute quickly became a nursery of great talents. Suffice it to say that among the young people there, as physics students just after World War II, one could find: Chamberlain, Chew, Garwin, Goldberger, Lazarus, Lee, Orear, Rosenbluth, Rosenfeld, Selove, Steinberger, Wolfenstein and Yang! This is an impressive list and more were to follow. Among Fermi's own achievements at that time now come:

vii) A theory of cosmic rays acceleration (1949).

viii) A theory of the π-meson as a bound state of a baryon and an antibaryon, (also in 1949).

ix) The experimental and theoretical study of π-nucleon scattering with the validation of charge independence, or isospin invariance, and the emerging of the famous 3-3 resonance. A first comprehensive look at pion interactions was the subject of his lecture series at the Varenna Summer School on pions and nucleons, which we already mentioned.

Fermi was both a theorist and an experimentalist. He was among the last ones achieving so impressively that very rare combination of research skills in nuclear and particle physics. Did one not say of him: *"For the theorists he was a great theorist and for the experimentalists, he was a great experimentalist"?* He was also a magnificent teacher leaving many monographs and lecture notes but first of all an extraordinary legacy in the minds of those who had the privilege to attend his lectures. Fermi was the trigger to success for the many young scientists who met him, and, let alone, for those who worked with him, first in Rome, in the late thirties, and later in Chicago, in the late forties and early fifties. Gian Carlo Wick, one of the best theorists of the last century, was Fermi's assistant in Rome in the thirties. He said: *"Fermi's advice over many years but, first of all, the example set by Fermi were my essential guidances as a young researcher."* Fermi's approach to new physics, with its powerful mixture of logic and pragmatic skill, as it shows clearly in some of his monographs, and in particular the one on nuclear (and particle) physics of 1949, is part of his great legacy. A reading of his Varenna notes allows almost one to see him at work.

Fermi died just before nuclear and particle physics split, each of the two fields becoming a rather specialised domain of study with particular sets of conferences and meetings. Nuclear physics specialised in the study of the structure of nuclei, seen as assemblies of nucleons. Particle physics devoted itself to the study of the inner structure of the nucleon.

By the late fifties it was already clear that the nucleon was an extended object (with a size of about 10^{-15} m —one fermi!) and with some inner structure, still to be explored.

The many new particles discovered in cosmic rays also showed that the list of fundamental particles could not be confined to the few involved in the structure of the atom.

Fermi could realise the complexity which all that was bringing to the study of the strong forces binding nucleons together. He said *"When the Yukawa theory was proposed (that the strong force corresponds to the exchange of π mesons between nucleons) there was a legitimate hope that the particles involved, protons, neutrons (the nucleons) and the π-mesons could be considered as actual elementary particles. This hope loses more and more of its foundation as new elementary particles are rapidly being discovered."* As we know today, these particles having strong interactions, which we now call hadrons, those known at the time of Fermi and a few hundreds more of others discovered since, are all made of quarks. The quarks are spin-(1/2) fermions. There are three quarks in a baryon (which is a fermion) and a quark and an antiquark in a meson (which is a boson).

In this essay on the scientific legacy of Fermi in particle physics we shall choose three topics among his many major contributions, which we listed under items iii), viii) and ix). They are respectively the theory of β-decay, the π-meson, or pion, seen as a nucleon-antinucleon bound state and, finally, the study of charge independence in the interaction between pions and nucleons. This is not exhaustive! These three contributions have deeply influenced the development of particle physics and they remain highly topical today.

2. – Nuclear β-decay and the structure of the weak forces.

Back in the late twenties, β-decay (the radioactive emission iof electrons) was still shrouded in mysteries. One just saw electrons emitted by nuclei that were changing their nature. Quantum theory was advanced enough that one did expect a rather sharply defined energy for the electron, corresponding to a transition between two well-defined nuclear quantum levels. The electron, much lighter than the nucleus, should take the overwhelming part of the energy released by the transition. However, a continuous spread of electron energies, extending from zero to the full energy released by the nuclear transition, was observed instead.

One had thought for a while that a nucleus of atomic number A and charge Z could be made of A protons and $A-Z$ electrons (called "nuclear" electrons) one of those electrons being simply liberated in the β-decay. In fact this picture was already embarrassing to accept from a quantum-mechanical point of view. It was indeed hard to think that these electrons could pre-exist in the nucleus. They are so light particles that, whereas they can be well localised within a "big" atom, they should not be localisable over such a small volume as that of an atomic nucleus. The quantum rule of thumb for the localisation of a particle consists in taking its Compton wavelength, which is proportional to the Planck constant and inversely proportional to the mass. It is between 10^{-12} and 10^{-13} m for an electron, over a hundred times larger than the dimensions of a nucleus!

The existence in the nucleus of A protons and $A-Z$ electrons was also not compatible with the relation between spin and statistics, as shown a few years earlier by F. Rasetti with a brilliant experiment on the observation of Raman spectra emitted by bi-atomic,

O_2 and N_2 molecules, performed in 1928. F. Rasetti was led to conclude that the nuclei $^{16}_{8}O$ and $^{14}_{7}N$ had both integer spin (0 and 1, respectively) and obeyed the Bose-Einstein statistics. On the other hand, in the theory of nuclear electrons, $^{14}_{7}N$ should be made of 14 protons and 7 electrons, a total of 21 spin-(1/2) particles, leading to half-integer spin and Fermi statistics, contrary to observation.

Be as it may, the most striking aspect of nuclear β-decay was that the electron emitted was not carrying the full energy released by the transition. Since one did not know much about the structure of the nucleus anyway, the apparent lack of energy conservation was seen as the most blatant problem. Energy conservation is a principle of physics that we have learned to associate with the invariance of physical laws under time translations. Energy should be conserved if phenomena have to develop in the same way, irrespectively of their absolute location in time! The energy released by the nuclear transition should therefore be found with the electron released in the decay. Nevertheless, even though invariance principles are parts of the "Constitution" of physics, the general framework for the formulation of the physical laws, they have to be challenged experimentally whenever needed. Was energy really conserved in β-decay, such a new and puzzling phenomenon? Bohr had doubts about it. Pauli, however, was among those who insisted that energy should be conserved and he introduced for that purpose a new particle, the neutrino. In β-decay, he said, both an electron and a neutrino are produced and they share in some variable and uneven way the energy released in the nuclear transition. What is dominated by chance is not the amount of energy released by the nucleus in the decay, which is always the same, but the way this energy is shared between the electron and the neutrino.

It must be said that the neutrino had to be a very special particle, to meet already known experimental facts. It should have no charge, a small mass (if any mass at all, as we would say today) and practically no interaction so as to be able to escape the detection techniques available at that time. It was bold of Pauli to propose such a new particle at a time when it was very unpopular to do so. It was expected that there should be very few fundamental particles. The look for simplicity was still in an economy of particles and not in an economy of principles as it is today. Indeed, at about the same time, Dirac barely dared to introduce the positron as a new (anti) particle beside the electron. He had even tried for a while, though unsuccessfully, to identify his positive particle with the proton.

In any case, the neutrino was proposed as a solution to the energy riddle in β-decay. The price to pay to save energy conservation was the existence of this hitherto unknown particle with such peculiar properties. Pauli first mentioned his idea in a letter to Heisenberg in 1930 and the idea sank in.

Pauli supported his idea also on the ground that, as we have seen, the existence in the nucleus of A protons and $A - Z$ electrons was not compatible with the relation between spin and statistics, as shown by the Rasetti experiment. Referring to this problem, Pauli in his letter referred to what his predecessor in charge, R. Peierls, said about it: *"it is like taxes, the less you speak of it, the better it is."* The emission of a neutrino together with the electron allowed reclassifying the constituents of the nucleus in such a way as to restore the spin-statistics relation. Assuming neutrinos to have spin 1/2, the nucleus would be made by Z protons and $A-Z$ complexes made by proton + electron + neutrino (in total, half-integer spin complexes). Thus A-even nuclei would always correspond to integer spin and therefore obey the Bose-Einstein statistics (the new picture gives a different statistics with respect to the old one in the case when $A-Z$ is odd, like $^{14}_{7}N$, and it reproduces correctly Rasetti's result). Something else obeying Fermi statistics had to be in the nucleus beside the electron, he insisted.

The neutrino was eventually discovered, by Reines *et al.*, but only in 1956, two years before Pauli's death. By that time the huge flux of neutrinos emitted by a nuclear reactor could create a few reactions in Reines' imposing (for that time) detector.

It is in that context that Fermi's theory was formulated. Pauli's hypothesis had become quickly known to Fermi, who took it up with great interest. Fermi indeed coined the name "neutrino" by opposition to the name "neutron" which had been used at first by Pauli, but had, as Fermi thought, to be "reserved" to the neutral partner to the proton, which had been discovered by Chadwick in 1932. The discovery of the neutron definitely clarified the problem associated with the composition of the nucleus, thereupon understood as an assembly of Z protons and $A-Z$ neutrons. The electrons of β-decay were indeed produced during the transition of a neutron into a proton accompanied, according to the hypothesis of Pauli, by a neutrino.

The neutron is to be identified with the "complex" proton+electron+neutrino we have mentioned before. The neutron is able to decay into this complex by a new interaction, distinct from the electromagnetic (atom binding) and the strong (nucleus binding) forces, which Fermi called the "weak interaction". This is the way it was in 1933.

Let us now say a few words at this stage about relativistic electrodynamics. The Coulomb law for the action between charges and the Biot-Savart law for the action between currents, merge into a single interaction between two currents which are now 4-vectors in relativistic mechanics. One says that the interaction has a vector nature. The quantum form of the electromagnetic interaction corresponds to the exchange of one photon (a vector particle) between the two currents which are associated with the presence and motion of the two interacting particles (two electrons, say). This is nowadays summarised by a Feynman diagram, which gives the first and leading contribution to the electron scattering process in a perturbative expansion in terms of the square of the electric charge. The electric charge measures the strength of the coupling of the photon to the electromagnetic current.

Through scattering, each charged particle (here the two electrons) changes its momentum under the emission or absorption of the photon but it does not change its nature. Quantum relativistic electrodynamics, as thus presented, became the model for Fermi's picture for β-decay.

Fermi considered that a current could emit an electron-neutrino pair in much the same way as the electromagnetic current emits a photon in Quantum ElectroDynamics. We shall however present the theory directly in its current-current version, using the Dirac sea language of the time whereby particle production is associated with a change of motion for a particle, as intuitively connected to a current.

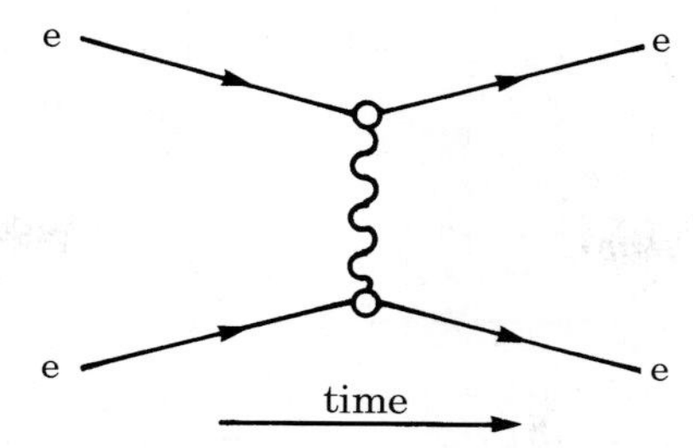

Fig. 1. – Electron-electron scattering with one photon exchange.

In order to avoid the problem raised by the presence of negative-energy states which appear in the Dirac electron theory, the vacuum (which acts as the lowest-energy state for the theory) has to be such that all negative-energy states are already filled. This is the so-called Dirac "sea" of negative-energy electrons. The Pauli exclusion principle, which forbids any two electrons from being in the same quantum state, then prevents positive-energy electrons from falling into negative energy states that are already all occupied by one electron.

One can extract such a negative-energy electron from the sea, providing its energy through photon absorption. The electron now appears as a positive-energy electron and one can naturally associate a current to that transition changing the motion of that electron. The operation, however, leaves a hole in the Dirac sea, which behaves as a positively charged particle, the positron, the antiparticle of the electron. One thus sees that the exchanged photon of fig. 1 can change the motion of a (positive energy) electron but also can create an electron-positron pair. These are but two views of the same dynamics.

In Fermi's theory, β-decay is modelled along the lines defined by electrodynamics. The currents associated with the initial and final particles are now allowed to change their nature. One current can be associated with a neutron turning into a proton and the other one to a neutrino turning into an electron, or, as we just saw when the initial neutrino is taken from the negative-energy sea of neutrinos, to the production of an electron-antineutrino pair. One needs a neutrino sea together with the electron sea, filling in both cases all the negative-energy states! One may say that a negative-energy neutrino is "kicked out of the sea" to a positive-energy state but also transformed into an electron at the same time. The hole left behaves as an antineutrino. The overall reaction can be written as

$$\mathrm{N} \to \mathrm{P} + \mathrm{e}^- + \bar{\nu}$$

What does the "kicking"? Something should be exchanged to transfer energy and momentum from the nucleon, N-P, to the electron-antineutrino system. This "something", unlike the photon in electrodynamics, must now carry one unit of electric charge. If it were also a vector particle, like the photon, the new particle would lead to a description of the neutron decay closely analogous to the electrodynamics picture of electron scattering (fig. 2).

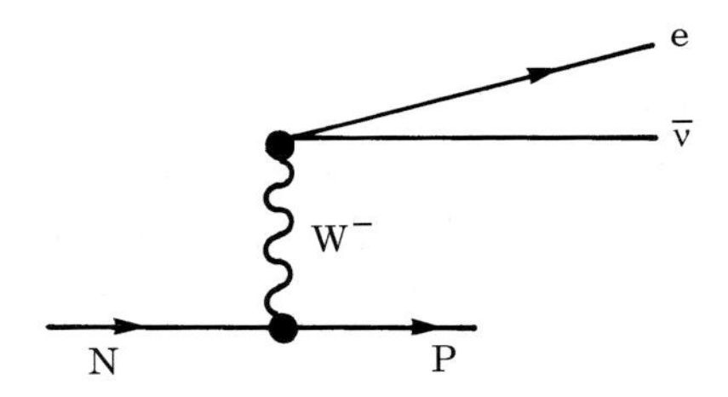

Fig. 2. – Neutron decay by one W exchange.

To economise in particles, Fermi did not introduce explicitly the W particle, which was however considered immediately after Fermi's theory (this possibility is already mentioned by Yukawa in his theory of nuclear forces of 1935). At any rate, the W has to be very massive in order to explain the short-range nature of the interaction. This exchange mechanism was to be eventually vindicated experimentally but only in the early eighties, with the discovery of the W meson, with a mass close to 90 times that of the proton! The range of such an interaction is then of the order of 10^{-18} m, which remained much beyond experimental possibilities for a long time.

In Fermi's theory one does not consider the nature of this intermediate particle but assumes that the two currents are directly interacting among themselves, in a point-like way, as if the particle exchanged had an infinite mass. As a result, the global coupling between the two currents has the dimension of an inverse mass squared. This coupling constant has ever since been known as the Fermi constant, with a value

$$\frac{G}{\sqrt{2}} \cong 10^{-5} M_p^{-2},$$

where M_p is the proton mass (the smallness of the Fermi constant justifies the name of weak interaction given to the forces responsible for the β-decay).

We have introduced a current associated with an electron-antineutrino pair in the Dirac sea language of the time, in order to stress the analogy between the electromagnetic interaction and β-decay, which is the essence of Fermi's theory. Fermi could not readily say all that in the early thirties. He indeed did not first distinguish an antineutrino from a neutrino but rather considered a vector built out of an electron and a neutrino wave functions which would globally produce something which could replace the photon emitted in electrodynamics. A Dirac current duly associated with an electron-antineutrino pair came only shortly afterwards. Fermi could also not use a Dirac current for the neutron turning into a proton, since relativistic electrodynamics was not yet extended to the proton. It was still considered a big puzzle that the proton had a magnetic moment different from the one predicted by the Dirac theory to which the electron was obeying so well. One can however but admire the insight of Fermi in his using only a nucleon density, something legitimate in view of the smallness of the energy release. And (this is in retrospect a masterpiece!) he defined this charge-changing density according to the isotopic spin operator of Heisenberg, which changes a neutron into a proton through a rotation in charge space, in much the same way as a spin operator flips a spin down into a spin up in ordinary space. With that definition for the nucleon current he had enough to define a coupling and relate it to the decay rate. He had enough to distinguish allowed transitions (with no change of nuclear spin) from forbidden ones. He used the allowed transitions to estimate the coupling. This involved much technical work with Coulomb wave functions for electrons emitted by heavy nuclei.

The introduction of this isospin current (once put in the proper relativistic form) can be seen in retrospect as a great insight. At first it was just a short-hand notation to express the transformation of a neutron into a proton. The connection with isospin however paves the way to the universality of the weak interaction with a unique weak coupling constant. The interaction introduced by Fermi does not "look" differently at different particles. It deals with the isospin current whether carried by a nucleon, a pion or any other particle, in much the same way as a photon couples to a charge, irrespectively of the particle which carries it. There is one coupling for the isospin current, irrespectively of the way it is built out of different particles.

We talked about the range of the interaction. It was Wick, who was at that time the assistant of Fermi in Rome, who had shown that, in relativistic quantum mechanics, the range of a force due to the exchange of a particle is inversely proportional to the mass of that particle. Quantum fluctuations allow for the appearance of this extra mass but only for a time inversely proportional to it. The range is then approximately given by that time multiplied by the speed of light. It corresponds to how far anything can be felt from the quantum fluctuation. This looks trivial today for a generation brought up with relativistic quantum mechanics but, at that time, even Bohr was impressed by the argument! The photon being massless, the electromagnetic interaction has an infinite range. The pion mass imposes the range of the nuclear force. It is of the order of 10^{-15} m, or one fermi. The heavy W mass gives the very small range of β-decay which, following Fermi's theory, was long taken as zero.

Wick also showed how Fermi's theory of β-decay, with electron emission, could be readily used to describe the emission of positrons, now with a proton turning into a neutron inside a nucleus and a neutrino being emitted together with the positron. This new form of β-decay had just been discovered by the Joliot-Curies. One can describe the process in the Dirac sea picture. A negative-energy electron in the sea is turned into a positive-energy neutrino leaving a hole behaving as a positron.

One may however now use the particle-antiparticle symmetry of quantum field theory that Majorana was by then strongly advocating within the Roman group. As Majorana was saying, one should forget about individual particles but rather consider quantum fields, particles, as they are observed, being the quantum excitation of those fields. In the framework of quantum field theory, the negative-energy problem disappears. There is no longer any need for a Dirac sea in the formulation of the theory. But there is now nothing like a well individualised particle! Particle-antiparticle pairs can be freely emitted or absorbed, just as photons. For instance, in the case of electron scattering off a proton, one cannot tell whether the outgoing electron is the initial one or a member of

TABLE I. – *Weak processes related to β-decay in a relativistic, quantum field theory (RQFT). Time reversal symmetry allows to interchange the direction of the reaction while the so-called "crossing symmetry", characteristic of a RQFT, relates different processes obtained one from the other by transforming particles (antiparticles) in the initial state into antiparticles (particles) in the final state. The last two processes are obtained under the additional hypothesis that weak interaction is described by the product $J_\mu J^\mu$, where J_μ is the total weak current.*

Reaction	Denomination
$\mathrm{N} \to \mathrm{P} + \mathrm{e}^- \overline{\nu}$	β^--decay
$\mathrm{P} \to \mathrm{N} + \mathrm{e}^+ \nu$	β^+-decay
$\mathrm{e}^- + \mathrm{P} \to \mathrm{N} + \nu$	K-capture
$\nu + \mathrm{N} \to \mathrm{P} + \mathrm{e}^-$	neutrino-nucleus scattering
$\overline{\nu} + \mathrm{P} \to \mathrm{N} + \mathrm{e}^+$	antineutrino-nucleus scattering
$\overline{\nu} + \mathrm{e}^- \to \mathrm{N} + \overline{\mathrm{P}}$	inelastic antineutrino-electron scattering
$\overline{\nu} + \mathrm{e}^- \to \overline{\nu} + \mathrm{e}^-$	elastic antineutrino-electron scattering
$\mathrm{N} + \mathrm{P} \to \mathrm{P} + \mathrm{N}$	nuclear weak force, gives rise to parity violations in nuclear levels

an electron-positron pair whose positron had annihilated against the incoming electron. But why should we be able to tell? The question has no experimental meaning since electrons are identical.

Wick also predicted "K-capture" whereby an electron (close to the nucleus, in the so-called K-shell) can be captured by a proton, which turns into a neutron as a neutrino is emitted.

In modern words one would say that all that results from the CPT invariance of quantum field theory, a global invariance property of the dynamics under the combined operation of C (changing particles into antipartides and vice versa), P (space inversion) and T (time reversal). One can freely exchange an ingoing particle into an outgoing antiparticle and vice versa. It is still the very same physical process, which is at work. This was, however, not yet part of the physicist's household tools in the mid-thirties and exploiting such properties and thus getting several processes out of Fermi's theory called for much insight (see table I).

Fermi first described his theory of β-decay to some of his colleagues during an "after ski" discussion at Val Gardena. He was probably not certain to have grasped the truth, as he indeed had, with the picture that he had put together in straight but subtle analogy with electrodynamics. His colleagues were, however, much interested and encouraging and he eventually published it but with the rather modest title of "Tentativo di una teoria dei raggi beta" (Tentative theory of beta rays). This was in 1933. This modest title may have to do with the fact that a first version of the paper had not been accepted by "Nature" on the ground that "it contained speculations too remote from reality to be of interest to the reader".

The vector coupling with its neutron-proton and electron-neutrino currents, as introduced by Fermi, has been the winning horse. Not ever since though, since, soon after people could master the quantum relativistic properties of the nucleons, and write a relativistic nucleon current, it was discovered that the point-like structure of β-decay introduced by Fermi could be realised with several different types of couplings. It could be between two vectors (two currents), as Fermi had written it, but also between two axial vectors (differing from a vector by the absence of a change of sign of the three space components under parity), or again between two scalars, two pseudoscalars or two tensors. It was shown that all forms could equally well reproduce the known data. They were known as V, A, S, P and T, respectively. The more detailed study of the β-decay spectra showed that it could be either S and/or T or V and/or A. It was noticed that electron-antineutrino angular correlations could distinguish between V and/or A, on the one hand, and S and /or T, on the other hand. Such experiments were very difficult and it is not too surprising that they led to definitely favour S and T over V and A for some time! Fermi could not witness the wandering away from the line that he had traced out.

At the end of the fifties, nuclear β-decay had become only one element of a large set of decay processes (muon decay, π decay, electron capture), which all appeared to be of a similar nature. All rates were tallying with what was expected from a coupling given by the Fermi constant. They globally represent different manifestations of the same weak interaction.

The strange particles (kaons, hyperons) were also seen to decay through the weak interaction that was violating the conservation of strangeness, respected by the strong interactions that create them. Some time later, in the mid-sixties, Cabibbo has shown how a universal approach could apply to all forms of the weak interaction, whether or not strangeness was changing in the process.

It was only by 1958 that the vector structure for β-decay, which was by then also recognised as that taken by all the different manifestations of the weak interactions, was fully ascertained. By that time, however, the weak interaction was known not to conserve parity and, instead of a pure vector, one had a mixture of vector and axial, which behaved in an opposite way under parity. For the leptons (electron and neutrino) and later for the quarks, the interaction was taking the simple symmetrical $V - A$ form, corresponding to a maximum violation of parity. If we neglect masses, we can say that particles enter weak interactions only through their negative-helicity component, that is when they spin to the left around their momentum, whereas antiparticles enter through their positive-helicity component. For a massless particle, positive- and negative-helicity states are not connected through Lorentz transformations and they can be considered separately. Helicity and not charge can be used to define the proton and the electron as particles (both coupled with negative helicity) and the antiproton and the positron as antiparticles.

By 1957 the data available on weak decay were already numerous but still contradictory. It called for the insight of Feynman and Gell-Mann and, independently, of Marshak and Sudarshan to claim $V - A$ as the truth and work out the dynamical consequences. They were right. All experiments then opposing $V - A$, and they were several at that time, were eventually shown to be in error.

One may mention for instance the story of the Michel (ρ) parameter the value of which characterises the shape of the electron spectrum in muon decay. If the interaction is V (or a mixture of V and A) the parameter ρ has to be $3/4$. The first reported value was

0! With time (and more accurate experiments!), the measured value eventually climbed monotonically, each new reported value being compatible with the preceding one granting the error bars, though somehow higher. It was only by 1958 that it had reached 3/4 to stabilise there. The V (or a mixture of V and A) coupling also specifies the rate for π electron-neutrino decay to be about a tenth of a thousands that of π muon-neutrino decay. For some time, nothing was found of the former mode, even below the predicted value. It was the first CERN experimental success to find this decay mode and that with the expected rate. It turned out that previous experiments were rejecting the few electrons seen as (overestimated) background.

It was an eventful and exciting saga all through the later part of the fifties, but eventually Fermi theory with its universal character, and now with its modern $V - A$ attire, was fully vindicated.

Fermi had quickly stopped working on his theory, which in retrospect appears as a great masterpiece. The neutron had been discovered and was becoming a tool for induced radioactivity. Fermi was going to be one of the great pioneers in this endeavour and, already by 1934, he published a paper with the title "Radioattività provocata da bombardamento di neutroni" (Radioactivity induced by neutron bombardment), with a systematic study for increasing atomic number values. Was Fermi the theorist turning into Fermi the experimentalist?

In 1934, after the results just mentioned, Rutherford actually wrote to him: *"I congratulate you on your successful escape from the spheres of theoretical physics! It seems that you have struck a good line to start with. You may be interested to know that Professor Dirac is also becoming an experimentalist. This seems to be a good omen for theoretical physics... "*. Rutherford was well known for such sweeping and provocative statements.

But we shall not touch here Fermi's extraordinary contribution to nuclear physics in the middle and late thirties. We shall instead briefly describe the future of the field of β-decay after 1958. Fermi left it soon but his legacy was felt for a long time. However, before doing so, we may reflect once again upon Fermi's insight. Fermi showed that β-decay, with the creation of an electron and an antineutrino, was an entirely new force, the forebearer of what is now referred to as the weak interaction. It silenced the view that the electron could come out of the neutron, seen for some time as a proton-electron bound state. As we already mentioned, this view does not resist our present quantum point of view but it prevailed for some time in a doomed effort to limit the number of elementary particles. Fermi's theory was the first one ever where massive "bona fide" particles (as opposed to light quanta) appear and disappear, as they naturally do in the framework of a quantum field theory. In modern parlance this was the first use of second quantized spin-(1/2) fields.

Let us now go back to 1958-59. With the $V - A$ structure being established, it was natural to expect that the interaction be mediated by a vector particle acting much as the photon for the electromagnetic interaction but carrying charge and having a very heavy mass in order to limit the range. It was named W, for weak. The W was discovered in 1983, a discovery which was rewarded by the Nobel Prize given to Rubbia and Van der Meer. The mass of the W is 81 GeV, and it was necessary to transform the CERN Super Proton Synchrotron into a proton-antiproton collider to achieve collision energies permitting its production. It was correctly estimated that the proton-antiproton collision energy should be about 6 times the W mass in order that its formation through quark-antiquark fusion should not be quenched by threshold effects. By that time the weak interaction had been put in a form which provided a unique structure for the weak and electromagnetic interactions. That form is the theory of Glashow, Salam and Weinberg, also rewarded by a Nobel Prize. The analogy between β-decay and electromagnetism heralded by Fermi is then brought to the ultimate point with a unique theory for the two interactions, and not only an analogy between the two. The theory, however, predicts a new form of weak interaction with no charge exchanged, a short-range coupling between neutral currents, mediated by a neutral vector particle, called Z.

The theory brings in parallel lepton pairs (e-ν_e; μ-ν_μ) and quark pairs. We have skipped the discovery of two types of neutrinos, one associated with the electron and one associated with the muon, since this dichotomy does not change the structure of the Fermi theory. This was, however, a very important discovery, made in the early sixties, for which Lederman, Schwartz and Steinberger received the Nobel Prize.

There was for long only a quark pair (u-d) to put in parallel with the first lepton pair and an "odd" strange quark. In this scheme, the interactions mediated by the Z should violate strangeness, like the other weak interactions. However, this was in stark conflict with the experimental facts, which showed a remarkable suppression of strangeness-changing processes in neutral weak interactions. For instance, the decay of a neutral kaon into a $\mu^+\mu^-$ pair is suppressed by several orders of magnitude with respect to the similar decay of a charged kaon in a $\mu\nu_\mu$ pair. Glashow, Iliopoulos and Maiani (1970) solved the problem with the introduction of another new particle, the charmed quark. With two doublets of quarks (u-d and c-s) in parallel to the two lepton doublets (ν_e-e and ν_μ-μ), strangeness may change only in the interactions mediated by the charged W, in line with the experimental observation.

The new form of "neutral current" interaction was discovered at CERN in 1973 by a group led by Lagarrigue using the big heavy liquid Gargamelle bubble chamber. The W and the Z (mediating the neutral current interaction) were then at bay. All that was needed, but it took 10 years to achieve, were high enough collision energies (several hundreds of GeV).

As we have said, the charmed quark and the strange quark make a new pair. It took a few more years for the charmed quark, with all its expected properties, to be discovered (1976). With the discovery of another lepton, the τ-lepton discovered at Stanford by Perl and coll., and then of the b-quark (Lederman and coll.) there was, however, no doubt any longer that a new pair of doublets was there, even though it took several years again to discover the top quark and the neutrino associated to the τ.

LEP (the Large Electron Positron collider), at CERN, was built as the best possible instrument to test the validity of the electroweak theory and also many aspects of the theory of the strong interactions at the quark level (Quantum ChromoDynamics). These theories altogether constitute the Standard Model of fundamental particles and fundamental interactions. LEP really vindicated the Standard Model with precise measurements of higher-order effects, testing the theory in great detail. LEP did not have enough energy to produce a top-antitop pair and the discovery of the top quark was made at the Fermilab collider. From the LEP results one could however predict the mass of the top quark (175 GeV) before it was observed.

LEP is now phased out to leave room for the Large Hadron Collider.

3. – Is the pion a fundamental particle?

We now turn to another important contribution of Fermi. This seminal work was done in Chicago in collaboration with C. N. Yang, in 1949. At that time theorists faced great difficulties in formulating testable predictions for strong interactions. The only solid tool available was perturbation theory. It worked remarkably well for the electromagnetic interaction, but failed miserably with the strong-coupling constant of meson theory. As Goldberger once said of this period, *"There was some excitement as the first experiments were being done. Then it was announced that the pion is a pseudoscalar and gloom again fell upon the theoreticians."* In the same paper, in the mid-fifties, Goldberger goes on discussing dispersion theory which, by the mid to late fifties, had already thrown some light on strong interactions.

Fermi's contribution shows once again his insight and his pragmatism. With the discovery of π-mesons, of strange mesons, of the strange baryons etc. the number of particles was simply growing too large to make it plausible that a fundamental field should be associated to each of them. Indeed, the proposal of the paper was that the pion might not be elementary after all, but rather a bound state of a nucleon and an antinucleon.

There was no possibility at the time to prove that fermion-antifermion interaction was attractive and fermion-fermion repulsive, let alone to perform an actual calculation of the binding energy. However, the idea would explain at least two mysteries. Assuming that the lowest bound state should correspond to the lowest possible orbital angular momentum, $L = 0$, the pion had to have negative parity, since the parities of the nucleon and of the antinucleon are opposite. The pions could also naturally exist in three charged states (π^+, π^-, π^0, a triplet of isospin) since these are the combinations obtained from nucleon-antinucleon pairs. One could thus make the economy of three fundamental particles. Pursuing the idea, one could expect an isoscalar state of spin zero and negative parity and also spin-one particles (spin triplet) in an isovector and an isoscalar versions. They are respectively the η, the ρ and the ω. One could anticipate the beginning of a rich spectroscopy of hadrons, which in fact started being discovered in the early sixties. The existence of the pion as a bound state does not impose their existence, but it makes it very likely.

In their effort to reduce the number of fundamental particles, Fermi and Yang were led to assume a point-like interaction (of the Fermi type) between the constituents of π-mesons, thus avoiding the introduction of the quanta of an inter-nucleon field. This hypothesis was abondoned later on, by Gell-Mann, in favour of a vector force. The elementary quanta of this field have been called "gluons", for their role of gluing together the elementary constituents. We know today that constituent interactions display a "hidden" symmetry, called colour, which plays a basic role in forbidding fermion-fermion and allowing fermion-antifermion bound states. As a consequence, a whole octet of gluons exist, related among themselves by the symmetry.

Sakata brought the Λ-hyperon into the picture, to carry the negative unit of strangeness, and extended the Fermi-Yang model to strange particles. The combinations $T\bar{T}$ (T denotes the generic element of the basic triplet P, N, Λ) could now describe the π-mesons, the K-mesons and their antiparticles, as well as two additional neutral particles, today identified with the η and the η', a total of 9 particles.

The thing that did not work, in the Sakata model, and which has much precluded the progress of this idea in the late fifties, was the structure of strange baryons, particles of spin-(1/2) which decay weakly into one nucleon and one pion, discovered in the meanwhile (sometimes also called hyperons, the word "baryon" including also P, N and Λ).

In the Sakata model, hyperons would be bound states of the type $TT\bar{T}$. However, among these combinations, there are states with positive strangeness, like, *e.g.*, PN$\bar{\Lambda}$, which have never been observed: all spin-(1/2) hyperons have been found to have negative strangeness. In addition, there was the "philosophical" problem that one could hardly see an objective difference between, *e.g.*, the proton —assumed to be elementary— and the allegedly composite pion. On the contrary, several elements pointed to the essential unity of particles sensitive to the strong interactions (collectively named "hadrons" by Gell-Mann) a concept which was being developed in the late fifties by G. Chew and collaborators, under the suggestive name of "nuclear democracy".

A crucial step forward was the Eightfold Way scheme of Gell-Mann and Ne'eman, based on the symmetry SU_3, the unitary transformations in a complex, three-dimensional space. The symmetry extends the isospin symmetry of Heisenberg (SU_2) to include strange particles. In the Fermi-Yang-Sakata model this symmetry is natural because there are three basic constituents, and the SU_3 transformations are nothing but unitary transformations of P, N and Λ among themselves (fermions are represented by complex fields, hence their three-dimensional space is complex). However, the basic idea was to put the symmetry on a more abstract basis: assume its existence and try to classify the particles according to SU_3 multiplets, without prejudices arising from what their composition would be, in terms of the Fermi-Sakata-Yang triplet. Surprisingly, it was found that mesons would fit in octet and singlet representations of SU_3 (like in the triplet model) and that the known baryons could be accommodated in octets and decuplets, the latter being an SU_3 representation never considered before. The octet includes P, N and Λ, at par with the hyperons and consistently with "nuclear democracy". The decuplet of spin-(3/2) resonances was incomplete at the time when the Eightfold Way was proposed. The discovery of the missing particle, the Ω, with exactly the predicted properties, convinced the physicists that they were on the right track! The situation was described by Gell-Mann in terms of a culinary analogy. In the French cuisine, to cook a pheasant one often puts two steaks around it. When the cooking is done, steaks are discarded (the P, N and Λ triplet) and the pheasant (the SU_3 symmetry) is kept.

However successful, the Eightfold Way did not answer the basic question: are all these particles, baryons and mesons, to be really considered elementary? The next step is due

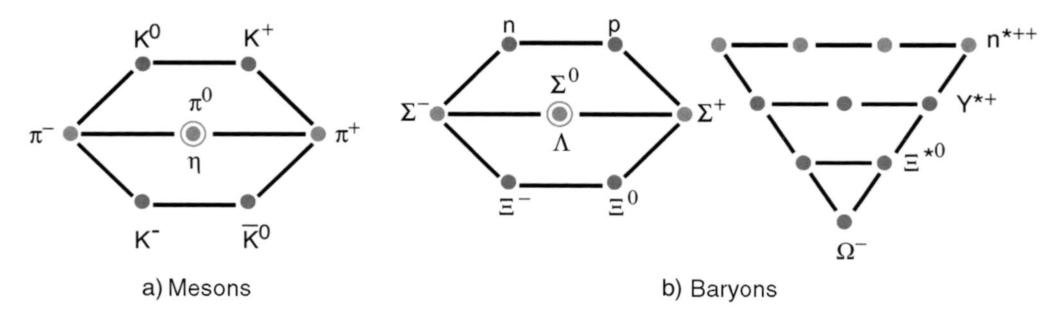

Fig. 3. – The distribution of the electric charge and strangeness for the lightest hadronic particles (mesons (a) and baryons (b)). The electric charge increases along the horizonatl direction and strangeness along the vertical direction, at steps of one unit. Except for proton and neutron, the value of the electric charge is indicated as a right superscript to the name of each particle. For mesons, positive and negative strangeness are distributed symmetrically with respect to the horizontal axis (assuming zero strangeness for π-mesons). On the other hand, assuming zero strangeness for proton and neutron, hyperons have all negative strangeness. The strangeness distribution for mesons is compatible with the Fermi-Yang-Sakata model, while the distribution for baryons is consistent only with the "eightfold way" scheme of Gell-Mann and Ne'emann and finds its explanation in the quark model of mesons and baryons (mesons $= \mathrm{q}\bar{\mathrm{q}}'$, baryons $= \mathrm{qq'q''}$ with $\mathrm{q} = \mathrm{u, d, s}$).

to Gell-Mann and Zweig who, independently discovered that the structure of mesons and baryons could be reproduced assuming them to be bound states of a fundamental triplet of spin-(1/2) fermions, quarks, denoted by u, d and s.

Mesons are quark-antiquark bound states (meson $= \mathrm{q}\bar{\mathrm{q}}$) while baryons are made by three quarks (baryon = qqq). The former statement requires the electric charge of u to be larger than one unit than the electric charge of d and s (similarly to the Fermi-Yang-Sakata triplet). However, to reproduce baryons, quarks have to have fractional electric charges (in terms of the proton charge), with $Q(\mathrm{u}) = +2/3$, $Q(\mathrm{d}) = Q(\mathrm{s}) = -1/3$. Fermi's view has now to be rephrased in terms of quarks instead of nucleons but the idea remains basically the same.

In the mid sixties the idea of quarks as constituents of the known hadrons was gaining currency. A detailed spectroscopy of the many particles known by then, in terms of quark bound states, was worked out by Dalitz, who had followed Fermi in Chicago and was then in Oxford, and by Gatto and his school in Florence.

In the mid fifties the idea of Fermi and Yang had been used by Goldberger and Treiman in their calculation of the decay rate of the charged pion, using the newly available technique of dispersion relations. The pion turns into a nucleon-antinucleon pair which, following Fermi theory, annihilates into a lepton pair. The universality of the Fermi coupling could naturally extend to π-meson decay.

However, and as just mentioned, it is only with the advent of the quark model that the Fermi picture found its completion. A π^+ is made of a u quark and of an anti d quark. Calculations at the quark level are still often impossible with good accuracy. Neverthe-

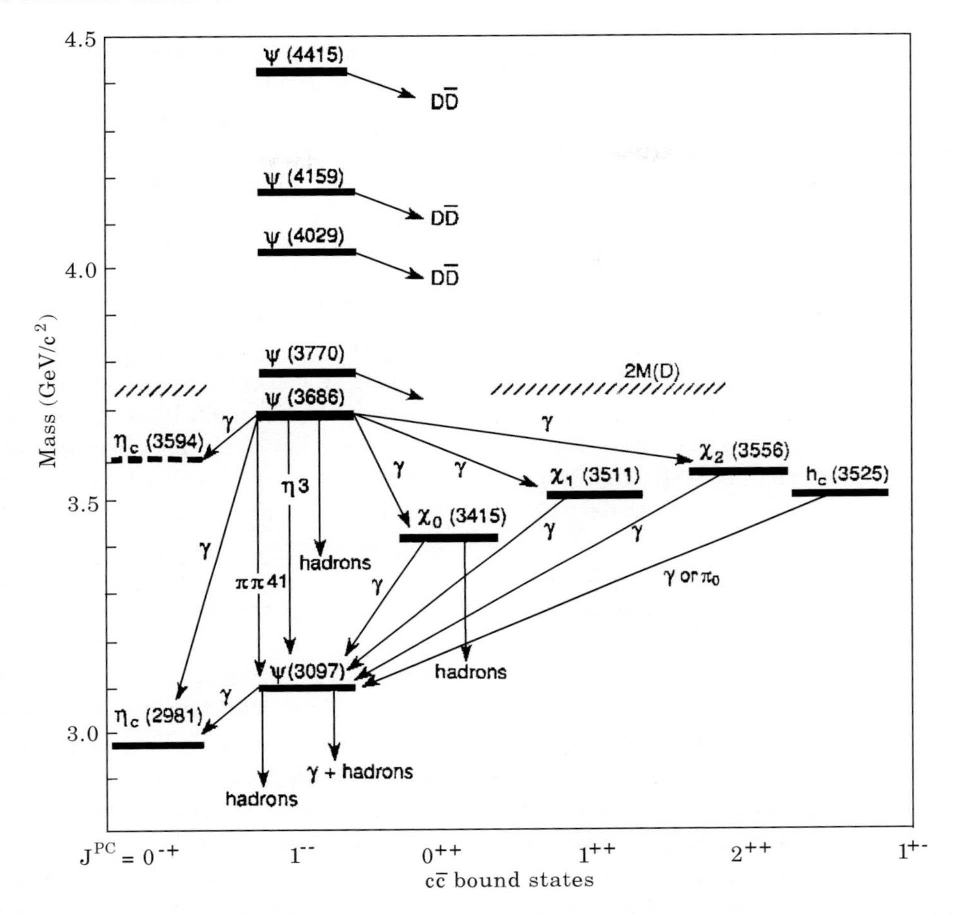

Fig. 4. – The charmonium state. Level structure and intensity of the radiative transitions between the different levels.

less, it works in some cases because of particular symmetry or kinematical configurations.

This quark structure of the pion can be actually used to calculate with accuracy the decay rate of the neutral pion into a photon pair. One can indeed show in this way that each quark has to exist under different varieties of "colour". Without the needed summation over "colours" one would underestimate the decay rate by a factor N^2, with N the number of colours. From this one derives that $N = 3$ is necessary.

The small mass of the u and d quarks involved in the π structure makes them highly relativistic and widely spread apart (over one fermi as they dress themselves with gluons and become more massive). A precise calculation of the properties of the bound state is still not possible in Quantum ChromoDynamics (QCD) the theory that describes the forces mediated by gluons and determined by the colour symmetry. This is however not the case for charmonium (fig. 4) a system made of a charmed quark and its antiquark.

In that case the quarks are very massive (1.5 GeV) and the bound system is very tiny and non-relativistic. A calculation of the energy levels of the bound state and of the radiative transition between them becomes then possible. One obtains a succession of levels and the system, once formed in an excited state, can, some of the times cascade down by photon emission before annihilating. This is the "atomic model" at the quark level, a vindication of Fermi's approach to meson structure.

4. – Pioneering work in meson scattering

The Yukawa theory had had many appealing features but a difficult start. A particle in the expected mass range had been discovered among the cosmic ray secondaries but it was weakly interacting. As demonstrated by Conversi, Pancini and Piccioni (a former Fermi student) in Rome, the muon seen in cosmic rays could not be the particle postulated by Yukawa, the pion, as mediator of the nuclear forces. Things clarified only after World War II. There existed actually two particles, the pion and the muon. The pion is produced in the interaction of cosmic rays and was eventually discovered by Powell and Occhialini but, if not readily absorbed in the atmosphere, it swiftly decays into a muon and a neutrino. With the pion established, with the properties expected from Yukawa theory, the study of nuclear interactions was ready for a new start and the more so that the new synchrocyclotrons could produce pions in great numbers. This was in particular the case for the one in Chicago used by Fermi. His extensive study of the π-nucleon interaction started in 1951 and this constituted the core of what he reported in his Varenna lectures.

Fermi had no time to write up his Varenna lecture notes, since he fell ill soon after returning to America. Sets of notes were written based on the tape recording of his lectures and edited by Feld, who had been following his class when he had arrived at Columbia. As Feld said *"It was with considerable misgiving that I agreed to edit this set of notes, based on the lectures of Enrico Fermi at Varenna...Many portions are, essentially word-for-word, in the original form. This is certainly not the form in which Fermi would have written them for publication, for his methods of oral and written presentation were very different. In their present form, however, they illustrate the unique qualities of Fermi as an expositor and teacher. In order to preserve this quality I have tried to minimise the changes from the original notes (which were based on tape recordings)...In working on these notes, I have been rewarded by frequently encountering sections which are so unique in their language and approach as to evoke for me, again, the picture of Fermi as he lectured during those lovely mornings in the beautiful setting of the Villa Monastero on Lake Como. It is to be hoped that these notes have succeeded, in some small measure, in capturing the spirit of those unforgettable lectures."*

As Feld says in his foreword: *"There was no foretaste of tragedy in the lectures themselves. Here was Fermi at the height of his powers, bringing order and simplicity out of confusion, finding connections between seemingly unrelated phenomena; wit and wisdom emerging from lips white, as usual, from contact with chalk, in that clear, resonant voice that had never lost the soft Italian vowel endings on a perfectly colloquial American delivery."*

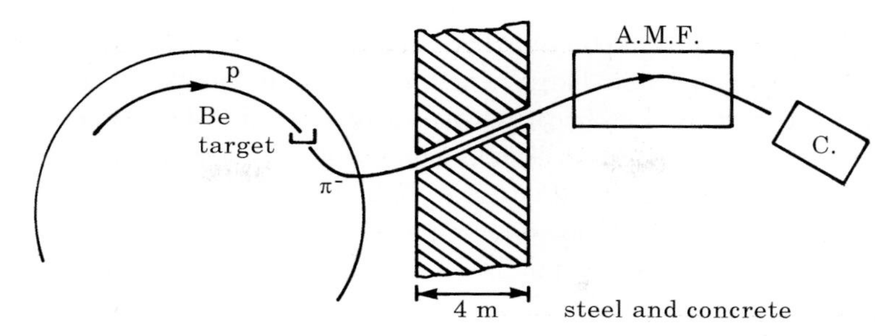

Fig. 5. – Fermi's schematic presentation of the set-up used to produce negative pions on the Chicago Synchrocyclotron. He has calculated the trajectories himself.

Fermi started on the theoretical side, explaining the basic of isotopic spin and its properties at describing the charge independence of nuclear forces. Heisenberg had introduced the concept of isotopic spin in 1932, immediately after the discovery of the neutron. The proton and neutron are thus seen as two states of the same particle, the nucleon, in much the same way as the spin-up and spin-down states can be taken as the two polarisation states of a spin-(1/2) particle. Rotations in charge space are put in exact parallel with rotations in ordinary space. This is still a matter of notations. However, as discussed later by Kemmer and Wigner, the charge independence property of the strong interaction, which was being recognised, corresponds to invariance under rotations in charge space. This is where actual physics comes into the picture. This invariance property (Isospin Symmetry) is not exact since electromagnetism distinguishes proton and neutron. Nevertheless, it is almost exact and a recognised property of the strong interaction among nucleons. Pauli matrices can be used to describe the charge states of nucleons, in much the same way as they allow one to deal easily with spin. In the two-nucleon case, one obtains a separation between the symmetric and the antisymmetric configurations, which are put in parallel with the triplet and singlet states of spin.

Fermi presents the formalism and shows how Isospin invariance corresponds to a symmetry property of the interactions among nucleons. He then continues extending the formalism to the π-nucleon states. He distinguishes the isospin-(3/2) from the isospin-(1/2) states, and relates them to processes where a π^+ scatters elastically on a proton (pure isospin 3/2) or a π^- scatters on a proton to yield either the same state or a neutron and a π^0 (a mixture of isotopic spin 3/2 and 1/2). The presentation of the formalism is then intertwined with experimental considerations on π-P scattering with pions produced from the synchrocyclotron and counter detection. Fermi had calculated himself their trajectories! (fig. 5). There are comments about the budding use of bubble chambers.

Fermi then introduces the partial waves amplitudes to be used in the phenomenological analysis that will transform the information in the data (angular dependence) into phase shifts for the different states of well-defined angular momentum, parity and isospin. The partial-wave analysis makes use of all conservation properties. The wave function for each of these states suffers only a phase shift (*i.e.* a multiplication by a phase factor $e^{i\delta_l}$) through the collision. Combining the differently phase shifted waves one reproduces the angular and charge distributions in the final state as they vary with incident energy. The energy of the pions was rather low and the S-wave ($l = 0$) could *a priori* have been considered as still dominant. But Fermi noted that "there is good evidence that the interaction in the P-state is abnormally strong so there is perhaps some point in including both S and P states in the analysis". As we know today, there is the famous 3-3 resonance (isospin 3/2-spin 3/2 and therefore at least P-wave) lurking there, even if the peak of the resonance could not be reached at the Chicago Synchrocyclotron (this particle is the non-strange member of the decuplet of resonances described in the previous section, see fig. 3).

Next comes a detailed study of the phase shifts and their interpretation in terms of the range of the interaction. This is a great piece in phenomenology.

Later on, in the study of the reaction where two nucleons bind into a deuteron as a pion is emitted, Fermi underlines the importance of the strong enhancement in the P-state, isospin-(3/2) state and he is led to assume that the emitting nucleon and the emitted pion are dominantly in a (3-3) state. The reader may feel that Fermi was procrastinating with the 3-3 resonance. Fermi knew about it, in particular from Bruckner, but he lectured as if he did not want to mention it as a reality, when others could have focused on it. This is most likely because Fermi knew too much about the ambiguities still present in the phase shift analysis. There was a strong enhancement in the isospin-(3/2) state and in the P-wave but one could not be certain yet that the corresponding phase shift was actually going through 90 degrees with an eventually falling partial-wave amplitude after the sharp rise. The peak, if any, was still beyond the limit of the experimental energy range available at Chicago. One may compare that with Gell-Mann's attitude with respect to Unitary Symmetry in the early sixties. He was apparently also procrastinating, slow to claim in print the validity of Unitary Symmetry because he knew so much about the data and had to convince himself that this was the only way to understand the physics.

Fermi, of course, knew about the 3-3 resonance which had been predicted in strong coupling model calculations which were however still relying on unjustified approximations. Nevertheless, throughout his notes he preferred to keep a strictly phenomenological attitude, simply noting the big increase of the isospin-(3/2) P-wave component in π-P scattering with increasing energy.

As Pais says in his book "Inward bound": *"When I met Fermi on the evening preceding the second Rochester conference (this was in 1952), he told me with enthusiasm of the work he and his collaborators were then engaged in at the 450 MeV Chicago synchrocyclotron: the pioneering pion-nucleon scattering experiments. He was particularly intrigued by indications that at their highest energies (about 140 MeV) the scattering appeared to proceed predominantly in the isospin-(3/2) state. When I asked him how he knew that, he replied that the cross-sections for the processes* $\pi+$p *elastic,* π-p *charge exchange and* π-p *elastic stood in ratios close to 9/2/1 and that these numbers follow from isospin considerations alone as long as isospin 3/2 dominates. I excused myself shortly afterward, Pais says, and went to my room to check this statement."* Fermi's report at the Rochester conference was received with great interest and triggered much activity using isospin symmetry in a number of processes. Since these symmetry considerations have nothing to do with perturbation theory, theorists could at long last propose valuable and valid relations.

The pion-nucleon cross-section was eventually found to peak at 180 MeV. A detailed analysis of the phase shift showed that the 3-3 phase shift was indeed going through 90 degrees. There was a genuine resonance in pion-nucleon scattering and others were to follow. One may then ask "what distinguishes this resonance from a new particle?" Is not everything happening as if a new particle was produced through the fusion of the pion and the nucleon that would then decay into a pion and a nucleon? That particle, the Δ, has a mass of 1230 MeV. Its lifetime is very short (about 10^{-23} second) and this corresponds to a rather large width (of the order of 110 MeV). Nevertheless strong interactions do not demand a longer time to manifest themselves. All what is needed is the time taken by light to cross a nucleon, which is typically one fermi wide. Such unstable particles cannot be considered as that special from the point of view of strong interactions. The Δ should therefore be considered just as fundamental as the nucleon. It has all the attributes of an unstable but still "elementary particle", namely a well-defined mass, spin, parity and isospin. We are now very familiar with this point of view. Nevertheless it took some years before the idea that there is no real difference between a resonance and an unstable particle became part of the physicist's thinking.

5. – The rise of Quantum Chromodynamics

When Fermi gave his Varenna lecture, there was still a strong current of thought to attach to the proton and the neutron an elementary label. There was some reluctance to introduce new elementary particles. A little later the Δ could be seen as a dynamical effect of the pion-nucleon interaction. It took a few years before it was realised that the Δ was actually as elementary (or non-elementary) as the proton. Whereas a Δ can be seen as a resonance in pion-nucleon scattering, the proton can equally be seen as a bound state in pion-Δ scattering. We touch here again what became in the sixties the concept of "nuclear democracy" and the basis for the Reciprocal Bootstrap, a very fruitful and interesting idea whereby all hadrons are dynamical effects resulting from the interactions of all hadrons.

By the early sixties the hadronic world just exploded. The advent of higher-energy accelerators (the Brookhaven Alternating Gradient Synchrotron and the CERN Proton Synchrotron, both reaching proton energy above 25 GeV) and the development of bubble chambers, particularly at Berkeley with the group of Louis Alvarez, led to the discovery of many new particles. Strong-interaction physics became dominated by these resonances. One could even forget about the detailed and patient phase shift analyses, which actually become very cumbersome when inelastic channels open up. It was only necessary to check that a peak in the cross-section could be associated with a well-defined set of quantum numbers (again mass, spin, parity and isospin) to enter it as a bona fide new particle on the expanding list of elementary particles. Models were developed to deal with this many-particle world, the latest step along that line (with an infinite number of resonances!)

being the Veneziano model of the late sixties. The catchword was "bootstrap", whereby particles exist as a result of the interaction of all other particles.

The quark approach, whereby mesons and baryons were composites of a small number of quarks, eventually won the day. However, as it often happens in physics, the bootstrap approach found a brilliant future along different lines. The Veneziano model had become the first step in what became known as the dual models. As these models could not incorporate the quark structure, they had eventually to be abandoned for a deeper study of the strong interaction, as thereafter pursued with success in the framework of Quantum ChromoDynamics. Yet, dual models eventually led to the impressive theoretical development known today as string theory. They no longer refer to the strong interaction and to the hadronic scale of 150 MeV but to the gravitational interaction and to the Planck scale of 10^{+19} GeV.

The quark approach turned out to be the right one. The Δ is now made of three quarks, just as the proton. But then comes a riddle. The Δ has isospin 3/2 and there is, therefore, a doubly charged Δ made of three identical quarks: $\Delta^{++} = \mathrm{uuu}$. To produce a spin-(3/2) state, the three quarks have to have their spin-(1/2) aligned. So far so good but should we not expect that the Δ; which is the lowest mass state of that type, be made of three quarks in an S-state? And one may then ask, what about Fermi's (once again his legacy) statistics? The three fermions should be in an antisymmetric state while everything in the quark construction of the Δ gives a symmetric configuration! The problem does not occur in such a blatant way with the nucleon since two of the spins have to be in an antisymmetric state, but for the Δ, it cannot be escaped. The answer lies with the "colour" of the quarks. As we said earlier, the neutral pion decay gives evidence for quarks to exist in three "colours". The same concept helps us here: the three quarks that build the Δ (and the nucleon) are in a fully antisymmetric state with respect to "colour". Combined with full symmetry in all other quantum states (quark type and spin) this allows a fully antisymmetric state, as required by Fermi statistics.

Colour helps us also in understanding why only certain configurations of quarks are bound. The forces associated to colour increase very strongly with the distance, so strongly that as we try to separate two quarks, quark-antiquark pairs are created by the interaction that screens the colour of the original particles. The QCD vacuum is "opaque" to colour thus preventing the free propagation of any isolated quark. This explains why free quarks have never been observed. But there is more to it. Han and Nambu first found that only "colour neutral" states can have a finite energy. Therefore we may understand in this way why the configurations $\mathrm{q\bar{q}}$ and qqq are bound, since they can form colour singlets, but simple configurations like qq, which cannot be colour neutral, are not bound.

The quest for particles corresponding to other colour singlet quark configurations (for example exotic mesons = $\mathrm{q\bar{q}q\bar{q}}$) or to states made solely of gluons is still going on.

6. – Chiral dynamics

Let us go back to the legacy of Fermi's work on π-nucleon scattering. The guide line is Isospin Symmetry, an *a priori* valid theoretical assumption for this scattering process, with a well-defined theoretical machinery which he recalls, showing how to check it in that particular case. The key message of Fermi was indeed the great use which one could make of isospin symmetry, as developed by Wigner, and this irrespectively of the details of the dynamics which were still unknown. The next big step along that line was the advent of the wider symmetry, SU_3, and of the Eightfold Way described previously. Its main architect, Gell-Mann, had been much influenced by Fermi as a young researcher, while in Chicago.

One can see the practical use of Unitary Symmetry as the continuation of Fermi's work with π-nucleon scattering, generalising the use of Isospin symmetry. Many relations imposed by the symmetry could be exploited independently of any detailed knowledge of the dynamics.

In isospin, SU_2, symmetry, the lowest representation (doublet) occurs in Nature. The nucleons illustrate it. In the exact symmetry limit, nucleons have to have the same mass, which is not far from reality, as the neutron-proton mass difference is only about 0.15% of the total mass. As we understand it today, Isospin Symmetry results from the fact that the quark dynamics, based on "colour" exchange according to Quantum ChromoDynamics, simply ignores the type ("flavour") of the quarks. With quark masses for the u and the d which are almost identical (a few MeV difference) when considered at the scale of 150 MeV, typical of hadronic size and hadron excitation levels, Isospin Symmetry should then be very good indeed. It would be perfect in a world where the u and d quark would have the same mass. The strange quark is more massive, with a mass of the order of 150 MeV. As a result Unitary Symmetry is much less accurate. There is no practical interest to consider a global SU_4 symmetry, including particles with a very massive charmed quark. We should rather use Isospin (and Unitary) Symmetry among charmed particles, made by the charmed quark in combination with lighter quarks or antiquarks.

It is interesting to consider a world where quark masses would be *exactly zero*. In this case, one could introduce a further symmetry, the so-called Chiral Symmetry, which transforms among themselves the (massless) quarks of negative and positive helicities and, separately those of negative helicities. We have noted already that the helicity of a massless particle is a conserved quantity and, being the same in all reference systems, can be used to characterise different particle states. If we restrict to u and d quarks, chiral symmetry extends the symmetry group SU_2 to $SU_2 \times SU_2$ (independent isospin transformations on the two sets of positive and negative helicity states).

Chiral Symmetry leads to very interesting developments. If simply valid as such in the real world, it would imply either a massless proton or it would bring the proton within a larger multiplet, together with particles with the same spin but with opposite parities. However (much alike the situation we met with positive strangeness states in the Sakata model) the opposite-parity partners of the proton simply do not exist!

There is another possibility, fortunately, namely that the symmetry is, as one says, *spontaneously broken* by a lack of symmetry of the vacuum. The study of this alternative has been of crucial importance to the understanding of the basic interactions and we may try to illustrate it.

One can consider the lowest energy state of the theory (the vacuum) as filled with a uniform density (a "condensate") of massless quark-antiquark pairs. Since particles in that condensate are formed of u and d quarks of opposite helicities, the presence of this condensate spoils the global chiral symmetry to leave only an explicit Isospin Symmetry. The symmetry is not lost however. As a consequence of the lack of invariance of the vacuum state, a set of massless particles must exist, called the Goldstone bosons. In scattering reactions, Goldstone particles are emitted and/or absorbed and the symmetry implies typical relations between the corresponding amplitudes. These relations replace the symmetry relations first explored by Fermi for isospin related processes and are equally significant to establish the symmetry.

When we add an explicit breaking of the chiral symmetry, in the form of a small quark mass, the Goldstone bosons become slightly massive, with the square of their masses being proportional to quark masses. In this way, departure from exact but spontaneously broken chiral symmetry is associated to the square of the pion mass, which indeed is a small quantity on the natural hadronic mass scale provided by the proton's mass. The square of the ratio of pion to proton mass is about 1/50.

The notion of a spontaneously broken chiral symmetry was established by the end of the sixties. The picture that emerged was that u and d quarks are very light indeed, with masses of the order of a few MeV, as determined from the pion mass. u and d quark mass differences are of the same order of their absolute values, so that chiral $SU_2 \times SU_2$ is almost as good as the classical SU_2, only less apparent because of the non-invariance of the vacuum!

Chiral $SU_3 \times SU_3$ is considerably worse, as the s-quark mass, of about 160 MeV, is much closer to the typical hadronic scale and symmetry-breaking effects tend to obscure the simple picture of a spontaneously broken symmetry. For the heavier quarks, chiral limit is too far to provide a useful description of their strong interactions (but the situation is different for the weak interactions, as discussed below).

The standing of this spontaneously broken symmetry turns out to be in a rather good shape. It no longer refers to a world where the nucleon mass would be zero, but to a world where the pion masses would vanish, which is not too far from reality. We shall stop here since we are already far away from pion-nucleon scattering but, with the massless quark-antiquark bound states in that condensate, we may be touching the very nature of the pion as first glanced at in the Fermi-Yang paper, about fifty years ago.

7. – The Standard Theory of particle interactions

The final outcome of the investigations described in this essay goes under the low-profile name of Standard Theory. In fact, the Standard Theory represents a most remarkable synthesis, able to describe all particle world —but gravity— in excellent agreement with the available experimental information. The Standard Theory was delineated at the

TABLE II. – *Fundamental Particles of the Standard Theory.*

Quarks and leptons		
$\begin{pmatrix} u_{r,g,b} \\ d_{r,g,b} \end{pmatrix}$	$\begin{pmatrix} \nu_e \\ e \end{pmatrix}_L$	$e_R,\ (u_{r,g,b})_R,\ (d_{r,g,b})_R$
$\begin{pmatrix} c_{r,g,b} \\ s_{r,g,b} \end{pmatrix}_L$	$\begin{pmatrix} \nu_\mu \\ \mu \end{pmatrix}_L$	$\mu_R,\ (c_{r,g,b})_R,\ (s_{r,g,b})_R$
$\begin{pmatrix} t_{r,g,b} \\ b_{r,g,b} \end{pmatrix}_L$	$\begin{pmatrix} \nu_\tau \\ \tau \end{pmatrix}_L$	$\tau_R,\ (t_{r,g,b})_R,\ (b_{r,g,b})_R$

Force related particles	
Gravitational:	graviton
Weak:	intermediate vector bosons, W^-, Z Higgs-Brout-Englert boson (?)
Electromagnetic:	photon
Strong:	gluons, $(g)^{r,g,b}_{r,g,b}$

Note 1: The indices L and R denote particles with different helicity (respectively left- and right-handed); for antiparticles, helicity is flipped (*e.g.* left-handed electron and right-handed positrons are represented by the same symbol). Quarks come in different "colours", conventionally denoted as red, green and blue (r, g, b). Gluons are identified by ordered pairs of colour indices; pairs with unequal indices are all allowed, but only 2 out of the 3 combinations with equal colour (rr, gg, bb) are physically allowed, giving a total of 8 gluons out of the 9 possible combinations.
Note 2: Known stable matter (nuclei, atoms) is made out of particles of the first generation. All particles above have been positively observed except for the Higgs boson (see text) and the graviton. Emission of gravitational waves, the classical counterpart of gravitons, has been indirectly observed by the secular change in the revolution frequency of certain binary stellar systems.

beginning of the '70s and was experimentally tested extensively throughout the remaining part of the century.

The fundamental particles of the Standard Theory are illustrated in table II.

Quarks of six different types (flavours) make up all the observed hadrons, with their colour neutral combinations, $q\bar{q}'$ and $qq'q''$. Strong inter-quark forces are mediated by gluons, vector particles associated to the colour symmetry. Colour forces become very strong at large distances, which allows colour neutral states only to be able to freely propagate in vacuum and explains why we cannot observe free quarks or free gluons. The same forces decrease in intensity at small distances, which explains why we do see quarks in high-energy, large-angle collisions.

Six lepton flavours appear in parallel to the quarks. Leptons do not carry colour. Therefore, like the electron, they are insensitive to the strong interactions. Neutrinos, being in addition electrically neutral, respond only to the weak forces.

In the making of the hadrons, each quark flavour participates with four equivalent states: particle and antiparticle, each with two possible helicities. On the other hand, weak forces act differently on the different states. This is why we have distinguished in the table between left-handed and right-handed particles.

Emission and absorption of charged intermediate vector bosons, $W^{\pm}$, occur only between particles in doublets. This is, of course, the final evolution of the Fermi idea that the basic weak transition is the proton-neutron transformation. At quark or lepton level, the basic processes are $u \longleftrightarrow dW^{+}$ and $\nu_e \longleftrightarrow e^{-}W^{+}$, for the first generation, and similar ones for the higher generations. Other allowed processes are obtained by moving particles on the two sides of the equations (when crossing a particle one should replace it with the corresponding antiparticle, *e.g.* $\bar{d} \longleftrightarrow \bar{u}W^{+}$).

For light quarks and leptons, the emitted W cannot leave the interaction region, because of energy conservation. Rather, it has to materialise into a fermion-antifermion pair (*e.g.* $W^{-} \longleftrightarrow \bar{\nu}_e e^{-}$) thus giving rise to Fermi-type weak decays, *e.g.*: $\mu \rightarrow \nu_\mu e \bar{\nu}_e$ or $c \rightarrow s e^{+} \nu_e$.

The top quark is heavy enough to decay into a b quark, its partner in the doublet, and a W, thus revealing the basic weak interaction.

The foundation of the unified theory of the weak and electromagnetic interaction lies in the symmetry of quarks and leptons under transformations which treat differently left-handed (L) and right-handed (R) states. In the terminology of the previous Section, it is a chiral symmetry, and it is therefore incompatible with quark and lepton masses, which are on the other hand required by the observation. The way out from this conflict is again the notion of spontaneous symmetry breaking.

The solution provided by the Standard Theory is radical and simple. A scalar field, the Higgs field, "condensates" in the vacuum. The Higgs field couples in a different way to L and R fermions, so that this condensate breaks the invariance of the vacuum under chiral symmetry. This is much alike the condensation of quark-antiquark pairs to explain the proton's mass, and similarly to this case, it provides the desired mass to the quarks and leptons. However, here there is an important departure with respect to the previous case. The electroweak symmetry applies to transformations which are space-time dependent (gauge transformations). It is this special invariance which leads to the existence of vector particles, the photon and the intermediate bosons W and Z, which in turn mediate the electroweak interactions.

We have mentioned, in the previous case, that the signal of spontaneous symmetry breaking is the appearance of an exactly massless particle, the Goldstone boson. In the case of a gauge symmetry, it was discovered by Higgs and, independently, by Brout and Englert, that there is no Goldstone boson, rather the vector particles corresponding to the symmetry transformations broken by the condensate, acquire a mass. This is precisely what is needed to generate a mass for the W and Z, leaving the photon massless (in correspondence to the exact conservation of the electric charge). The Higgs field does not disappear completely, however, but it leaves a physical particle, the Higgs boson, coupled to the particles of the theory in a prescribed way.

The Higgs-Brout-Englert mechanism has been a key discovery to formulate a theory of the electroweak interactions, which is at the same time theoretically consistent with the gauge symmetry and in agreement with the observed particle masses. This explains the interest to find the Higgs boson, the real sign that we are indeed on the right track.

The Higgs boson is actively searched at the present accelerators. As of today, no convincing signal of a Higgs boson has been found with LEP, which implies that its mass has to be larger than about 114 GeV.

APPENDIX

The Fermi maximal accelerator ([1])

A well known story in all Physics Departments in Italy, was that Fermi had considered a maximal proton accelerator with a ring going around the Earth.

This story emerged again recently at a CERN seminar, when F. Gianotti showed the logo of the Very Large Hadron Collider project, fig. 6.

A brief search carried out at the Chicago University Library gave interesting results.

The figure is in fact fig. 6 of the talk "What can we learn with High Energy Accelerators?", that Enrico Fermi presented at the American Physical Society, on January 29, 1954. Fermi was then leaving the APS Chair which he had taken during 1953. The University of Chicago library has short personal notes that Fermi wrote for the talk as well as the slides of the figures.

With his typical insight, Fermi considered from the outset a proton accelerator with the Earth's radius as the maximum reachable. Assuming a magnetic field of 2 Tesla, this gives an energy: $E_{\text{Max}} \approx 5 \times 10^{15}\,\text{eV}$. It is the energy of the cosmic rays around the "knee", the most energetic cosmic rays that can be accelerated by the galactic magnetic clouds, according to Fermi's ideas developed in the very same years. By extrapolating with the available Livingstone plots of energy or money *vs.* time, Fermi concluded that this energy could be reached in the year 1994, at a cost of about 170 B$.

What to do with the high energy? Fermi underlines the difficulty of looking into a *"very, very cloudy crystal ball"*. He mentions the observation of antinucleons, the puzzle of the long lifetime of strange particles (high angular momentum barrier? associated production *"at present more probable"*?), the need for precision measurements. But also the possibility of *"a lucky break, or theoretical leap, or more probably a combination of hard work, ingenuity and a little bit of good luck"*.

All that and much more did in fact happen from the fifties until now in High Energy Particle Physics. Progress is exemplified in fig. 7, by the chart of what are now considered the elementary constituents of matter, the three generations of quarks and leptons.

It must be said that Fermi's successors did not fare badly at all. Transforming back to fixed target energy, the Tevatron (2 TeV in the c.o.m.) has reached $2 \cdot 10^{15}\,\text{eV}$ in 1987. LEP and HERA have explored about the same energy range with the electron-positron probe (something unthinkable at Fermi's times) and the electron-proton probe.

The LHC ($B \approx 9$ Tesla) will reach $1 \times 10^{17}\,\text{eV}$ in 2006 (50 times E_{Max}) at an all-out cost of about 5 B$. If the VLHC which is being considered today at FermiLab or the

([1]) Extracted from a talk given at the Conference: *Neutrino Telescopes 2001*, Venezia, 9 March 2001).

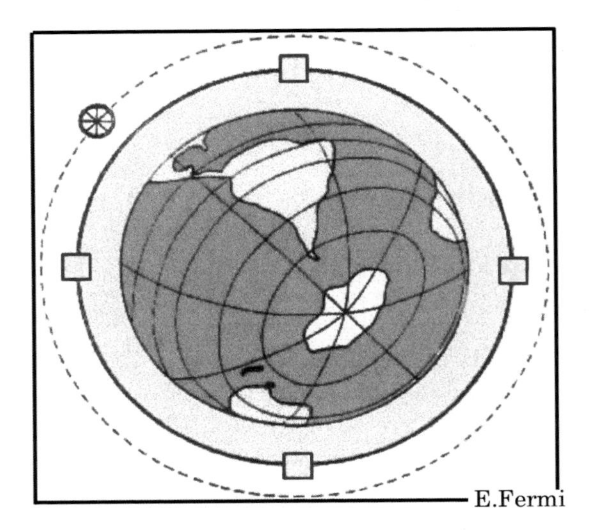

Fig. 6. – The logo of the Very Large Hadron Collider project at FermiLab.

Eloisatron proposed by INFN will materialise, with a c.o.m. energy of 100 TeV and a corresponding fixed target energy around 10^{19} eV, mankind will have been able to produce collisions at an energy equivalent to that of the highest energy cosmic rays that

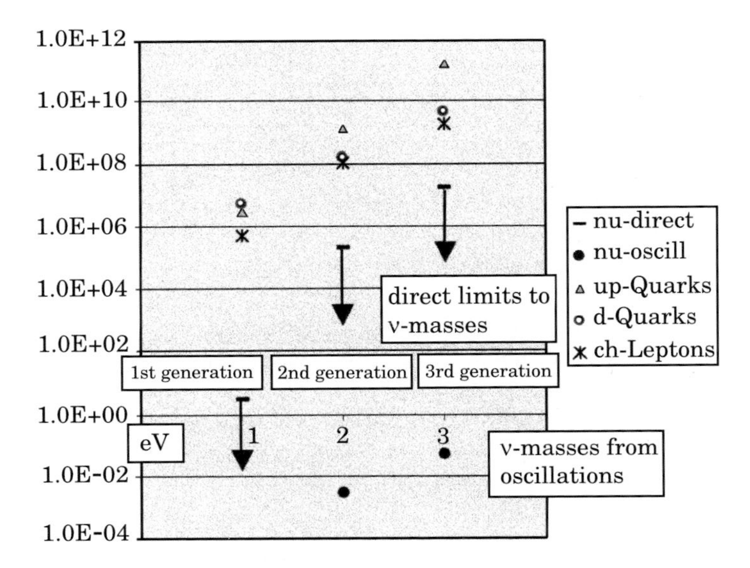

Fig. 7. – The lepton and quark mass spectrum. Upper bounds to neutrino masses are taken from beta decay spectra; estimates of ν_μ and ν_τ, masses are from solar and atmospheric neutrino oscillations.

can originate from nearby galaxies, that is those below the GZK cut-off due to the onset of hadron photoproduction in the scattering of cosmic ray protons off the microwave cosmic background photons.

The key to high energy and relatively low cost (very low indeed, compared to Fermi's extrapolation) is of course technological innovation, above all the invention of the colliders: electron-positron (AdA, Frascati, 1962), proton-proton (ISR, CERN, 1971), proton-antiproton (S$p\bar{p}$S, CERN, 1981). We must do all we can to keep innovation going on in the particle accelerator world.

What's next in Particle Physics? In brief, the Standard theory gives an accurate picture of the elementary particle interactions, but it leaves several open questions. The Higgs boson has not yet been observed. The naturalness of Standard Theory parameters requires a change at a critical energy around 1 TeV. Neutrino masses are exceptionally small, on the scale of other constituents masses, and may indicate new interactions at very high energy. The first two issues call for accelerator searches in the High Energy Frontier, at the Tevatron, the LHC and beyond. Neutrino masses are investigated via the phenomenon of neutrino oscillations, probably seen in solar and atmospheric neutrinos but to be confirmed with long-base line neutrino beams and dedicated neutrino factories, as well as more advanced non-accelerator facilities (SNO, Borexino, Kamland).

In addition, the merging of the Standard Theory with gravity requires additional (curved) space-dimensions. Recent speculations raise the possibility that the new dimensions may turn up already in the TeV region, again an objective for high energy accelerators.

We cannot say precisely what, but there should a lot to be found, with "a lucky break, or theoretical leap, or more probably a combination of hard work, ingenuity and a little bit of good luck".

* * *

I would like to thank FABIOLA GIANOTTI for calling my attention to the VLHC logo and the Chicago University Library for making available the unpublished notes and figures of the Fermi APS Seminar. ADRIENNE KOLB, MARK OREGLIA and JAMES PILCHER have made the reconstruction of the story possible, with their prompt interest.

About the Authors

MAURICE JACOB, a theoretician in the field of particle physics, has been a member of the CERN Theory Division for 31 years, and chaired that division in the '80. He worked also at Brookhaven, SLAC and Fermilab. He retired in 1998 after spending the last 5 years at CERN as advisor to the Director General for the Members States. He has been president of the French Physical Society in the mid '80s and of the European Physical Society at the beginning of the '90s. He has been editor of the journal "Physics Letters" and "Physical Reports". He chaired the Fundamental Physics Group of ESA, there, he has also been a member of the "Space Science Advisory Committee" in the second half of the '90s. He is known for his helicity formalism (with G.C. Wick) and for his jet study in hadron collisions.

LUCIANO MAIANI was born in Rome on July 16, 1941; after taking his degree in physics, he has been researcher at the "Istituto Superiore di Sanità" in Rome, and began his career in theoretical physics with the group of R. Gatto at the University of Florence. He has been post PhD Fellow at Harvard University and Visiting Professor at Ecole Normale Supérieure in Paris and at CERN (the European Organization for Nuclear Research) in Geneva, where he has been Director General until the end of 2003. He is full Professor of Theoretical Physics at "La Sapienza" University in Rome since 1976, and has been President of INFN (Istituto Nazionale di Fisica Nucleare) in 1993-1998. He published more than one hundred scientific papers in the field of theoretical particle physics; together with Glashow and Iliopoulos he theoretically predicted the "charm" particle, that has been fundamental for the formulation of the unified theory of electroweak interaction. He is a member of "Accademia dei Lincei" "Accademia Nazionale delle Scienze detta dei XL" and of the Russian Academy of Sciences. He received a honorary degree in Physics by the Universities of Aix-Marseille and St. Petersburg, by the Slovenian Academy of Sciences and Warsaw University. He was awarded the Matteucci Medal by "Accademia Nazionale delle Scienze detta dei XL" and won the J. Sakurai Prize of the American Physical Society".

Enrico Fermis's contribution to non-linear systems: The influence of an unpublished article

Massimo Falcioni and Angelo Vulpiani

1. – Introduction

Among Fermi's scientific articles "Studies of non-linear problems", written in collaboration with J. Pasta and S. Ulam, is considered as being particularly important [1]. Following a common procedure among the people working in dynamical systems we shall refer to this work using the acronym FPU. The article, written as an internal report of the Los Alamos Laboratories, was completed in May 1955, after Fermi's death, but appeared for the first time in 1965, as contribution N. 266 to "Note e Memorie" (Collected Papers), an anthology of Fermi's writings edited among others by Emilio Segrè and published jointly by the Accademia dei Lincei and the University of Chicago [2].

FPU was decisive in the development of many important fields of research, for instance, the studies of integrable systems and dynamic chaos. In his comment on FPU, written for the publication of "Note e Memorie", S. Ulam reports Fermi's opinions on the great relevance of the results and, more generally, on:

a) the importance of a systematic study of non-linear systems

b) the necessity of using numerical simulations as a research tool complementary to theoretical studies or laboratory experiments.

However, we should emphasize that the importance of the work was not immediately recognized by the physics community, so much so that only three lines are devoted to FPU in the only scientific biography about Fermi, written by E. Segrè [3].

In this chapter we shall discuss FPU in some detail within the framework of those problems it originated from, namely the foundations of statistical mechanics, and the new research fields which, following its lead, have developed since the sixties (non-linear mechanics, chaos, integrable systems and numerical simulations).

The layout of this chapter is as follows: in sect. **2** we discuss the general issue of the foundations of statistical mechanics and the ergodic problem. Section **3** is devoted to a presentation of the state of the art of the ergodic problem and, more generally, of the "thermalization" of non-linear systems before FPU. In sect. **4** we will discuss the FPU results. Section **5** will dwell on some developments resulting from this work: chaos in Hamiltonian systems, integrable systems (solitons) and numerical simulations (molecular dynamics).

2. – A brief survey of the foundations of statistical mechanics

Macroscopic systems contain a very large (on the order of Avogadro's number) number of particles; this implies the practical necessity of a statistical description, using "statistical ensembles", *i.e.* more precisely, probability distributions in phase space.

Denoting with $\mathbf{q}_i$ and $\mathbf{p}_i$ the position and momentum vectors of the i-th particle, the state of an N-particle system at time t is described by a vector $\Gamma(t) \equiv (\mathbf{q}_1(t), \ldots, \mathbf{q}_N(t), \mathbf{p}_1(t), \ldots, \mathbf{p}_N(t))$ in a $6N$-dimensional space which is called phase space. The system observables are described by the functions $A(\Gamma)$, defined in phase space. The particles follow the deterministic laws of classical mechanics (the quantum case does not change the problem significantly), hence $\Gamma(t)$ evolves in phase space according to Hamilton equations. In the following we shall always assume that the Hamiltonian does not depend explicitly on time, hence the energy will be conserved during the motion, which will occur on a constant-energy hypersurface. If $V(\{\mathbf{q}_j\})$ is the interaction potential between particles, the Hamiltonian can be written as

$$H = \sum_{i=1}^{N} \frac{|\mathbf{p}_i|^2}{2m} + V(\{\mathbf{q}_j\}) \,, \tag{1}$$

and the evolution equations are

$$\begin{aligned} \mathrm{d}\mathbf{q}_i/\mathrm{d}t &= \partial H/\partial \mathbf{p}_i = \mathbf{p}_i/m \,, \\ \mathrm{d}\mathbf{p}_i/\mathrm{d}t &= -\partial H/\partial \mathbf{q}_i = -\partial V/\partial \mathbf{q}_i \,, \end{aligned} \tag{2}$$

with $i = 1, \ldots, N$. Here it is important to note that the macroscopic observations time scale is much larger than the microscopic dynamics time scale, over which molecular changes take place. This means that an experimental measurement is actually the result of a single observation during which the system goes through a very large number of different microscopic states. If the measurement refers to the $A(\Gamma)$ observable, it must be compared with an average performed on a very long time (from the microscopic point of view):

$$\overline{A}(t_0, T) = \frac{1}{T} \int_{t_0}^{t_0+T} A(\Gamma(t))\mathrm{d}t \,. \tag{3}$$

The calculation of the time average $\overline{A}$, in principle, requires both the knowledge of the complete microscopic state of the system at a given instant and the determination of its trajectory. This is evidently an impossible requirement so that, if $\overline{A}$ depended too strongly on the initial state, not even statistical predictions could be made, even neglecting the difficulty of integrating the system (2). The ergodic hypothesis makes it possible to overcome this obstacle [4]. It essentially assumes that any energy hypersurface is totally accessible to any motion with that energy; in other words, a constant-energy hypersurface cannot be subdivided into (measurable) regions each containing complete motions, *i.e.* invariant under time evolution (if this condition is satisfied the hypersurface is called *metrically non-decomposable or metrically transitive*). Furthermore, for any trajectory the average permanence time in a given region is proportional to the region volume. If the preceding conditions, which make up the core of the ergodic hypothesis, are satisfied, it follows that for sufficiently large T the average (3) depends only on the system energy and hence it has the same value for all evolutions with the same energy; furthermore, this shared value can be obtained from an average of $A(\Gamma)$ in which all, and only, the states with the given energy contribute equally. The uniform probability density on the constant-energy surface defines the microcanonical measure, or microcanonical ensemble; if $P_{\rm mc}(\Gamma)$ is such density, from the ergodic hypothesis it follows that

$$\overline{A} \equiv \lim_{T\to\infty} \frac{1}{T} \int_{t_0}^{t_0+T} A(\Gamma(t))\mathrm{d}t = \int A(\Gamma) P_{\rm mc}(\Gamma)\mathrm{d}\Gamma \equiv \langle A \rangle \,. \tag{4}$$

We note that the validity of such an equation eliminates the necessity of determining an initial state of the system and solving the motion equations. Whether eq. (4) is valid or not, *i.e.* whether it is possible to substitute the temporal average with an average in phase space constitutes the ergodic problem. If an isolated system can be described by the micrononical ensemble, it is not difficult to show that a system in thermal contact with a reservoir can be well described by the canonical ensemble, so that the demonstration of the validity of (4) provides the dynamical justification for the introduction of statistical ensembles. From the above considerations about the metrical non-decomposability of constant-energy surfaces, it becomes clear that the study of the validity of (4) and the search for constants of the motion other than energy are closely related. If, in fact, first integrals, besides energy, existed, the system would be non-ergodic since it would not be metrically transitive: choosing one of the first integrals as the observable A, one would have $\overline{A} = A(\Gamma(0))$, which depends on the initial condition $\Gamma(0)$ hence, in general it would be different from $\langle A \rangle$.

The ergodic problem originates (together with the ergodic hypothesis) from L. Boltzmann's ideas about statistical mechanics and has since been studied in general mathematical terms mainly by J. von Neumann and D. Birkoff [5]. We shall not discuss the details of the ergodic theory in all its generality, instead, we shall limit our discussion to some of its aspects in connection with statistical mechanics and systems with many degrees of freedom.

There exists an approach (its major proponents are A. I. Khinchin [6] and L. D. Landau [7]) which considers the whole issue of ergodicity as being substantially irrelevant when dealing with statistical mechanics, since the ergodic hypothesis would not in fact be necessary to justify (4) for physically relevant observables. This approach is based on the following facts:

a) in the systems which are of interest to statistical mechanics the number of degrees of freedom is very large;

b) the interesting statistical mechanics issue is the validity of (4) for just the thermodynamically relevant observables.

The conclusions of this approach to the ergodic problem can be summarized in the following theorem [6, 8]:
If A can be expressed as the sum of N functions, each depending on the single-particle variables ($A = \sum_{i=1}^{N} f(\mathbf{q}_i, \mathbf{p}_i)$), then the relative measure (i.e. the probability with respect to the microcanonical measure $P_{\rm mc}$) of the constant-energy hypersurface points, for which the inequality

$$\left| \frac{\overline{A}}{\langle A \rangle} - 1 \right| > kN^{-1/4} \tag{5}$$

(where k is a constant independent of N) holds, is a negligible quantity on the order of $N^{-1/4}$.
Essentially, in the limit $N \to \infty$, (4) is valid (apart from a region in phase space which becomes even smaller with increasing N) for an interesting class of functions, independently of the dynamics details.

The requirement of ergodicity (the validity of (4) even in its weaker form, *i.e.* for particular observables and in the $N \to \infty$ limit) is not by itself sufficient to insure a statistical "good behavior" in a Hamiltonian system described by eq. (2). A further reasonable requirement is that the times necessary to reach the statistical equilibrium (*i.e.* those times which are long from the microscopic point of view so that the temporal average (3) is close to the ensemble average) still be short enough from the macroscopic point of view. It is expected that in "reasonable" times the system loses memory of its initial state and this can be obtained if it can be ensured that two phase space trajectories which start from nearby states rapidly diverge; such behavior is typical of chaotic systems. In these case evolutions are extremely sensitive to the initial state.

A possible characterization of the approach to equilibrium can be given by the concept of "mixing" [5], which was proposed initially by W. Gibbs. An intuitive idea of how equilibrium is approached can be given by considering an ensemble of identical systems, mental copies of the same system, whose representative points (the different Γ's) at a given instant are concentrated in a small region F of the constant-energy hypersurface. It is natural to expect (in agreement with Gibbs) that after some time the evolution will lead to a practically uniform distribution on the hypersurface, *i.e.* the microcanonical measure. The rapidity of the redistribution gives us the idea of the equilibrium relaxation times. More strictly speaking, the mixing can be defined as follows: a dynamical system, *i.e.* a deterministic evolution rule $\Gamma(0) \to \Gamma(t) = S^t\Gamma(0)$ (in our case the evolution is given by Hamilton's equations), is defined as mixing with respect to a probability distribution $P(\Gamma)$ (in our case the microcanonical distribution) which is invariant under time evolution, if for every couple of phase space regions G and F, we have

$$\mu\left(G \cap S^t F\right) \stackrel{t\to\infty}{\longrightarrow} \mu\left(G\right)\mu\left(F\right), \tag{6}$$

where S^tF indicates the region into which F has evolved at time t, and $\mu(G) = \int_G P(\Gamma)\mathrm{d}\Gamma$ is the relative weight of region G. The meaning of the mixing property as defined by (6) can be intuitively understood if referred to the example we just described: the fraction of the systems which at time $t = 0$ are in phase space region F and which at time t are contained in G tends to a constant value, proportional to the volume of region G. The mixing time, *i.e.* the time necessary to reach the limit (6), is the time necessary to lose memory of the initial condition and hence reach equilibrium. The aforementioned considerations imply that the mixing must take place within reasonable times, in order to be physically significant. Unlike ergodicity, the requirement for mixing is not simplified in the $N \to \infty$ limit.

3. – Some results from analytical mechanics and their connection to the ergodic problem

The subject of ergodicity is entangled with the problem of the existence of non-trivial first integrals (*i.e.* conserved quantities) in Hamiltonian systems: this is a typical problem in celestial mechanics. Given a Hamiltonian $H(\mathbf{q}, \mathbf{p})$, with $\mathbf{q}, \mathbf{p} \in \mathbb{R}^N$, if there exists a canonical transform from the variable $(\mathbf{q}, \mathbf{p})$ into the action angle variables $(\mathbf{I}, \phi)$, such that the Hamiltonian depends only on the action $\mathbf{I}$:

$$H = H_0(\mathbf{I}), \tag{7}$$

then the system is called integrable. In this case the temporal evolution of the system can be written simply:

$$\begin{cases} I_i(t) &= I_i(0), \\ \phi_i(t) &= \phi_i(0) + \omega_i(\mathbf{I}(0))\,t, \end{cases} \tag{8}$$

where $\omega_i = \partial H_0/\partial I_i$ and $i = 1, \ldots, N$. Note that in an integrable system there are N independent first integrals, since all the actions I_i are conserved and the motions evolve on N-dimensional tori. The Solar System provides an important example, if the planetary interactions are neglected: in this limit we are led to the two-body problem (Sun-Planet) for which the integrability can be easily shown.

It is fairly natural to study the problem of perturbation on (7), *i.e.* study the Hamiltonian

$$H(\mathbf{I},\,\phi) = H_0(\mathbf{I}) + \epsilon H_1(\mathbf{I},\,\phi)\,. \tag{9}$$

For the Solar System, this would imply an accounting for the interactions between planets, leading to $\epsilon \approx 10^{-3}$, which is the ratio between the mass of Jupiter (the largest planet), and the Sun. Do the perturbed system (9) trajectories result to be "close" to those of the integrable system (7)? Does the introduction of the $\epsilon H_1(\mathbf{I},\,\phi)$ term still allow for the existence of integrals of the motion besides the energy?

These questions are of obvious interest to celestial mechanics, and are also relevant to the ergodic problem, as mentioned before. Curiously in statistical mechanics and in celestial mechanics there are contrasting expectations, or, better, wishes: in statistical mechanics one would wish for "irregular" dynamical behavior, in order to justify the ergodic hypothesis; conversely, in celestial mechanics "regular" behavior is desired so that accurate previsions could be made.

In a very important work of 1890, H. Poincaré [9] showed that generally a system like (9), with $\epsilon \neq 0$, does not allow analytic first integrals, besides energy. Such result seems discouraging for celestial mechanics; we shall see however, in the following sections, that this conclusion is not obvious. In 1923 the young Fermi [10] first generalizes Poincarè's theorem, showing that a hypersurface which contains all the trajectories originating from its points must coincide with a constant-energy hypersurface. He then shows that for any two arbitrarily small regions of a hypersurface H = constant a trajectory which crosses both always exists. From these results Fermi argued that Hamiltonian systems (apart from the integrable ones, which must be considered atypical) in general, are ergodic, as soon as $\epsilon \neq 0$; this conclusion has been generally accepted (certainly by the physics community).

4. – An unexpected result

Following Fermi's 1923 work the ergodicity problem seemed, at least to physicists, to have been essentially solved: even in the absence of a rigorous demonstration, there was a vast consensus that non-existence theorems for regular first integrals implied ergodicity. In the thirties the ergodic problem thus became a subject mainly studied by mathematicians, who tackled it in a general and abstract fashion, with little interest for its connections with statistical mechanics.

After the war Fermi continued to visit periodically Los Alamos Laboratories and, as Ulam wrote in the introduction to FPU for "Note e Memorie", quickly became interested in computers development and in their use for scientific research. The use of the computer which Fermi had in mind was extremely original (this shall be discussed in sect. **5**), and went well beyond the traditional use made of it during the forties (*i.e.* roughly as a large calculator for operations involving big many-digit numbers). Fermi and Ulam thought about a project for the systematic investigation, using the available computers, of the dynamical behavior of some non-linar systems. Such a program was successively pursued by S. Ulam and J. Pasta, and, more broadly speaking, by a new generation of researchers who, beginning in the sixties and seventies, started to use numerical simulation as a systematic investigation technique for non-linear problems.

FPU studied the dynamical evolution of a chain of $N+2$ identical particles of mass m, connected by non-linear springs (*i.e.* Hooke's law is not exactly valid). The Hamiltonian of the system is

$$H = \sum_{i=0}^{N} \left[\frac{p_i^2}{2m} + \frac{K}{2} \left(q_{i+1} - q_i \right)^2 + \frac{\epsilon}{r} \left(q_{i+1} - q_i \right)^r \right] \tag{10}$$

with $q_0 = q_{N+1} = p_0 = p_{N+1} = 0$ and $r = 3$ or $r = 4$. For $\epsilon = 0$ the system is integrable, since it is equivalent to N independent harmonic oscillators. In fact, using the normal modes:

$$a_k = \sqrt{\frac{2}{N+1}} \sum_i q_i \sin \frac{i\,k\,\pi}{N+1} \qquad (k = 1, \ldots, N)\,, \tag{11}$$

the system reduces to N non-interacting harmonic oscillators, with angular frequencies

$$\omega_k = 2\sqrt{\frac{K}{m}} \sin \frac{k\,\pi}{2(N+1)} \tag{12}$$

and energies

$$E_k = \frac{1}{2} \left(\dot{a}_k^2 + \omega_k^2 a_k^2 \right) . \tag{13}$$

The E_k's, in this case, are clearly constants of the motion and turn out to be proportional to the action variables $E_k = \omega_k I_k$. The Hamiltonian (10) is, therefore, a typical example of a perturbed integrable system. For small values of ϵ it is not difficult to compute all the thermodynamically relevant quantities in the framework of statistical mechanics, *i.e.* averaging over a statistical ensemble (*e.g.* the canonical or microcanonical ensembles). In particular, it can be shown that

$$\langle E_k \rangle \simeq \frac{E_{\text{tot}}}{N}\,, \tag{14}$$

which becomes exact for $\epsilon = 0$. Equation (14) is one way of writing the famous equipartition law. It must be noted that equipartition is valid for $\epsilon = 0$, or small; however, $\overline{E_k}$ (*i.e.* the average calculated along the trajectory) can coincide with $\langle E_k \rangle$, only if ϵ is different from zero, so that the normal modes will interact, hence losing memory of the initial conditions.

What happens if an initial condition is chosen in which all the energy is concentrated in a few normal modes, for instance $E_1(0) \neq 0$ and $E_k(0) = 0$ for $k = 2, \ldots, N$? Before

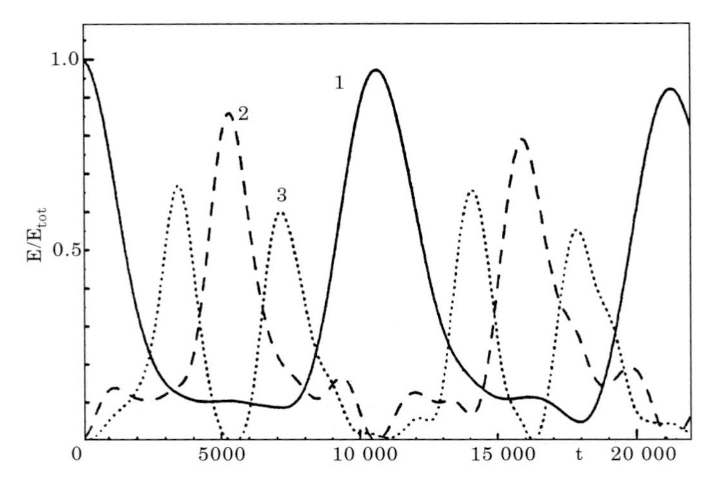

Fig. 1. – $E_1(t)$, $E_2(t)$ and $E_3(t)$ according to FPU with $N = 32$, $r = 3$, $\epsilon = 0.1$ and $\mathcal{E} \simeq 0.07$. This figure, as all the others, is taken from ref. [16].

FPU the general expectation would have been (on the basis of the discussion of sect. **3**) that the first normal mode would have progressively transferred energy to the others and that, after some relaxation time, every $E_k(t)$ would have fluctuated around the common average value given by (14). Even though there is no specific evidence (Ulam does not mention it explicitly), it is reasonable to expect that Fermi shared this expectation. Probably Fermi was interested in a numerical simulation, not so much to verify his "demonstration" of the ergodic hypothesis, as to investigate the thermalization times, *i.e.* the times necessary for the system to go from a non-equilibrium state (all energy concentrated in only one mode), to the equipartition situation expected by statistical mechanics. In FPU a numerical simulation was performed with $N = 16, 32, 64, \epsilon \neq 0$ and all energy concentrated initially in the first normal mode. Unexpectedly, no tendency towards equipartition was observed, even for very long times. In other words, what was observed was a violation of ergodicity and mixing.

In fig. 1 we show the time behavior of the quantity E_k/E_{tot}, for several k's, for $N = 32$ and $r = 3$. Instead of a loss of memory of the initial condition, we see an almost periodic mode: after a long time, the system reverts practically back to the initial state. The non-equipartition of energy can be clearly observed in fig. 2, which shows the average energy of mode k, as a function of the observation time T:

$$\overline{E_k}(T) = \frac{1}{T}\int_0^T E_k(t)\mathrm{d}t\,, \qquad \text{with} \quad k = 1, \dots, N\,. \tag{15}$$

One might wonder whether the results in figs. 1 and 2 could be due to spurious numerical effects caused, for instance, by a too short simulation time, or peculiarities of the specific system being studied. In fig. 3 we show E_k/E_{tot} at long times: here the absence of mixing is clear still. As to the non-peculiarity of the non-ergodic behavior at low energy we refer

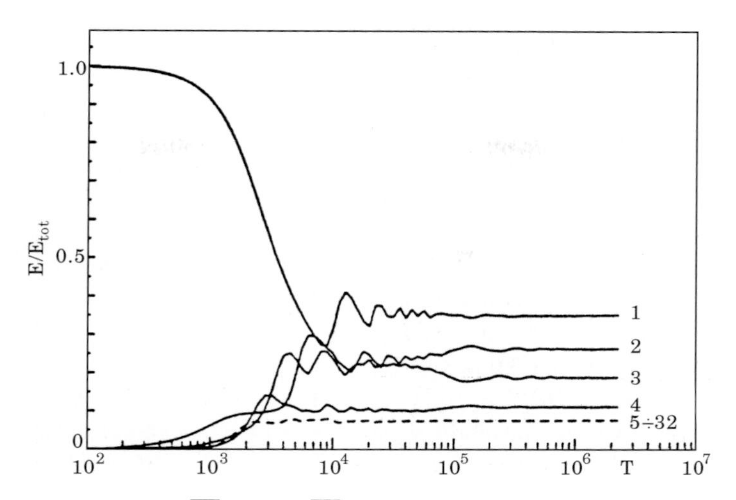

Fig. 2. – Temporal averages $\overline{E_1}(T), \ldots, \overline{E_4}(T)$ (continuous lines from top to bottom) and $\sum_{k=5}^{32} \overline{E_k}(T)$ (dashed line) under the same conditions as in fig. 1.

to the discussion in the following chapter. The FPU results strongly contrasted with expectations, and Fermi himself, writing to Ulam, said that he was surprised and declared that they were dealing with an important discovery which unambiguously showed that the prevalent opinion on the generality of mixing and thermalization properties of non-linear systems, might not always be justified. Fermi had been invited to give a prestigious lecture (the Gibbs Lecture) at the 1955 conference of the American Mathematical Society, and had decided to talk specifically on the FPU results. Because of the illness which caused his death, he was never able to give the lecture.

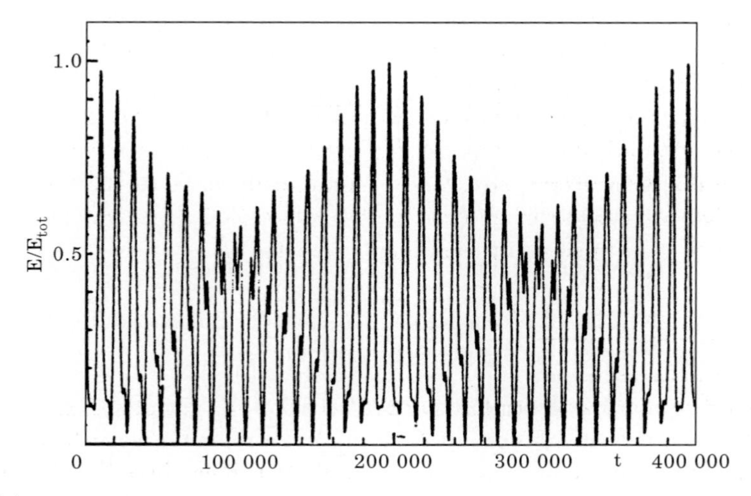

Fig. 3. – $E_1(t)$, same conditions as fig. 1.

The solution to the problem, which is an important milestone in modern mathematical physics, had been (paradoxically) already found by the soviet mathematician A. N. Kolmogorov [11] in 1954, a year before the writing of the Los Alamos report. This fact was unbeknown to the authors of FPU.

5. – Some modern developments

5·1. *The KAM theorem and the foundations of statistical mechanics.* – In 1954 A. N. Kolmogorov proposed (without detailed proof, but clearly expressing the basic idea) an important theorem, which was subsequently completed by V. Arnold [12] and J. Moser [13]. The theorem, now known as KAM, reads:

Given a Hamiltonian $H(\mathbf{I}, \phi) = H_0(\mathbf{I}) + \epsilon H_1(\mathbf{I}, \phi)$, *with* $H_0(\mathbf{I})$ *sufficiently regular and* $\det|\partial^2 H_0(\mathbf{I})/\partial I_i \partial I_j| \neq 0$, *if* ϵ *is small, then on the constant-energy surface invariant tori survive (called KAM tori, resulting from a small deformation of those present for* $\epsilon = 0$*) in a region whose measure tends to* 1 *when* $\epsilon \to 0$.

The KAM theorem might seem obvious, if the theorems on the non-existence of non-trivial first integrals were not known. Actually instead, as a result of these theorems the existence of the KAM tori is a rather subtle and strongly counterintuitive fact. In fact for every value (even small) of ϵ, some tori of the unperturbed system, the so-called resonant ones, are totally destroyed, and this forbids analytic first integrals. In spite of that, for small ϵ most tori survive, even if slightly deformed; thus the perturbed system (at least for "non-pathological" initial conditions) behaves similarly to the integrable one. It is certain that the authors of FPU were not aware of Kolmogorov's theorem. The FPU results can be seen as a "verification" of KAM and above all, of its physical significance, *i.e.* of the fact that the tori survive for physically significant values of the non-linearity parameter ϵ.

Let us reconsider the simulation described in sect. **4**: Izrailev and Chirikov [14] first had noted that for high values of ϵ (when KAM does not apply), there is good statistical behavior, as can be seen in fig. 4. The energy, initially concentrated only in the lowest frequency normal mode, can be seen to spread equally on all normal modes, in agreement with equilibrium statistical mechanics. For a given number of particles N the following scenario essentially holds [15, 16]:

> For a given energy density $\mathcal{E} = E/N$ there is a threshold ϵ_c for the strength of the perturbation such that

a) if $\epsilon < \epsilon_c$, the KAM tori dominate and the system does not follow equipartition;

b) if $\epsilon > \epsilon_c$, the KAM tori are negligible, the system follows equipartition and there is agreement with standard statistical mechanics.

It is easy to realise that if the value of the perturbation ϵ is given, as it happens in actual physical situations, the energy density could play the role of a control parameter and a threshold value $\mathcal{E}_c$ would exist which separates regular from chaotic behavior.

From the physical point of view several questions arise:

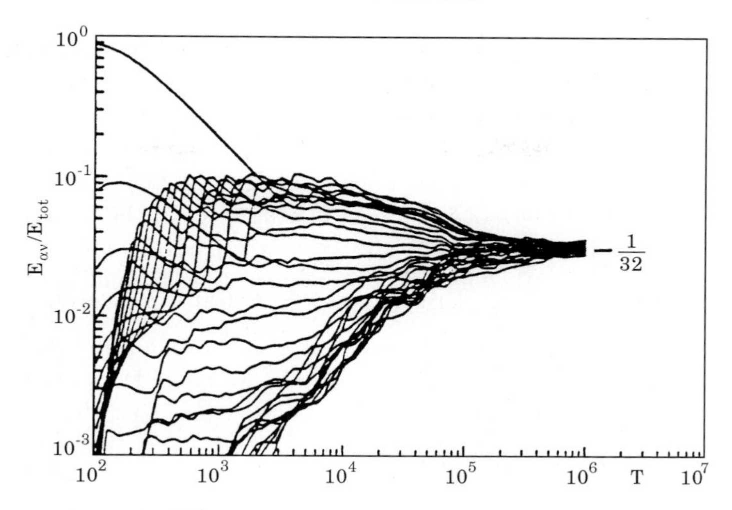

Fig. 4. – Temporal averages $\overline{E_k}(T)$, $k = 1, \ldots, 32$ in FPU with $N = 32$, $r = 3$, $\epsilon = 0.1$ and $\mathcal{E} = 1.2$.

1) whether this behavior (regular for small non-linearities and chaotic for large ones) is peculiar to the FPU Hamiltonian;

2) what is the dependence of ϵ_c on N (at fixed $\mathcal{E}$) or, equivalently, what is the dependence of $\mathcal{E}_c$ on N (at fixed ϵ);

3) which are the characteristic times of the equipartition process.

Point 1) is clear: the mechanism of the transition to chaos for increasing ϵ is standard for all systems which (as FPU) are obtained perturbing harmonic systems. Furthermore, this behavior is present not only in one-dimensional lattices, but also in the multidimensional ones, for instance in Lennard-Jones two-dimensional systems at low energy, where the Hamiltonian could be written in form (9), *i.e.* a harmonic part plus an anharmonic perturbation [17]. As to points 2) and 3), which are the most important for statistical mechanics, things are not well established. The dependence of ϵ_c (or equivalently, of $\mathcal{E}_c$ at fixed ϵ) on N is obviously very important: if $\epsilon_c \to 0$ when $N \to \infty$, the traditional point of view, *i.e.* that preceding FPU, would be re-established. On the contrary, if ϵ_c were not dependent on N, there would be a big discrepancy with the results predicted by equilibrium statistical mechanics. Detailed numerical simulations and analytic computations have been performed on points 2) and 3). In spite of the great efforts, due to technical and numerical difficulties, there still is no general agreement [15, 16, 18]. A detailed analysis will not be presented now, we shall discuss instead some recent works, which shows that indeed some relevant issues have still not been solved.

Casetti *et al.* [19] show clearly that in FPU with cubic non-linearity ($r = 3$), for an energy density smaller than $\mathcal{E}_c = E_c/N \sim 1/N^2$, the dynamics are very regular, even in the $N \to \infty$ limit, with solitonic behavior (see the following section), in agreement with the Zabusky and Kruskal interpretation. Above this threshold the system has good statistical behavior. However, the time $\tau_R(\mathcal{E})$ necessary to reach the equipartition, starting from a far-from-equilibrium initial condition (for instance all the energy is concentrated in a few normal modes) may be very long: $\tau_R \sim \mathcal{E}^{-3}$. Similar results for quartic ($r = 4$) non-linearities have been obtained by De Luca *et al.* [20]. More precisely, τ_R might also depend on the number of degrees of freedom N; for instance, if the initially excited normal modes are always those between k_1 and k_2 (with fixed k_1 and k_2), on increasing N we have $\tau_R \sim N^{1/2}\mathcal{E}^{-1}$ [21]. Since Hamiltonian systems do not have an attractor, the choice of initial conditions (particularly for $N \gg 1$) is technically very difficult since it may not have such a trivial influence, even on a qualitative level, on the relaxation to statistical equilibrium. Without going into details we quote that, even starting from initial conditions which are typical of statistical equilibrium, partially regular behavior is observed also above the stochasticity threshold ($\mathcal{E} > \mathcal{E}_c$) [22]. We note also that the scenario turned out to be much more complex than originally expected. Even if the system turns out to be chaotic (*i.e.* with positive Ljupanov exponent) and most KAM tori are destroyed, the automatic validity of ordinary statistical mechanics is, in fact, not obtained, at least over long but finite times [23].

5'2. *Solitons.* – We shall briefly discuss the impact of FPU on another research subject. In the sixties N. J. Zabusky and M. D. Kruskal [24] developed the idea that the regular behavior of the system associated with Hamiltonian (10) could be attributed to some solutions, called solitons, of a partial differential equation for which the Hamilton equations, obtainable from Hamiltonian (10), are a discrete approximation.

The equations which govern the temporal evolution of the FPU system, for $r = 3$, are

$$m\,\frac{\mathrm{d}^2 q_n}{\mathrm{d}t^2} = f(q_{n+1} - q_n) - f(q_n - q_{n-1})\,, \tag{16}$$

where $f(y) = Ky + \epsilon y^3$. Assuming that $q_n(t)$ is the value of a spatially continuous variable, the field $\psi(x,t)$ at $n\Delta x$, where Δx is the spacing of the lattice with which we approximate a given continuous interval, it is easy to write a partial differential equation for $\psi(x,t)$:

$$\frac{1}{c^2}\frac{\partial^2 \psi}{\partial t^2} = \left(a + 2\,g\,\frac{\partial \psi}{\partial x}\right)\frac{\partial^2 \psi}{\partial x^2}\,, \tag{17}$$

where, for $\Delta x \to 0$, we have assumed

$$\frac{(\Delta x)^2}{m} \to c\,, \qquad \frac{K}{(\Delta x)^2} \to a\,, \qquad \epsilon(\Delta x) \to g\,. \tag{18}$$

The solutions to (17) can be shown [25] to develop spatial discontinuities after a finite time $t_c \sim 1/(gc\psi_0)$, where ψ_0 in the maximum field amplitude at $t = 0$.

We now look for a solution to (16) which in the continuum limit will be slowly varying with t if $x - ct$ is fixed. Let us introduce the variables

$$\xi = \frac{\tilde{\epsilon}}{\Delta x}(x - ct) , \qquad \tau = \frac{\tilde{\epsilon}^3}{\Delta x} ct , \qquad \psi = \tilde{\epsilon} u , \tag{19}$$

where $\tilde{\epsilon}$ is a (small) parameter connected to the rate of variation. Assuming that $\tilde{\epsilon} \to 0$, when $\Delta x \to 0$, we obtain for the variable $v = \partial u / \partial \xi$ the equation

$$\frac{\partial v}{\partial \tau} + \epsilon v \frac{\partial v}{\partial \xi} + \frac{1}{24} \frac{\partial^3 v}{\partial \xi^3} = 0 . \tag{20}$$

Equation (20) is one of the possible ways of writing the Korteweg-de Vries (KdV) equation, dating back to 1895, when it was formulated to describe the propagation of surface waves in shallow water. Equation (20) can be shown to admit a "solitonic wave" solution of the type $v = F(\xi - V\tau)$, where V is a constant related to the wave velocity within the frame of reference where eq. (20) is valid and $F(z)$ is a function which is localized in a finite region and which decays at great values of $|\xi - V\tau|$. Solitary waves have been considered for a long time as being a mere mathematical curiosity of little physical relevance. After the work of Zabusky and Kruskal on vibrations in anharmonic crystals and plasma waves, solitonic properties have turned out to be fundamental in many phenomena, such as Bloch wall motion in magnetic crystals and propagation of magnetic flux in Josephson junctions [25].

The original Zabusky and Kruskal explanation of the regularity of the FPU system in terms of solitary waves originating from the KdV equation, however, is not totally convincing. In fact the passage from eq. (16), with variables (q_n) defined on a discrete set of points, to an equation with variables ($\psi(x)$) defined over a continuous set of points is very delicate, since similar assumptions can lead to very different systems. For instance, an equation such as (17) can be obtained which develops singularities in a finite time, or such as (20) which has an extremely regular behaviour.

Independently of technical details, it is a fact that Zabusky and Kruskal started from the Fermi, Pasta and Ulam results, and the study of equations with solitonic solutions has become an important chapter of mathematical physics.

5·3. *The role of simulations in physics.* – The FPU work has pioneered computer simulations, as one of the very first numerical experiments. FPU was preceded by the work of Metropolis *et al.* [26] on Monte Carlo (MC) techniques for studying equations of state of liquids, which should be considered to be the first case of numerical simulation. Even before FPU and MC there had been examples of numerical analyses, even on a large scale, and of the use of computers in physics (in particular, in military research at Los Alamos during the war). However, FPU and MC have signalled a notable change in the ways the computer was used. From the use of the computer as a fast adding machine destined to grind over thousands of trivial operations, we passed to the computer as an instrument of experimentation.

Fermi and collaborators did not use the computer to obtain numerical details within the context of a well founded and understood theory; they instead performed a true *gedankenexperiment*, verifying conjectures (which turned out to be wrong in the specific case) to try and shed light on a problem which was not well understood. This new way of using the computer is still with us, and we can state that numerical simulation has become a new branch of physics; together with theoretical and experimental physics, we speak now also of a computational physics [27, 28]. The computer is not only useful for studying a given phenomenon, it can also, in a way, create the phenomenon through modelling. Simulation is somewhat like an experiment "*in vitro*", in which one can choose those facets of a given phenomenon which are (or are deemed) relevant, bringing to the extreme the Galilean objective of "difalcare gli impedimenti" ([1]), which is not always possible in a real experiment [27, 29].

It would take too long to discuss in detail all the subjects included in computational physics: molecular dynamics, Monte Carlo methods, numerical fluidodynamics, ...

It is interesting that, independently of technical aspects, the growth in the use of the computer should not be attributed only to the increase in computing power, but also to the evolution of algorithms and to the introduction of new and suggestive graphics.

We might state, exaggerating a bit, that most physicists were convinced of the importance of non-linear systems not so much because of the significance of the Kolmogorov, Arnold and Moser theorems, but rather because of the suggestive power of the drawings obtained from the computer simulations of the pioneers of chaos: Fermi, Pasta, Ulam, Chirikov, Lorenz, Hénon.

The systematic use of the computer has favoured the rebirth of entire research sectors: the most significant examples are perhaps turbulence and chaotic dynamical systems, which had been marginal and relegated among the engineering applications (turbulence) or were considered as being more esoteric branches of mathematics (dynamical systems).

* * *

We thank G. B. BACHELET, M. CASARTELLI, G. CICCOTTI and S. RUFFO for a critical reading of the manuscript.

REFERENCES

[1] FERMI E., PASTA J. and ULAM S., *Los Alamos Sci. Lab. Rep.* LA-1940 (1955).
[2] FERMI E., *Note e Memorie (Collected Papers)* (Accademia Nazionale dei Lincei and The University of Chicago Press) 1962, 1965.
[3] SEGRÈ E., *Enrico Fermi Physicist* (The University of Chicago Press) 1970.
[4] CERCIGNANI C., *Ludwig Boltzmann e la Meccanica Statistica* (La Goliardica Pavese, Pavia) 1997.
[5] ARNOLD V. I. and AVEZ A., *Ergodic Problems of Classical Mechanics* (Benjamin, New York) 1968.
[6] KHINCHIN A. J., *Mathematical Foundation of Statistical Mechanics* (Dover, New York) 1949.

([1]) Removing the obstacles [T.N.].

[7] Landau L. D. and Lifschitz E. M., *Statistical Physics* (Pergamon Press) 1969.
[8] Truesdell C., *Ergodic Theory in Classical Statistical Mechanics* in *Proceedings of the International School of Physics Enrico Fermi, Course XVI*, edited by Caldirola P. (Academic Press, London) 1961.
[9] Poincaré H., *Acta Math.*, **13** (1890) 1.
[10] Fermi E., *Phys. Zeits.*, **24** (1923) 261; translated into Italian in Fermi E., *Nuovo Cimento*, **26** (1923) 105; **25** (1923) 267.
[11] Kolmogorov A. N., *Dokl. Akad. Nauk SSSR*, **98** (1954) 527.
[12] Arnold V. I., *Russ. Math. Surv.*, **18** (1963) 9.
[13] Moser J. K., *Nachr. Akad. Wiss. Göttingen Math. Phys. kl.*, **2** (1962) 1.
[14] Izrailev F. M. and Chirikov B. V., *Dokl. Akad. Nauk SSSR*, **166** (1966) 57.
[15] Bocchieri P., Scotti A., Bearzi B. and Loinger A., *Phys. Rev. A*, **2** (1970) 2013; Casartelli M., Casati G., Diana E., Galgani L. and Scotti A., *Theor. Math. Phys.*, **29** (1976) 205; Livi R., Pettini M., Ruffo S., Sparpaglione M. and Vulpiani A., *Phys. Rev. A*, **31** (1985) 1039.
[16] Benettin G., in *Molecular-Dynamics Simulation of Statistical-Mechanical Systems*, edited by Ciccotti G. and Hoover W. G. (North-Holland, Amsterdam) 1986.
[17] Benettin G. and Tenenbaum A., *Phys. Rev. A*, **28** (1983) 3020.
[18] Kantz H., *Physica D*, **39** (1989) 322; Kantz H., Livi R. and Ruffo S., *J. Stat. Phys.*, **76** (1994) 627.
[19] Casetti L., Cerruti-Sola M., Pettini M. and Cohen E. G. D., *Phys. Rev. E*, **55** (1997) 6566.
[20] De Luca J., Lichtenberg A. J. and Ruffo S., *Phys. Rev. E*, **60** (1999) 3781.
[21] Ruffo S., in *Chance in Physics: Foundations and Perspectives*, edited by Bricmont J. *et al.* (Springer-Verlag, Berlin) 2000.
[22] Alabiso C. and Casartelli M., *J. Phys. A*, **33** (2000) 831.
[23] Livi R., Pettini M., Ruffo S. and Vulpiani A., *J. Stat. Phys.*, **48** (1987) 539.
[24] Zabusky N. J. and Kruskal M. D., *Phys. Rev. Lett.*, **15** (1965) 240; Zabusky N. J., in *Nonlinear Partial Differential Equations*, edited by Ames W. (New York) 1967.
[25] Cercignani C., *Riv. Nuovo Cimento*, **7** (1977) 429.
[26] Metropolis N., Rosenbluth A. W., Rosenbluth M. N., Teller A. H. and Teller E., *J. Chem. Phys.*, **21** (1953) 1087.
[27] Wilson K., *La Recherche*, **14**, No. 146 (1983) 1004.
[28] Ciccotti G., Frenkel D. and McDonald I. R. (Editors), *Simulation of Liquids and Solids* (North-Holland, Amsterdam) 1987.
[29] Livi R., Pettini M., Ruffo S. and Vulpiani A., *Giornale di Fisica*, **26**, No. 4 (1985) 285.

About the Authors

Massimo Falcioni, born in 1950 in Rome, is presently a researcher at the Physics Department of the Rome University "La Sapienza". He has worked on particle physics; his more recent interests are on chaotic dynamical systems.

Angelo Vulpiani, born in 1954 in Borgorose (Rieti), is presently Full Professor of Theoretical Physics at the Physics Department of Rome University "La Sapienza". His major interests are statistical mechanics of disordered systems, turbulence and chaotic dynamical systems.

Fermi's last lessons

Renato Angelo Ricci

"In working on these notes, I have been rewarded by frequently encountering sections which are so unique in their language and approach as to evoke for me, again, the picture of Fermi as he lectured, during those lovely mornings in the beautiful setting of the Villa Monastero on Lake Como."

The excerpt above is taken from B. T. Feld's presentation of the publication, "Lectures on Pions and Nucleons" for the Proceedings of the International School of Physics held in Varenna. These lessons were taught by Enrico Fermi [1] for the II course on "Questioni sulla rivelazione delle particelle elementari" (Questions about the detection of elementary particles), directed by G. Puppi in 1954 just a few months before Fermi's death.

The lectures, compiled by the students of the course and edited by Feld were reprinted and translated into Italian by the Italian Physical Society in 1983, as part of the 30th anniversary celebration of the school named after him one year after Fermi's death [2]. "*... This is certainly not the form in which* —continues Feld— *Fermi would have written them for publication, for his methods of oral and of written presentation were very different. In their present form, however, they illustrate (subject to the limitations of those who transcribed and edited them) the unique quality of Fermi as an expositor and a teacher.*"

There could not have been a better way of presenting not only these lessons, which were the last, but also those which were taught at Chicago which preceded these, for example the course on "Nuclear Physics " from 1949-1950, published in the well known book "Nuclear Physics" (University of Chicago Press) based on the notes compiled by J. Orear, A. H. Rosenfeld and R. A. Schluter. And also the handwritten notes on Thermodynamics and Statistics and Quantum Mechanics Courses which Fermi taught respectively in 1951-1952 and in 1954. These were published by the University of Chicago Press, with a preface by E. Segrè, the former in 1966 and the latter in 1961.

As for Varenna School lectures, in the Nuclear Physics Course we rely on notes taken and reproduced by others (in this specific case by Orear, Rosenfeld and Schluter, whose preface will be cited in the following), while for the other two books the notes are Fermi's own and make up the final version of his presentation as continuously modified and improved during the course of many years of teaching. Finally special mention must be made of the volume "Elementary Particles", which Fermi himself worked on. This was published in 1951 by Yale University where he taught 6 lectures for students and experimental physicists. These lectures are the contents of the book and, as we shall see, Fermi believed that they were essential (we would say fundamental) in understanding subjects which are often specialized or very theoretical.

In addition to this, Fermi at this time also gave a cycle of 6 lessons for the general public. These lessons were probably linked to those of the Italian Atomic Physics conferences. Mention of this will be made later.

Thus, with great respect and admiration we shall discuss Enrico Fermi's "cultural heritage", which is equally as great as his fundamental discoveries and inventions. Fermi, undoubtedly, was gifted with the great ability of being able to explain things in a very clear and precise manner which together with his ingenious theoretical intuitions and his notable experimental aptitude, made him a teacher who had the rare gift of being able to reinterpret and present subjects in a way which was very beneficial for those who were learning.

In the preface of "Notes on Quantum Mechanics", Emilio Segrè explains this ability of Fermi's very well. In the hope that these notes are appreciated especially by young physicists of future generations *"...who have never come in direct contact with Fermi, and for whom he must be a little more than a name among the great scientists of the century..."* and that it would be very useful for them to have readily available *"a notebook on such an important topic as quantum mechanics written for them by such a master in his own hand..."*, he adds *"...they cannot be construed in any way as the final presentation of quantum mechanics by Fermi, such as he could have given in a more elaborate text. Heisenberg, Pauli, Dirac, de Broglie, Jordan, Kramers, to mention only some of the creators of quantum mechanics, have all presented their own versions of quantum mechanics in books which are justly famous. The notes by Fermi are not to be compared in any way with these texts. They are written in a spirit and for a purpose completely different ... Fermi in the last ten or fifteen years of his life scarcely read any book. He kept abreast of scientific developments mainly by hearing the results of investigations and reconstructing them on his own. It is practically certain that he did not consult any text of quantum mechanics while compiling these notes. ... the notes were clearly prepared only for the lectures and ... their distribution beyond the class group was not intended by the author."*

This article will deal, for the most part, with these "lessons". In any case, it would be best, at least, to cite two other series of Fermi's didactic scientific works, for example "Conferenze di Fisica Atomica"(Conferences on Atomic Physics) [5] given in 1950 in Rome and Milan and, a little before this, the course on Neutron Physics [4] given at Los Alamos in 1945.

The 9 conferences on Atomic Physics, 6 in Rome and 3 in Milan, were given by Fermi when he was in Italy during the fall of 1950 after an absence of 11 years. He had been invited by the Donegani Foundation. These conferences were recorded and then edited in sequence by Sebastiano Sciuti and Lucio Mezzetti ("Le particelle elementari" (The elementary particles)), Ettore Pancini ("Teorie sull'origine degli elementi" (Theory on the origin of the elements)), Nestore B. Cacciapuoti ("La ricerca di un'attrazione tra elettrone e neutrone" (The research on the attraction between the electron and the neutron), Mario Ageno ("Orbite nucleari" (Nuclear Orbits)), Giuseppe Morpurgo ("Nuovi sviluppi dell'elettrodinamica quantistica" (New Development in Quantum Electrodynamics)), Carlo Salvetti ("Il neutrone e analogie ottiche nelle proprietà dei neutroni" (The Neutron and optical analogies in neutron properties) and Piero Caldirola (Il monopolo di Dirac" (The Dirac Monopole).

These lessons were attended, as you could imagine, and as G. Castelnuovo confirms in his presentation of the volume edited by the Accademia dei Lincei, *"... by many people who crowded into the classroom of the Institute of Physics in Rome and the hall of the Montecatini Company in Milan"* and truly can be considered as being admirable lessons on the evolution of modern Physics and several aspects of fundamental importance, as can be deduced from the titles of the contributions.

The 30 lessons of the Neutron Physics Course, given by Fermi in the fall of 1945 as part of the program of the University of Los Alamos (a few taught by R. F. Christy and E. Segrè (when Fermi could not be there)) interest us for two reasons:

The first is that these lessons were a real course designed for young physicists (Segrè mentioned about 30 undergraduate and graduates) who took notes which were compiled, as were others of Fermi's, without his revision. We are speaking about a full immersion teaching situation, which was represented by the closed community of Los Alamos during the concluding period of the war.

This course on Neutron Physics, on the other hand, which was compiled by I. Halpern and revised by B. T. Feld is also a sampling of Fermi's didactic style and a large part of the material which later appears in more conventional texts, for example, the already cited course on Nuclear Physics, is found here.

The second reason is that this series of lessons deals with original information from that period and which was also secret information. The first part, about Neutron Physics, with no reference to chain reactions, was immediately declassified and freely circulated, whereas the second part was declassified and published in 1962. It will be mentioned again in connection with the Nuclear Physics text.

This short introduction has allowed us to reiterate an essential concept that justifies this task of attempting to present Fermi's last lessons. Paraphrasing Segrè we would say that *"It is only because we know his great interest in teaching that we think is not irrelevant to his memory to publish the notes for the benefit of other students."*

1. – Fermi's courses from 1945 to 1950

1·1. *"Nuclear Physics"* (fig. 1). – The text —now a classic— of "Nuclear Physics" in the editions compiled by Orear, Rosenfeld and Schluter (in 1949 and 1950) is a complete course which ranges from properties of atomic nuclei to interaction of radiations with

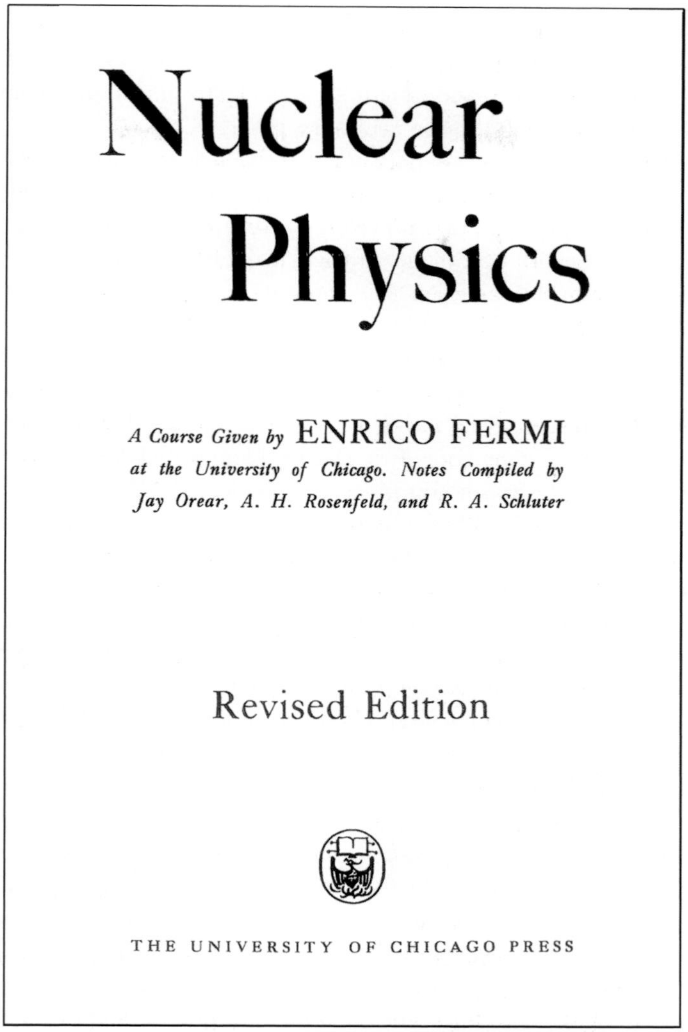

Nuclear
Physics

A Course Given by ENRICO FERMI
at the University of Chicago. Notes Compiled by
Jay Orear, A. H. Rosenfeld, and R. A. Schluter

Revised Edition

THE UNIVERSITY OF CHICAGO PRESS

Fig. 1. – Front page of the book on Enrico Fermi's Nuclear Physics course (1949-50).

matter and radioactive decays, to nuclear forces and meson theories with particular attention devoted to neutron and cosmic rays physics. This is a corner stone, not only because of its content and relative explanations which are still so up-to-date, but also because it dates back to a crucial period, the one which lays the ground for and precedes the "conceptual" separation between nuclear physics understood as physics of nuclei and the physics of elementary particles also known as "subnuclear". As V. Weisskopf noted in 1960, this separation is a prelude to the distinction between "intensive" physics (typically reductionist such as the physics of particles) and "extensive" physics (holistic, which favours the typical complexity of condensed matter),while nuclear physics is the watershed. This would explain why, from the mid 50's, an evolution which would separate the study of structural properties and dynamics of nuclei from that of elementary particles and their primary interactions became necessary [8].

During the last ten years, this separation has been less marked, either due to the specific introduction of subnucleonic degrees of freedom in the study of nuclear behaviour, or because of the use of the nuclear framework to decipher fundamental questions such as the quark structure and the problems of their deconfinement.

And, yet Fermi, who had foreseen this separation (the subnuclear structures were limited to the mesonic field at that time), in his Nuclear Physics course of 1949 presented the nuclear static and dynamic description in a unified manner, including the α, β, γ radioactive decays, in the framework of nuclear forces considered as nucleon-nucleon interactions, already foreseeing the mesonic theory (exchange forces) and dedicating an entire chapter to meson phenomenology (the problem will later be taken up more in detail during the course in Varenna on pions and nucleons).

The course was then extended to include a chapter on cosmic rays, which is a prelude to a more specific treatment of elementary particles.

Some significant examples are:

In presenting the exchange nature of nuclear forces, Fermi explains:

"From electrostatics we know that two particles attract or repel one another according to Coulomb's law. For a classical treatment we say that this force arises from the potential field $\phi = e/r$ of one of the particles. However if we wish to take into account the corpuscular nature of light, we can describe this interaction by saying that one particle "emits" a photon which is subsequently absorbed by the other."

Simple and illuminating.

And he continues:

"Analogously, the interaction of two nucleons can be partially (note the "partially" and recall the residual interactions, in addition to those of the mean field) *interpreted by the picture of one nucleon "emitting" a quantum which is promptly absorbed by the second nucleon. These quanta are called mesons, and we shall call them π-mesons in this discussion. The reason for this nomenclature is that we know experimentally that nucleons interact more strongly with the π's than with the μ's. If we are going to attribute nuclear forces to one sort of meson, we might as well call it a π."* Better than this!

At any rate, the whole section on the ensemble properties of the nuclei (Chapt. 1) and those about radioactive decays (Chapts. 3, 4 and 5) as well as the chapter (8) on nuclear reactions is already a "classical" nuclear physics course.

The first part essentially goes back to the liquid drop model (nuclear species, semiempirical mass formula, binding energy, isobaric behaviour, electric and magnetic momenta, and neutron excess) and refers to the chapter on nuclear reactions, where the model for the compound nucleus is introduced (in particular, with an explicit reference to resonance phenomenology in the n, γ cross-section) and the statistical gas model (which is Fermi's model anyway) precursor of the shell model of the nucleus.

"Various models of the nucleus —says Fermi— *emphasize various different features*

of the nucleus. No single simple model explains all nuclear properties. (We still say the same thing today, despite all the progress which has been made in the overall microscopic description of the nuclear structures.) *We shall consider the statistical or gas model, then the liquid drop model applied to fission, and finally the nuclear shell model."*

I am still convinced that this course is exemplary (chapt. 8) for explaining Fermi's gas model using nuclear temperature, Fermi's energy and density of energy levels concepts.

The shell model, which Fermi calls *"orbit model of the nucleus"* is phenomenologically introduced using a clear justification of the nuclear orbits, a subject not readily understood by students, in view of the "compactness" of the nuclear structure which among other things, legitimates the concept of the drop model, as explained in the following:

"This model describes the nucleus in terms of nucleon orbits somewhat like the description of the atom in terms of electron orbits. The orbit picture is valid if collisions are rare enough so that a nucleon may travel at least across the nucleus without collision. This requirement seems at first not to be fulfilled in a nucleus, for at $\approx$ 20 MeV the n-p scattering cross-section is of the order of 0.3 barn, and, for the known density of nucleons, the mean free path is only about 1/3 or so the radius of the nucleus. However, there are two factors which this calculation ignores, and this makes the orbit picture not so untenable.

1) When one nucleon passes another, it passes through a potential well. If the nucleon is constantly passing other closely spaced nucleons, the wells may be so closely spaced so as to blend together to form a roughly uniform potential.

2) The nucleus is a degenerate system in which the lowest energy states are, for the most part, filled. A collision can occur between nucleons only if the collision results in transferring both the nucleons to empty states. The Pauli exclusion principle prevents two nucleons of the same kind in the same state."

This is a very clear illustration of the validity of the quasi-free motion of a nucleon within the nucleus (extension of the mean free path caused by the Pauli principle and a first approximation of absorption in the mean potential (central) of the interactions between the "orbiting nucleon" and the rest of the nucleons).

Another very significant example is the illustration of the "spin-orbit" coupling introduced by M. G. Mayer to explain the magic numbers for closing the major nuclear shells which are empirically found (N, $Z = 2, 8$, 20, 50, 82, 126, to be compared with those derived from quantum calculations for the rectangular potential well: 2, 8, 20, 40, 70, 112, 168).

"Suppose that spin-orbit coupling splits the energy levels corresponding to different J values (note $J = l + s$)*, that is, 1g splits into $1g_{9/2}$ and $1g_{7/2}$. Assume that the level with larger J is more stable, i.e. lies lower. This assumption is not contrary to any known facts about the nucleus.* (Note the "Galilean" statement.) *Then the former closed shell number 40, for example, must be altered as follows* (fig. 2, taken from the original manuscript):

The importance of the spin-orbit coupling will be discussed later on when referring to the Varenna course.

We should now like to digress into the subject of neutron physics, a substantial com-

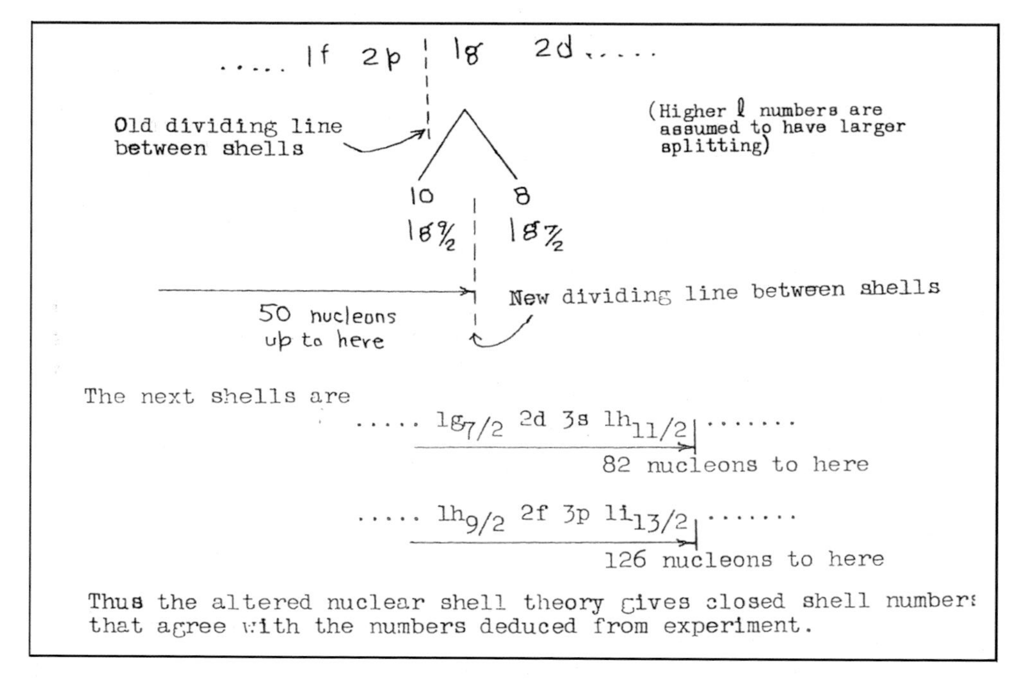

Fig. 2.

ponent of Fermi's didactic work. In the Nuclear Physics Course, as already mentioned, Fermi devoted an entire chapter (9) to neutrons, using the Course of Neutron Physics held at Los Alamos in 1945 as a basis. This latter course was obviously more detailed, since it contains, to the benefit of the students, a part with more general notions about nuclear physics, such as the isotope chart and nuclear models used for neutron reactions.

As to the remaining subjects all you have to do is to leaf through either of these texts, starting from natural neutron sources (radioactive) and artificial (reactions induced by cyclotron-accelerated particles) and devoting attention to neutron collisions, to scattering and diffusion theory and, in particular, to the slowing-down and distribution of slow neutrons in materials and ending with nuclear fission. It has already been mentioned that in 1945 for secrecy reasons this last subject had not been fully dealt with; in the Nuclear Physics course the Theory of Chain Reactions and Fermi's article published in the January 10th, 1947 issue of "Science" on "Elementary Theory of Chain-reacting Pile", which makes reference to a report presented on June 21st, 1946 at the American Physical Society Conference, are included.

A copy of the first page of this article reproduced from the 1949 text is shown in fig. 3.

We shall now discuss briefly the part which deals more specifically with properties and interactions of charged particles, in particular, electrons, and cosmic rays.

In chapt. 2 we shall find a classical treatment of the radiation-matter interaction

Elementary Theory of the Chain-reacting Pile

(Reprinted by permission from Science Jan. 10, 1947)

Enrico Fermi
Institute for Nuclear Studies, University of Chicago

THE RESULTS AND THE METHODS DISCUSSED in the following outline of the theory of a chain-reacting pile working with natural uranium and graphite have been obtained partly independently and partly in collaboration by many people who participated in the early development work on the chain reaction. Very important contributions to the theoretical ideas were given by Szilard and Wigner. Many physicists contributed experimental results that helped to lead the way, among them, H. L. Anderson and W. H. Zinn, first at Columbia University and later at the Metallurgical Laboratory of the University of Chicago; R. R. Wilson and E. Creutz, at Princeton; and Allison, Whitaker, and V. C. Wilson, at the University of Chicago. The production of the chain reaction was finally achieved in the Metallurgical Laboratory directed by A. H. Compton.

ABSORPTION AND PRODUCTION OF NEUTRONS IN A PILE

We consider a mass, "the pile," containing uranium spread in some suitable arrangement throughout a block of graphite. Whenever a fission takes place in this system, an average number (ν) of neutrons is emitted with a continuous distribution of energy of the order of magnitude of 1,000,000 EV. After a neutron is emitted, its energy decreases by elastic collisions with the atoms of carbon and to some extent also by inelastic collisions with the uranium atoms. In the majority of cases the neutrons will be slowed down to thermal energies. This process requires about 100 collisions with carbon atoms. After the energy of the neutron is reduced to thermal value, the neutron keeps on diffusing until it is finally absorbed. In several cases, however, it will happen that the neutron is absorbed before the slowing-down process is completed.

The neutron may be absorbed by either the carbon or the uranium. The absorption cross-section of carbon for neutrons of thermal energy is quite small, its value being approximately $.005 \times 10^{-24}$ cm.[1] For graphite of density 1.6, this corresponds to a mean free path for absorption of about 25 m.[1] It is believed that the absorption cross-section follows the $1/v$ law, and consequently the absorption cross-section, which is already quite small at thermal energies, becomes practically negligible for neutrons of higher energy. It is therefore a sufficiently good approximation to assume that absorption by carbon during the slowing-down process can be neglected.

The absorption of a neutron by uranium may lead either to fission or to absorption by a (n, γ) process. We shall refer to this last possibility as the process of resonance absorption. The relative importance of fission and resonance absorption in the different energy intervals is not the same. In this respect we can consider roughly three intervals:

(1) Neutrons with energy above the fission threshold of U^{238}—We can call these conventionally "fast neutrons." For fast neutrons the most important absorption process is fission, which normally takes place in the abundant isotope U^{238}. Resonance absorption is smaller but not negligible.

(2) Neutrons of energy below the fission threshold of U^{238} and above thermal energy—We shall refer to these neutrons as "epithermal neutrons." For epithermal neutrons the most important absorption process is the resonance capture. The cross-section for this process as a function of energy is quite irregular and presents a large number of resonance maxima that can be fairly well represented by the Breit-Wigner theory. In practical cases the resonance absorption becomes important for neutron energy below about 10,000 EV and increases as the energy of the neutrons decreases.

(3) Neutrons having thermal agitation energy or "thermal neutrons"—For thermal neutrons both the resonance and fission absorption processes are important. In this energy range both cross-sections follow approximately the $1/v$ law, and therefore their relative importance becomes practically independent of the energy. Let σ_f and σ_r be the cross-sections for fission and resonance absorption for neutrons of energy kT, and η be the average number of neutrons emitted when a thermal neutron is absorbed by uranium. Then η differs from ν, since only the fraction $\sigma_f/(\sigma_r + \sigma_f)$ of all the thermal neutrons absorbed by uranium produces a fission. It is, therefore,

$$\eta = \nu\sigma_f/(\sigma_f + \sigma_r). \tag{1}$$

The preceding discussion leads one to conclude that only a fraction of the original fast neutrons produced will end up by producing a fission process. For systems of finite size, further losses of neutrons will be expected by leakage outside the pile.

Limiting ourselves for the present to systems of practically infinite dimensions, we shall call P the probability that a fast neutron ultimately is absorbed by the fission

This paper, presented at the Chicago meeting of the American Physical Society, June 21, 1946, is based on work performed under Contract No. W-7401-eng-37 with the Manhattan District at the Metallurgical Laboratory, University of Chicago.

Fig. 3. – First page of the article by Fermi on "The theory of the atomic pile" (1947).

(energy loss, absorption, polarization, ionization, scattering, photoelectric absorption, Compton scattering, pair formation) which follows a well known pattern typical of the old university courses on Advanced Physics which today can be found here and there in more specialized courses.

The last chapter of the book on Cosmic Rays (chapt. 10) is an admirable synthesis of a subject which was just beginning to be taught in universities, and which was opening up a field of investigation full of potential for future prospects, since it was essentially the beginning of elementary particle physics and of its splitting from nuclear physics (many of us remember it well, as do many standard texts, such as L. Janossy's "Cosmic Rays" of 1949 or D. J. X. Montgomery's "Cosmic Ray Physics" of 1949).

"This field is expanding —says Fermi— *very rapidly. Many facts are known but the present theories to explaining them are mostly tentative. For brevity we shall take many liberties and talk as though both fact and theory were better established than is the case."*

It is, in any case, the chapter on mesons (7) which, even if brief and synthetic, opens the door of the new physics. This aspect will become clearer when we discuss the course taught at Varenna in 1954.

We report here Fermi's introduction to the properties of mesons known from experiments: *"In this section we shall discuss briefly some of the facts known about mesons, and summarize them in a table..."* (the table on p. 133 of that text reports the properties of $\pi^{\pm}$, π^{o} pions and $\mu^{\pm}$ muons known at that time).

"By mesons we mean unstable particles of mass greater than that of the electron, less than that of the nucleon. ...

Mesons were postulated by Yukawa in 1935, and soon thereafter μ-mesons (they will be called "μ's" or muons from here on) were observed as secondary particles in cosmic radiation. In 1948 π-mesons (π's or pions) were created artificially by bombarding various targets in the Berkeley cyclotron. During 1949-50, overwhelming evidence has been found for the existence of a neutral pion π^0."

The distinction which Fermi makes between pions and muons is already clear, the first being essentially the Yukawa particle, subject to strong interactions, while the second is subject to weak interactions, even if both are considered within the frame of nuclear exchange forces.

These comments are the forerunners of the drafting, this time by Fermi himself, of the 1950 text on Elementary Particles [9].

1·2. *"Elementary Particles" (1950-51).* – In April 1950 Fermi gave a series of 6 lessons at Yale University for the public and 6 for physics students. These latter lessons were compiled and expanded by Fermi and are the core of his famous volume "Elementary Particles" published by Yale University Press in 1951 [9] (see fig. 4).

In the preface Fermi himself states: *"Many of the theoretical papers on the subject of elementary particles and of their interactions are very difficult reading except for a small, highly specialized group of theoretical physicists. This book is not written for that group. It attempts instead to make accessible to a larger number of students and, I hope, a large fraction of experimental physicists some of the most significant results of the field*

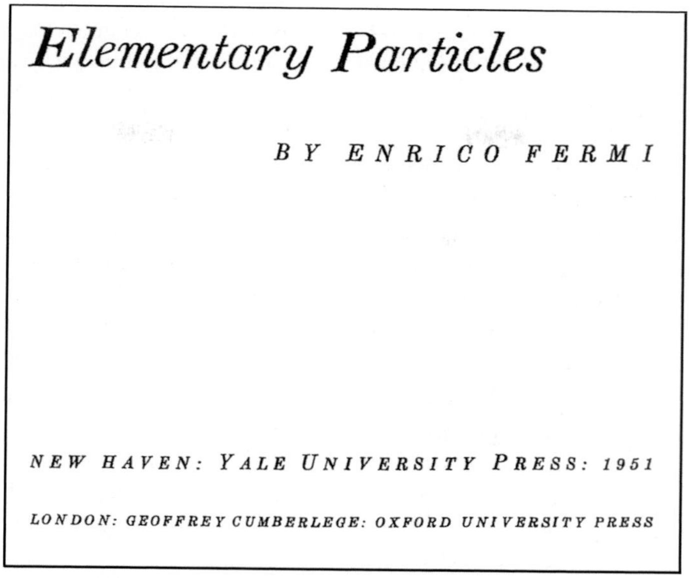

Elementary Particles

BY ENRICO FERMI

NEW HAVEN: YALE UNIVERSITY PRESS: 1951

LONDON: GEOFFREY CUMBERLEGE: OXFORD UNIVERSITY PRESS

Fig. 4. – Cover of the book by Enrico Fermi on "Elementary Particles" (1951).

theories of elementary particles that can be understood, at least in a semi-quantitative way, without excessive mathematical apparatus."

There is no doubt that, other than being very clear in its purpose, such a "treatise" was to become a very useful and much consulted text. On the other hand, this is not only a fundamental didactic tool (you only need to cite the chapters on quantum and field interactions beginning from the basic example of the electromagnetic field, the appendix about the second quantization, the measurability of the fields, the relativistic invariance and the relationship between the interaction constants) but it is also a lesson which foresees problems which at that time were only speculations: the fact that these "speculations" of Fermi's were well founded and brilliant is illustrated, for example, by the chapter on "Pions, Nucleons and Antinucleons" and more specifically on the annihilation of the antinucleons (still to be discovered). Fermi wrote:
"All the current theories of electrically charged particles have a symmetry property according to which for each particle a counterpart with the opposite charge and otherwise similar properties exists. This is true in particular of the Dirac electron theory which was established before the discovery of the positron. In most discussions about nucleons these particles are supposed to obey a Dirac-like equation. If this assumption is correct, negative protons must exist and also anti-neutrons. The anti-proton, here indicated by $\bar{P}$*, has the mass of the proton and has negative charge and magnetic moment equal and opposite to that of the proton. The anti-neutron, indicated by* $\bar{N}$*, has the mass of the neutron, no charge, and magnetic moment equal and opposite to that of the ordinary neutron.*

Since no experimental evidence has been found for the existence of these two particles we cannot be too sure that they really exist. It is interesting, nevertheless, to speculate as to what their properties are likely to be. In this discussion the somewhat similar case of the behaviour of electrons and positrons may be taken as a guide."

Fermi then proceeds to discuss the annihilation process with a release of the $2Mc^2$ energy and introduces, since absorbed neutrons have to be accounted for, not only the interaction of the nucleus with the electromagnetic field (emission of 2 photons with equal and opposite momenta) but also the one caused by the nucleon-pion coupling, in which case 2 pions with equal and opposite momenta are emitted.

This is the more general case of the nucleon-antinucleon annihilation which Fermi clearly and simply demonstrates to be more probable than electromagnetic decay "*...primarily because the coupling constant* e_2, *...*" relative to the expression of the probability, inverse of the average lifetime, of the pionic decay

$$\frac{1}{\tau_y} = \frac{e_2^4 n}{16\pi M^2 c^3},$$

"*...is much larger than the electromagnetic coupling constant e...*" which yields

$$\frac{1}{\tau_{\text{em}}} = \frac{\pi e^4 m}{N^2 c^3}.$$

"Indeed, it is probable that the rate of annihilation with pion emission is even faster than indicated ..., since the total energy available in the annihilation process is sufficient to produce more than two pions so that other processes could be operative, leading to a higher over-all probability of transition."

Going on to the numbers —as Fermi used to do to make the theoretical statements more concrete— assuming for the nucleonic density $n = 7 \cdot 10^{37}$ (the one inside the nucleus) we find $1/\tau_{\text{em}} = 1.5 \cdot 10^{17}$ and $1/\tau_y = 1.5 \cdot 10^{20}$, 1000 times greater. "*... this value* —explains Fermi— *is probably an underestimate. From it would follow, for example, that a negative proton traversing a nucleus of diameter* 10^{-12} *cm with velocity comparable to c would have a probability of less than 1 per cent of being annihilated."*

It is the multiple pion production which greatly increases this estimate. This is more or less what for example experiments with LEAR at CERN using antiprotons later demonstrated.

In the text examined so far a summary can be found of the evolution initiated as a result of the specificity of particle physics, which is illustrated by Fermi in the "Atomic Physics Conferences" [4], already cited. The 1st and 2nd conferences as already mentioned were compiled by Morpurgo and Mezzetti and deal with particle physics. Fermi's clearness and foresight are illustrated first and foremost in the introduction of the "elementarity" concept and in the critical description of the problem of the proliferation of particles considered as being elementary.

We shall cite (translating from the original Italian text):

"The subject I'll speak about today is "elementary particles": but if you ask me what is meant by elementary particles, I would be stumped because the term "elementary" is to be understood in a rather relative way with respect to what we know.... We could generally say that, at every scientific stage, those particles whose structure we know nothing about, are called elementary and they could thus be thought of as being dots .... Along these lines... we could also make another comment about the number of particles."

Here Fermi gives the example of the chemistry of the atom considered as being elementary in order to compare it with other various types of atoms, whose number increases more and more.

Thus:

"... the confidence... concerning the elementarity of the atom would gradually disappear since a large number of particles would be in contradiction to the very concept of elementariness. Also for the elementary particles (Fermi lists 9 of them, including the proton, neutron and π meson, besides the electron, the muon, and the photon, then adding the neutrino) *which are now known, we have a situation which is not very different...."*

It was a well-known issue that the number of elementary particles had been a problem for "reductionist" physics for a long time. The present scheme is based on 3 quark and lepton families (6×3 colours $+ 6$) with relative antiparticles (48 particles), the most elementary components of matter. To these must be added the particles which mediate the field interactions (the photon for the electromagnetic interaction, the $W^{\pm}Z^0$ intermediate bosons for the weak interaction, 8 gluons for the strong interaction plus eventually the quantum for the gravitational interaction (for a total of 61).

The problem of the unification of the forces, which during Fermi's last few years, was still "*in fieri*" necessitates, in addition to this, a further step, that is Higgs' particle.

Apart from the further evolutions (symmetry, supersymmetry, theory of strings) maybe we could imagine what would have been Fermi's role today in helping "students and experimental physicists" understand this impressive development of frontier physics.

2. – The last lessons at Chicago

2·1. *Thermodynamics and Statistics, Quantum Mechanics.* – The years between 1950-54, are highlighted by Fermi's teaching at the University of Chicago, with particular attention to the lessons on Thermodynamics, Statistics and Quantum Mechanics which, as has already been mentioned, have been "certified" by Fermi's handwritten notes.

Without going into too much detail on this, we should like to mention some particular cases which exemplify not only his clear presentation, but his extraordinary ability and capability of synthesizing and grasping the essential aspects without, however, sacrificing the sense and completeness of the subject. The beginning of the notes on the III principle of thermodynamics is reported in fig. 5: any comment would be very disrespectful, as would be any translation from English.

Of great interest are also the notes on "Matter under unusual conditions".

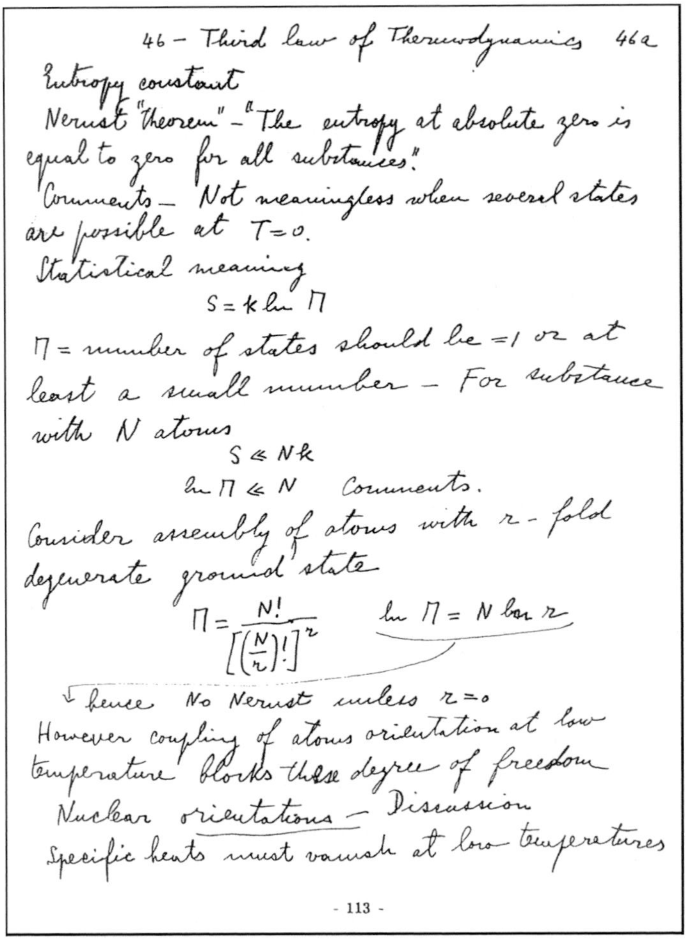

46 – Third law of Thermodynamics 46a

Entropy constant

Nernst "theorem" – "The entropy at absolute zero is equal to zero for all substances".

Comments – Not meaningless when several states are possible at $T=0$.

Statistical meaning

$$S = k \ln \Pi$$

Π = number of states should be $=1$ or at least a small number – For substance with N atoms

$$S \ll Nk$$

$\ln \Pi \ll N$ Comments.

Consider assembly of atoms with r-fold degenerate ground state

$$\Pi = \frac{N!}{\left[\left(\frac{N}{r}\right)!\right]^r} \qquad \ln \Pi = N \log r$$

hence No Nernst unless $r=0$

However coupling of atoms orientation at low temperature blocks these degree of freedom

Nuclear orientations – Discussion

Specific heats must vanish at low temperatures

- 113 -

Fig. 5. – Fermi's handwritten notes on "III Principle of Thermodynamics" (course taught in Chicago in 1951-1952).

Think about how, nowadays, such a subject is so popular considering the research and knowledge which has been acquired about phase transitions and the equations of state of systems, not only on an atomic-molecular level but also on a nuclear and particle level. All you have to do is to cite the problems of the nuclear, hadronic and quark-gluon plasma phases investigated by means of reactions with relativistic heavy ions.

The temperature-pressure diagram illustrated by Fermi on the page reproduced in fig. 6, which goes from ordinary condensed matter through the gas of electrons and of electrons and protons to the neutron gas varying temperatures and pressures to unusual limits, is a precursor of the phase diagrams which today describe the phase transitions of the nuclear matter up to the extreme conditions of the deconfinement of the quark and gluon plasma.

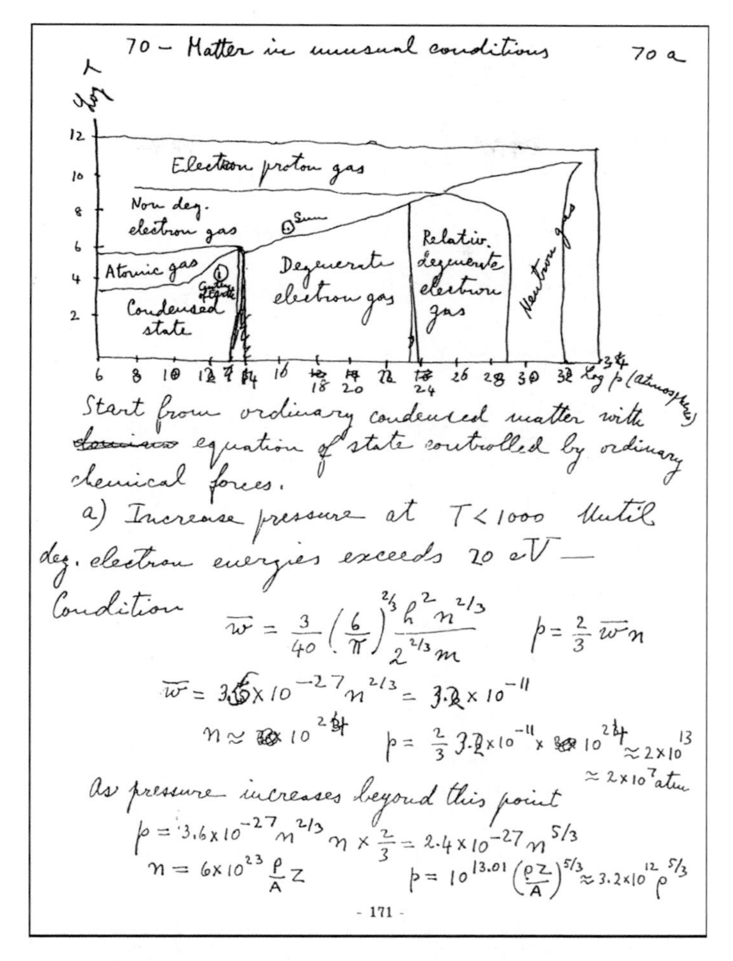

Fig. 6. – E. Fermi's handwritten notes on "Matter under Unusual Conditions".

Not to speak of the problems connected with magnetic confinement of the plasma of ions in the controlled thermonuclear fusion processes and those related to astrophysics.

In connection with these lessons on thermodynamics, it would be useful to cite B. T. Feld who, in his introduction to "Lectures on pions and nucleons" (paper no. 270 in "Note e Memorie") puts it this way:

"Fermi's greatness as a teacher stemmed from the fact that he made little, if any, distinction, between teaching and research. His famous power, of finding the simple and clear physical model for understanding a seemingly complicated idea, can be seen as well in his papers as in his lectures. his ability of recalling, for the solution of some new problem, an appropriate example from no matter which field in which the problem had already been solved, he applied in reverse in his course; the examples employed in

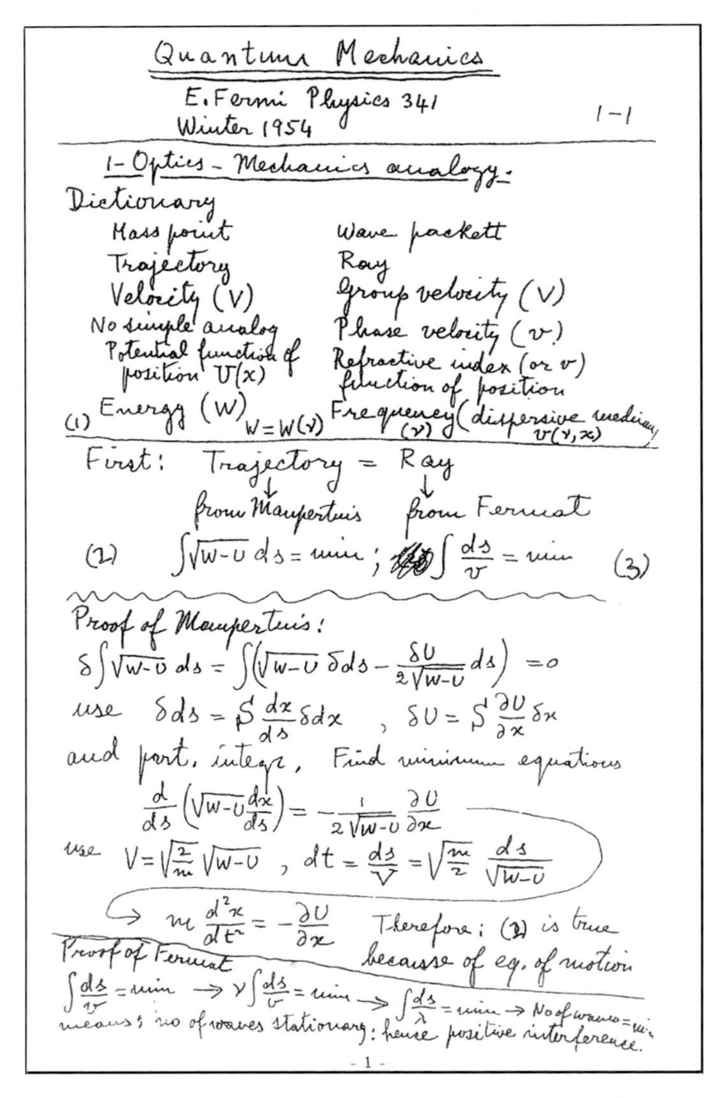
Quantum Mechanics

E. Fermi Physics 341
Winter 1954

1-1

1- Optics - Mechanics analogy.

Dictionary

Mass point	Wave packett
Trajectory	Ray
Velocity (V)	Group velocity (V)
No simple analog	Phase velocity (v)
Potential function of position $U(x)$	Refractive index (or v) function of position
(1) Energy (W) $W=W(\nu)$	Frequency (ν) (dispersive medium $v(\nu,x)$)

First: Trajectory = Ray

from Maupertuis — from Fermat

(2) $\int\sqrt{W-U}\,ds = \min$; $\int \frac{ds}{v} = \min$ (3)

Proof of Maupertuis:

$$\delta\int\sqrt{W-U}\,ds = \int\left(\sqrt{W-U}\,\delta ds - \frac{\delta U}{2\sqrt{W-U}}\,ds\right) = 0$$

use $\delta ds = \sum \frac{dx}{ds}\delta dx$, $\delta U = \sum \frac{\partial U}{\partial x}\delta x$

and part. integr., Find minimum equations

$$\frac{d}{ds}\left(\sqrt{W-U}\frac{dx}{ds}\right) = -\frac{1}{2\sqrt{W-U}}\frac{\partial U}{\partial x}$$

use $V=\sqrt{\frac{2}{m}}\sqrt{W-U}$, $dt = \frac{ds}{V} = \sqrt{\frac{m}{2}}\frac{ds}{\sqrt{W-U}}$

$\rightarrow m\frac{d^2x}{dt^2} = -\frac{\partial U}{\partial x}$ Therefore: (2) is true because of eq. of motion

Proof of Fermat

$\int\frac{ds}{v} = \min \rightarrow \nu\int\frac{ds}{v} = \min \rightarrow \int\frac{ds}{\lambda} = \min \rightarrow$ No of waves = min

means: no of waves stationary: hence positive interference.

- 1 -

Fig. 7. – Handwritten notes by E. Fermi on "Quantum Mechanics" (course held in Chicago in 1954).

his course in Thermodynamics were frequently exciting excursions along the frontiers of nuclear physics."

In connection with this course it would be worth mentioning the short book "Termodinamica" (Thermodynamics) by Fermi, edited by P. Borighieri in 1958, which many of us have consulted and used because of the many precious suggestions found in it for "Introductory Physics" taught at the University. This text was based on lessons from the course "Thermodynamics", taught by Fermi at Columbia University in 1936.

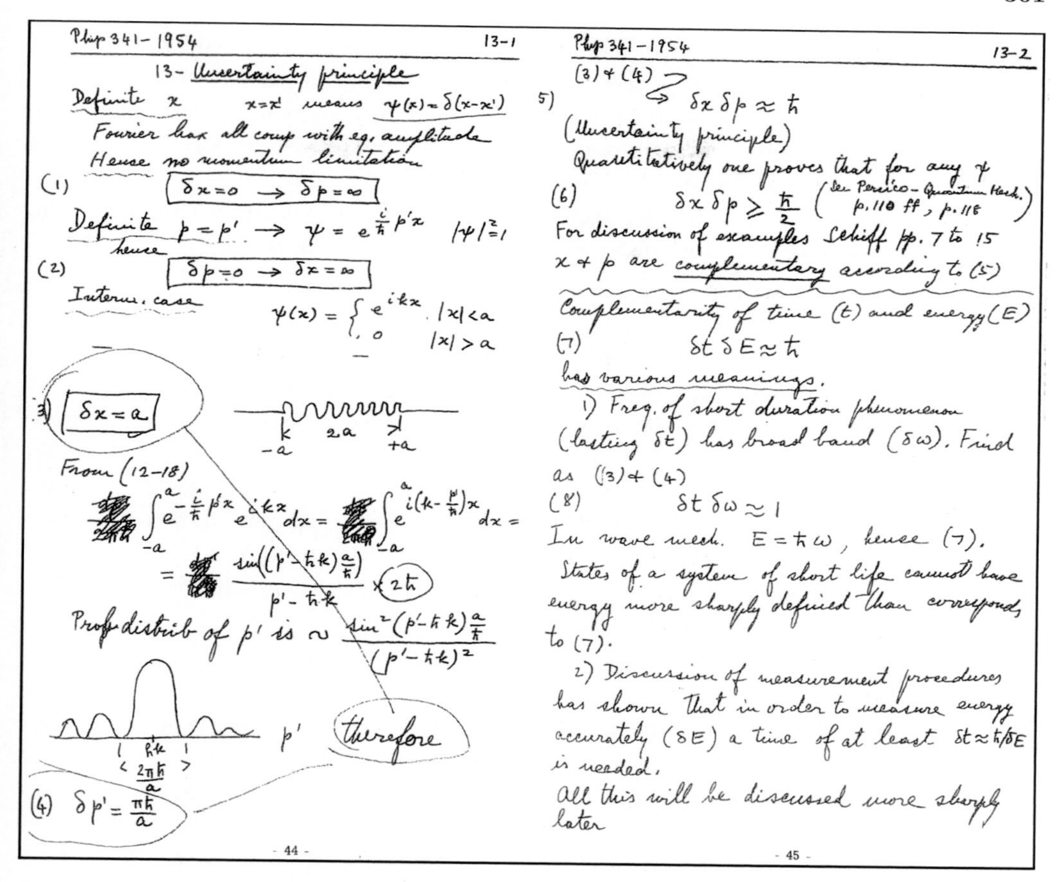

Php 341 – 1954 13-1

13 – Uncertainty principle

Definite x $\quad x=x'$ means $\psi(x)=\delta(x-x')$

Fourier has all comp with eq. amplitude

Hence no momentum limitation

(1) $\delta x = 0 \rightarrow \delta p = \infty$

Definite $p=p' \rightarrow \psi = e^{\frac{i}{\hbar}p'x}$ $\quad |\psi|^2=1$

hence

(2) $\delta p = 0 \rightarrow \delta x = \infty$

Interm. case $\quad \psi(x) = \begin{cases} e^{ikx} & |x|<a \\ 0 & |x|>a \end{cases}$

(3) $\delta x = a$ $\quad -a \quad 2a \quad +a$

From (12-18)

$$\int_{-a}^{a} e^{-\frac{i}{\hbar}p'x} e^{ikx} dx = \int_{-a}^{a} e^{i(k-\frac{p'}{\hbar})x} dx =$$

$$= \frac{\sin\left((p'-\hbar k)\frac{a}{\hbar}\right)}{p'-\hbar k} \times 2\hbar$$

Prob distrib of p' is $\sim \frac{\sin^2(p'-\hbar k)\frac{a}{\hbar}}{(p'-\hbar k)^2}$

p' therefore

$\hbar k$ $\quad \frac{2\pi\hbar}{a}$

(4) $\delta p' = \frac{\pi\hbar}{a}$

- 44 -

Php 341 – 1954 13-2

(3) + (4) $\rightarrow$

5) $\delta x\, \delta p \approx \hbar$

(Uncertainty principle)

Quantitatively one proves that for any ψ

(6) $\delta x\, \delta p \geq \frac{\hbar}{2}$ (See Persico – Quantum Mech. p. 110 ff, p. 118)

For discussion of examples Schiff pp. 7 to 15

x & p are complementary according to (5)

Complementarity of time (t) and energy (E)

(7) $\delta t\, \delta E \approx \hbar$

has various meanings.

1) Freq. of short duration phenomenon (lasting δt) has broad band ($\delta\omega$). Find as (3) + (4)

(8) $\delta t\, \delta\omega \approx 1$

In wave mech. $E = \hbar\omega$, hence (7).

States of a system of short life cannot have energy more sharply defined than correspond, to (7).

2) Discussion of measurement procedures has shown that in order to measure energy accurately (δE) a time of at least $\delta t \approx \hbar/\delta E$ is needed.

All this will be discussed more sharply later

- 45 -

Fig. 8. – Handwritten notes by E. Fermi on Quantum Mechanics about the Uncertainty Principle.

Let us now look at the lessons on Quantum Mechanics. As Emilio Segrè, in his preface, reminds us, the handwritten notes, as already mentioned, were written in 1954, less than a year before his untimely death. Recalling how Fermi, in his early years in Rome illustrated Schroedinger and Dirac's work, which had just been published, to his students in private seminars, *"in more familiar form "*, Segrè believes that comments and notes compiled by these students from the University of Rome, Columbia and Chicago do probably exist.

But it is in 1954 when *"... Fermi again gave a course in quantum mechanics at the University of Chicago. This time, however, he prepared the notes for the students himself by writing the outlines of the lectures on duplicator master sheets and delivering copies to the students in advance of each lecture."*

In fig. 7 the introductory page of the notes on the analogy between optics and mechanics on the basis of the particle-wave duality with the corresponding references to the theorems of Maupertuis and Fermat is reported.

From here we go on to the introduction of the time dependent Schroedinger equation and to the WKB (Wentzel, Kramers, Brillouin) method using as examples the one-dimensional and the linear oscillator problem. Central forces and the hydrogen atom are illustrated by demonstrating in a simple way the separation between relative motion and centre of mass motion, in analogy with classical mechanics.

A fine example that clearly illustrates this is that of the Uncertainty Principle. The descriptive pages, which need no further comments, are shown in fig. 8.

3. – The Last Lecture. Varenna 1954

In the volume published by SIF (Società Italiana di Fisica) in 1984 [2] (fig. 9) dedicated to the inauguration of the 1983 Courses, which I chaired as President, and the aforementioned 30th anniversary celebration of the International School of Physics, founded by Giovanni Polvani and named after Enrico Fermi, we can find the talks given by Gianni Puppi, Antonio Rostagni and Gilberto Bernardini who had taken part in these events.

G. Puppi (who was the first director of the school in 1953) writes:

"The School was founded at a time when there was not a great proliferation of schools... and at the right moment for establishing a name for itself.

During the 1st year everything was centred on detection techniques of elementary particles and it was complementary to the courses on cosmic rays and a preview of the physics of pions, which would be the subject of the courses for the following year. I should say that it went fairly well, so much so that SIF repeated the experiment the following year. For purposes of continuity I was again asked to direct and organize the 2nd course. This course was a memorable event....a series of heavenly conjunctions created a special charisma around it. Objectively speaking, at that moment a sort of "summa theologica" of all that was known about pion physics was made. This was to become a milestone for a long time thereafter.

Then Fermi's death shed a particular light on this second year and, when speaking of the Scuola di Varenna, the second year and Enrico Fermi come to mind..."

During this 2nd year Fermi taught "Lectures on Pions and Nucleons" [1], which we have already discussed. These lectures, translated into Italian and found in the above cited 30th anniversary issue [2] (see fig. 9), begin with the concept of isotopic spin and the description of the two-nucleon system, which involve not only the production of mesons and scattering, but also more specific nuclear questions such as those resulting from the diffusion of polarized nucleons by the nuclear potential (distinguishing between distribution of particle density in light and heavy nuclei). A typical example is the spin-orbit coupling, as evidenced in the discussion of the Nuclear Physics course.

I cite:

"What we have learned from the polarization effects is the following: there is good evidence that the l-s coupling of conventional nuclear physics persists with essentially the same strength at higher energies; this can presumably be tied to the primary origin of the l-s coupling, already present in a latent form in the nucleon-nucleon forces".

SOCIETÀ ITALIANA DI FISICA

CELEBRAZIONE DEL TRENTENNALE
DELLA
SCUOLA INTERNAZIONALE DI FISICA
"ENRICO FERMI"

VARENNA, LAGO DI COMO,
VILLA MONASTERO
FOUNDED IN 1953

Editrice Compositori - Bologna

Fig. 9. – Cover of the volume published by SIF for the 30th anniversary of the founding of the Varenna School. Fermi's course ("Lectures on Pions and Nucleons/Lezioni su Pioni e Nucleoni", Varenna 1954) can be found here.

As to the more specific part about the "summa of pion physics" Gilberto Bernardini in his comments during the 30th anniversary celebration of Varenna says:

"*... The subject of the 2nd year course at Varenna, still under Puppi's leadership, was 'The detection of elementary particles and their interactions'. Fermi and Heisenberg taught the first 2 lessons of that course; Fermi's was on 'Pions and Nucleons'.*

I feel he spontaneously included the concepts of quantum electrodynamics in his dis-

cussion on the interactions between nucleons by pions. Moreover, on this subject, he had written an article in 1932. This article was to become such an important point of reference in the future, that recently Pontecorvo referred to it as the 'Roman Bible'.

I think, for example, that for this correlation he had already said, among other things, that in 1938 G. C. Wick already had shown that within the nuclear dimension limits pions could be 'virtually' present only if the mass was at least 200 times that of the electron.

He also stated that just as the nucleon interaction should be attributed to pseudo-charges which, emitting some pions, limited the momenta and energies, similarly do the real or virtual photons among electrical charges. And speaking of charges and the related interaction constants, he compared the classical electrodynamics constant

$$\alpha = \frac{e^2}{4\pi\hbar c} = \frac{1}{137} ,$$

to which corresponds the Coulomb energy of 2 charges at a distance r

$$H(r) = -\frac{e^2}{4\pi r} ,$$

with the nucleon one

$$\frac{g^2}{4\pi\hbar c} \simeq 10 ,$$

which is present in the Yukawa potential

$$H(r) = -\frac{g^2}{4\pi r} \exp\left[-kr\right] ,$$

where $k = m_\pi \, c/h$, m_π *is the mass of the pion."*

Actually, also Fermi's didactic "illuminations" are innumerable. It is interesting to recall [8] an anecdote told by A. D. Bromley in his opening speech at the International Conference of Nuclear Physics in Florence in 1983. Bromley, speaking about his meeting with Fermi during the 50's at a conference in Rochester at which Fermi himself had presented the results of a work done with his students from Chicago on muon scattering, and in particular on their newly discovered resonance, that is the Δ resonance (excited nucleon), made the following statement:

"Knowing that I was a nuclear physicist... Fermi commented: you also should take this into consideration in nuclear physics..."—Bromley added— *"...Fermi, as usual, was right."*

Today we know that, for example, to explain the deuteron photoabsorption cross-sections it is necessary to include not only the mesonic exchange effects, but also the resonance Δ.

During the Varenna 2nd Course many were impressed not only by Fermi's lectures but also by the atmosphere of the course itself. A photograph of the group is shown in fig. 10. In addition to Fermi, who is in the center of the 1st row, there are Rostagni,

Fig. 10. – Photograph of the group of participants at Varenna School in 1954.

Borsellino, Caldirola, G. Bernardini, Puppi, Occhialini, Conversi, B. Rossi, Heisenberg, Steinberger, ... and others who today realize how fortunate they had been to have had such a group of people as teachers.

Once again as an example of the way a subject of physics is didactically introduced we shall cite the pages from these lessons which describe the isotopic spin:

"The isotopic spin notation was invented by Heisenberg in the early 30's almost immediately after the discovery of the neutron. The neutron appeared to be a particle with properties similar to those of the proton, and the idea was that they could both be described as different states of the same particle. Thus we can say that a particular nucleon, which I shall indicate by the symbol $\mathcal{N}$ can have two forms p or n, proton or neutron. The idea at this stage is purely formal and it could be adapted, though probably not fruitfully, to distinguish any two objects. As time went on, the fruitfulness of the notation became apparent, because the properties of nucleons are such that they make the notation more valuable than a purely formal device. However, just for a short time, let me pursue the purely formal consequences of this notation. We are here presented with what is usually called a dichotomic variable, i.e. *a variable that can take on essentially two values. If we represent, as is usual, a function of a dichotomic varaible by a vertical slot containing the two values a and b,* $f = \begin{vmatrix} a \\ b \end{vmatrix}$*, then this is a function which for the first of the variables takes the value a and for the second takes the value b.*

There is in Physics a wide amount of experience as to the behaviour of these functions of dichotomic variables. They were encountered for the first time when Pauli worked out the theory of the spin $\frac{1}{2}$*, which is also a dichotomic variable. The state may be specified by saying that the spin is 'up' or 'down'. There are certain standard linear operators that* *operate on variables of this type, and they are essentially the Pauli operators. We may consider adding the unity operator* $\begin{vmatrix} 1 & 0 \\ 0 & 1 \end{vmatrix}$ *to the three Pauli operators*

$$\begin{vmatrix} 0 & 1 \\ 1 & 0 \end{vmatrix}, \quad \begin{vmatrix} 0 & -i \\ i & 0 \end{vmatrix}, \quad \begin{vmatrix} 1 & 0 \\ 0 & -1 \end{vmatrix}.$$

We have then a list of four operators which are linear and have the following property. They and their linear combinations are all the possible linear operators on any function of a dichotomic variable. Whether this is a spin, or a variable which tells us whether the particle is a proton or a neutron, makes no difference whatsoever. We shall define

$$2\tau_1 = \begin{vmatrix} 0 & 1 \\ 1 & 0 \end{vmatrix}; \quad 2\tau_2 = \begin{vmatrix} 0 & -i \\ i & 0 \end{vmatrix}; \quad 2\tau_3 = \begin{vmatrix} 1 & 0 \\ 0 & -1 \end{vmatrix}.$$

The factors 2 here introduced will prove convenient later.

What is, for instance, the significance of the operator $2\tau_1$*, when applied to a proton* p*? The function which represents the proton state, is in the notation*

$$p = \begin{vmatrix} 1 \\ 0 \end{vmatrix}.$$

Similarly we shall introduce another function $n = \begin{vmatrix} 0 \\ 1 \end{vmatrix}$*, which defines the neutron. Now,*

$$\begin{vmatrix} 0 & 1 \\ 1 & 0 \end{vmatrix} p = \begin{vmatrix} 0 & 1 \\ 1 & 0 \end{vmatrix} \begin{vmatrix} 1 \\ 0 \end{vmatrix} = \begin{vmatrix} 0 \\ 1 \end{vmatrix} = n$$

means a linear operation with coefficients indicated by the square matrix $\begin{vmatrix} 0 & 1 \\ 1 & 0 \end{vmatrix}$ *applied to the function* $\begin{vmatrix} 1 \\ 0 \end{vmatrix}$ *and this yields* $\begin{vmatrix} 0 \\ 1 \end{vmatrix}$*. So we see that* $2\tau_1$ *changes* p *into* n *and, as can be shown, also changes* n *into* p*.* $2\tau_1$ *is thus the operator that interchanges a proton and a neutron. Similarly one can show the operational significance of the others."*

Fermi is here referring to two-nucleon system.

These concepts, as well as others, have been extended to different fields and the clear and precise illustration of this was Fermi's great prerogative. These lectures at Varenna have left an indelible mark.

4. – Conclusions

Concluding these remarks, without having even illustrated in the best way possible the lessons which Fermi had taught at the end of his life, when most of his time was dedicated to teaching, would do injustice to his memory. Therefore, we shall conclude with some references which evidence his activity as a great teacher.

Emilio Segrè said [10], with reference to Fermi's work at the Institute of Nuclear Studies in Chicago, beginning at the end of 1945:

"The Institute of Nuclear Studies was more oriented towards research than towards teaching. No diplomas were given, but Fermi, as were most of the others, was also a full university professor. At the Institute Fermi taught regular normal courses in physics, thermodynamics, statistical mechanics, nuclear physics, quantum mechanics, and solid state physics. He also insisted upon teaching an elementary course in introductory physics.

The carefully prepared notes used in these courses were the distillation of a long didactic and research experience. Some of these compiled notes, those on quantum mechanics for example, were continuously revised and corrected from year to year."

At this point it is worth mentioning what C. N. Yang (together with T. D. Lee the pioneer of the theory of parity non-conservation), a student of Fermi's, said almost immediately after his arrival in the United States from China (from the introduction to paper no. 239 in "Note e Memorie" by E. Segrè):

"As is well known, Fermi gave extremely lucid lectures. In a fashion that is characteristic of him, for each topic he always started from the beginning, treated simple examples and avoided as much as possible 'formalisms'. (He used to joke that complicated formalism was for the 'high priests'). The very simplicity of his reasoning conveyed the impression of effortlessness. But this impression is false: The simplicity was the result of careful preparation and of deliberate weighing of different alternatives of presentation.

... The fact that Fermi had kept over the years detailed notes on diverse subjects in physics ranging from the purely theoretical to the purely experimental, from such simple problems as the best coordinates to use for the three-body problem to such deep subjects as general relativity, was an important lesson to all of us. We learned that that *was physics. We learned that physics should not be a specialist's subject, physics is to be built from the ground up, brick by brick, layer by layer. We learned that abstractions come* after *detailed foundation work,* not *before. We also learned in these lectures of Fermi's delight in, rather than aversion to, simple numerical computations with a desk computer."*

In addition to this, Segrè also reminded us that Fermi had thought of a project for "his old age retirement " that of writing a book *"... that would contain all the difficult points of physics that are too often glossed over, by such phrases as 'it is well known'. I think that he was serious about this, because he started collecting critical questions, and even asked me to jot down seemingly elementary questions that I felt I did not really understand. This book would have been a great lesson to physicists, and may possibly have become the all-time best-seller in physics. Unfortunately, however, he did not have the time even to start it."*

In the absence of this we shall only say that at this time when we speak and discuss so much about ways of teaching, the example of this great Teacher's true and serious dedication to teaching, considering his stature as a scientist, remains universal.

Universal is also the man Fermi, despite some odd and unforgivable instances and even some contentious cases. I shall not dwell upon some despicable attempts to change the name of some lyceums in our country which had been named after him.

Instead I shall recall an episode in which I was personally involved. I was asked in extremis to fill in a great gap in a book edited by R. Cortina, for the series "Scienza e Idee"(Science and Ideas), of which Giulio Giorello,an old friend of mine, was the editor. The book, by Ernst Peter Fischer, "Aristotle, Einstein and the others", was being translated from German and Giorello asked me to look through it; upon doing this we both discovered that among "the others" —who go from Avicenna to Copernicus, to Galilei, Kepler, Descartes, Newton, Lavoisier, Faraday, Darwin, Maxwell, Mendel, Boltzmann, Marie Curie, Lise Meitner, Barbara McClintock, Niels Bohr, Pauling, Von Neumann, Max Delbruck, R. Feynman— Fermi was missing. I was asked to contribute a chapter as an appendix to the Italian edition, which was entitled "Il grande navigatore" (The great navigator) [11]. I shall cite a section which I had hoped would illustrate, as a way of trying to make amends, his genial understanding of all physics phenomena.

"This mental attitude of Fermi's of perceiving theory as the basis of phenomenology, even when using the most sophisticated mathematical methods, without ever sacrificing clarity, is the sign of a genial ability in describing the laws of matter in a truly 'Galilean' perspective."

The mark of an unavoidable destiny remains associated with Fermi. Such is the text on the stone of imperial porphyry dedicated to him in the classroom of Villa Monastero (where he taught for the last time):

HIC

ANIMO TOT INTER RERUM MIRA PACATO

ARCANA NATURAE PRIMORDIA

INTRA ATOMOS VOLVENTIA

DOCTORUM COETUI POSTREMUM APERUI

MEUM UNDE NOMEN IAM IMMORTALE FECERAM

(HERE, THE SOUL PACIFIED BY SUCH BEAUTIFUL SURROUNDINGS, FOR THE LAST TIME I OPENED, TO A THRONG OF MEN OF SCIENCE, THE DOOR TO THE MOST HIDDEN AND REMOTE ELEMENTS WHICH MOVE INSIDE THE ATOMS, WHEREOF I HAD ALREADY RENDERED MY NAME IMMORTAL).

APPENDIX

Fermi in Varenna

Fermi's stay at Varenna in 1954 was memorable not only for his already mentioned wonderful lectures, but also for his contributions during the discussions and subsequent analysis of the future of Italian Physics. Giorgio Salvini, who took part in these discussions, in particular those about the Frascati electrosynchrotron project, gives an exemplary though synthetic picture of it in the introduction to this book. There, among other things, Fermi's advice on electronic calculations in addition to his comments on the physics of accelerators can be found.

Actually that "memorable" course at Villa Monastero was not only culturally meaningful, since it was held at a very decisive moment of the "rebirth" of Italian physics as well as of the development of European physics; it was also a "corner stone" in terms of concrete proposals for the organization of research. All you have to do is look at the table of contents of the course, published in "Supplement to Nuovo Cimento", Vol. II (1955) issue no. 1, to realize the importance and specificity of the subjects, considering the quantity of subjects dealt with. The index of the course which I feel is worth including here (see the following page) is self-explanatory, no further comments are thus necessary.

You can see how, along with the frontier elementary particle physics, which however is linked to fundamental physics, not only subnuclear but also nuclear —suffice it to look, not only at Fermi's lectures , but also at all the questions connected with photoproduction processes— and the origin of cosmic rays, the foundations for the construction of the accelerators were laid: CERN as well as the English and French machines plus the Frascati Laboratories electrosynchrotron in Italy would be offsprings of this.

Thus, it can be said that the Varenna School of '54 was the cultural milestone for a new "internationalization" of physics on the European level.

At this point it might be appropriate to mention that the INFN (Istituto Nazionale di Fisica Nucleare) was set up in 1951 and reorganized in 1952. Its Board of Directors in 1953 decided to build an Electrosynchrotron of 1000 MeV. In 1954 the Accelerator Section with Giorgio Salvini as its director, was established in Pisa and 3 years later, in July 1957, this sector was transferred to Rome becoming the National Laboratory of Frascati. There the Electrosynchrotron was built and began to function in 1959. Salvini's lecture together with those of Enrico Persico at Varenna were thus in essence a preannouncement of the first great physics experimental endeavours that were to take place in Italy. In a similar way Edoardo Amaldi's lecture about the creation of CERN and related projects and those of J. B. Adams, T. G. Pickavance dealing with the 600 MeV Synchrotron project clarified just what the European Physics Frontier prospects were.

The following year the course entitled "Questioni di struttura nucleare e dei processi nucleari alle basse energie"(Questions about nuclear structure and low energy nuclear processes) directed by Carlo Salvetti, was held in Varenna. Among the partecipants were A. Bohr, D. Brink, J. Horowitz, I. I. Rabi, A. M. Weinberg, A. De Shalit in addition to M. Cini and S. Fubini. The prospects which were opening up also in the field of low energy nuclear physics can once again be traced back to Varenna and to Enrico Fermi's influence. Suffice it to remember that during those years research in nuclear physics was initiated at the Sicilian Center of Nuclear Physics in Catania with the Van de Graaff 2.5 MV accelerator (under the direction of Renato Ricamo) and at the Physics Institute in Padova with the design and installation of the 5 MeV ion accelerator under the direction of Antonio Rostagni.

These 2 initiatives were to then branch out into the establishment of the INFN "Laboratori Nazionali di Legnaro" (1968) and "Laboratori Nazionali del Sud" (1975). All of this resulted in an overall recognition of nuclear physics on an international level.

Fermi and nuclear physics

I feel that it would be of interest to add some additional comments to what has already been written about the chapter on the evolution of nuclear physics from the 30's until now which was excellently edited by Ugo Amaldi. I will now reiterate two statements which I myself had already made at the Accademia dei Lincei in 1992 during the 50th anniversary celebration of the first nuclear fission reactor. The first is related to Fermi's preliminary statement about the composition of the nucleus as it appears in the already cited text on Nuclear Physics: *"All nuclei are composed of Z protons + N neutrons. The mass number, A, is given by $A = Z + N$. Examine an isotope chart, such as at the back of this book, and notice that the stable elements lie along a curve starting out with $N/Z = 1$, ending with $N/Z = 1.6$. Nuclei with common Z are called isotopes, those with common A are called isobars, and those with common N, isotones."*

This statement is still valid but it is also a yardstick for measuring the progress made in nuclear physics: just think that at the time of Fermi and Segrè the "nuclear map" contained about 800 nuclear species, while today more than 2500 of them are known thanks above all to the reactions produced by heavy ions. And also that the synthesis of "exotic" nuclei far away from the stability valleys which have large values of isospins or superheavy nuclei (we are at $Z = 118$) is not only a question of chemistry. New structural problems about the nuclear systems are being posed.

The second remark is about the necessity, more or less evident, of introducing degrees of subnucleonic freedom (quarks interacting by gluon exchange) and of being convinced that the *"... (ultimate?) aim of nuclear physics is that of connecting the known phenomena of the nuclear medium to quarks and gluons and to the corresponding theory, quantum chromodynamics ..."*

Nevertheless today the greater part of the structural properties and a good part of the nuclear dynamics can be described by systems of interacting nucleons using "exchange forces" (as was done in Fermi's times), introducing as degrees of subnucleonic freedom only those of the mesons (π, ρ...). And this is true no matter what specific model for extracting "an effective" interaction among nucleons has been considered: shell model, collective model, interacting boson model and so on ....

On the other hand, it can be seen that the picture of the atomic nucleus which emerges, after 50 years of intensive study, is still complex, considering the fact that it represents a quantum system ranging from a few to many bodies. The study of many-body systems is a subject which pulls together the efforts of almost all branches of contemporary physics dealing with quantum systems, from macroscopic phenomena in condensed matter through molecular and atomic structures up to nuclei and elementary particles. Nuclear physics has a special role in this common effort. The reason for this is that it deals with systems which are complex enough to exhibit a variety of collective and symmetry phenomena which, however, are still sufficiently elementary to give rise to spectra exhibiting well-defined quantum states which can be studied in a very detailed manner. Thus particular symmetries such as those of the bosonic states of two nucleons (fermions) described by the "interacting boson" models can be used.

I should now like to make one last comment. The basic nuclear processes, such as those related to neutron interactions are still of current interest.

For example, we know that the residence time of slow neutrons within the nucleus (as was shown by the experiments which Fermi's group carried out in Rome and which is 10^{-18} s) exceeds the transit time (10^{-22} s) by such a large factor because the compound nucleus continues to be a crucial concept in the analysis of neutron reactions. Thus, the coexistence of independent particle motion with collective many-body behaviour of the compound system appears to be a fairly expected property in a great variety of quantum systems.

These systems, at zero temperature, can be liquid or solid (classical systems) or quantum liquids (the Fermi systems) which exhibit degrees of freedom of independent particle. The question is rooted in the major problem of different phases of nuclear matter: once again we see that the significance of the ideas which originated during Fermi's times and which he himself foresaw have a renewed meaning today (*"c'è qualcosa di nuovo, anzi d'antico..."* (there's something new, that is to say antique...)) in the science of today and not only in nuclear physics.

As Fermi himself says at the end of the famous article on the theory of β decay:

"Solo un ulteriore sviluppo della teoria, come un aumento nella precisione dei dati sperimentali, potrà indicare quale modificazione sarà necessaria." (Only further developments in the theory, as well as more precise experimental data, will be able to tell what modifications have to be made.)

REFERENCES

[1] FERMI E., *Lectures on pions and nucleons*, in *Proceedings of the International School of Physics, Varenna 1954, Course II*, edited by PUPPI G.; Supplemento to Vol. II, Serie X of *Nuovo Cimento*, **1** (1955) p. 17.

[2] Società Italiana di Fisica, *Celebrazione del trentennale della Scuola Internazionale di Fisica "E. Fermi"* (Editrice Compositori, Bologna) 1984.

[3] *Nuclear Physics*, A course given by E. Fermi at the University of Chicago. Notes compiled by J. Orear, A. H. Rosenfeld and R. A. Schluter (The University of Chicago Press). Reprinted in 1950.

[4] FERMI E., *Notes on Thermodynamics and Statisctics* (The University of Chicago Press) 1966.

[5] FERMI E., *Notes on Quantum Mechanics* (The University of Chicago Press) 1961.

[6] FERMI E., *Conferenze di Fisica Atomica* in *Note e Memorie (Collected papers)*, vol. II (Accademia Nazionale dei Lincei, The University of Chicago Press) 1965, p. 684.

[7] FERMI E., *A course in neutron physics*, compiled by J. Halpern in *Note e Memorie*, vol. II (Accademia dei Lincei, The University of Chicago Press) 1965, p. 440,

[8] RICCI R. A., *Nuclear Physics at the Fermi time and today*, Symposium in honor of E. Fermi, 1992 (Accademia Nazionale dei Lincei, Roma) 1993.
[9] FERMI E., *Elementary Particles* (Yale University Press, New Haven) 1951.
[10] SEGRÈ E., *Enrico Fermi Physicist*, (The University of Chicago Press) 1970.
[11] RICCI R. A., Appendix to E. P. FISCHER, *Aristotele, Einstein e gli altri*, Italian edition (R. Cortina Editore) 1997, pp. 403-416.

About the Author

RENATO ANGELO RICCI, alumnus of the Scuola Normale Superiore of Pisa and Paris, now Professor Emeritus of the University of Padua. Honorary president, and former President, of the Italian Physical Society, was President of the European Physical Society, and Vice-President of INFN; he has directed nuclear physics research in Turin, Amsterdam, Naples, Orsay, Rio de Janeiro, Yale, Florence and Padua; in particular he was a founder of the National Legnaro Laboratories where he supervised the construction of the first Italian heavy ion accelerator. He has contributed to the discovery of new nuclear species, to the determination of fundamental laws of nuclear spectroscopy and has promoted the Italian participation in the research at CERN on phase transitions in nuclear matter. He won the 2000 Somaini Prize in Physics.

Enrico Fermi's scientific work

Luisa Bonolis

"On Enrico Fermi's death many of his friends and admirers wished to commemorate him in some permanent fashion. It was immediately apparent that the greatest and most durable monument to Fermi was his own work, a monument that will last as long as the pursuit of scientific knowledge."

From the very beginning of the Preface the reader can understand how all of Fermi's fascinating story can be found, in the original documents, in "Enrico Fermi, Note e Memorie" (Collected Papers) jointly published by the Accademia Nazionale dei Lincei and the University of Chicago Press. Volume I (1962), edited by E. Amaldi, E. Persico, F. Rasetti and E. Segrè, covers Fermi's years in Italy from 1921 to 1939 before his emigration to the U.S. Volume II (1965), edited by E. Amaldi, H. L. Anderson, E. Persico, E. Segrè and A. Wattenberg, is made up of the remainder of the papers, and covers the years in the United States from 1938 to his premature death in 1954. (From now on they will be abbreviated as FNM I and FNM II.) All the editors of the first volume are physicists who were close friends and collaborators of Fermi's from the beginning of his scientific activity: Edoardo Amaldi, Enrico Persico, Franco Rasetti, and Emilio Segrè; Enrico Persico had even known Fermi since he was fourteen. Their scientific life had begun that very moment their personal relationship was struck up; the same must have happened with Herbert Anderson and Albert Wattenberg, who added their contribution to the editing of the second volume. Almost every paper is preceded by an illuminating introduction written by someone who had collaborated with Fermi at the moment, or who was in a position to reconstruct the circumstances in which the paper had been written. The long acquaintance of all the editors with Enrico Fermi has added to these volumes really unique features; the reader has actually the privilege of becoming acquainted with Fermi's work through the lively recollections of those who shared with him a very incredible scientific adventure. This account is based mainly on this two-volume collection. I have also freely used Emilio Segrè's biography, as a first-hand source of events and recollections, with a most precious insight into Fermi's style of working, and into his scientific thought.

1901-1917

"When I first met him he was fourteen years old, I noted with surprise that I had a schoolmate who was not only 'strong in science' (as one used to say in scholastic terms), but who was also endowed with a type of intelligence entirely different from that of other boys whom I knew and whom I considered good students and intelligent" [1]. Enrico Persico recalled the early times of his friendship with Fermi, which started shortly after the tragic death of Giulio, Fermi's beloved brother, and practically his only playmate during childhood. The two boys enjoyed building mechanical toys, small electric motors and model airplanes. When Fermi was ten, his parents enrolled him at the *Umberto I* Gymnasium-Lyceum, where he quickly developed a keen interest in physics and mathematics. According to Emilio Segrè, the ten-year-old Fermi was already trying to understand "what was meant by the statement that the equation $x^2 + y^2 = r^2$ represents a circle" [2]. For the rest of his life he was to consider this one of his greatest intellectual challenges.

In 1915, when Enrico Persico and Enrico Fermi became friends, they shared a great interest in physics, and began to do experiments such as determining with great precision the acceleration of gravity in Rome, the density of Acqua Marcia (tap water flowing from an ancient Roman aqueduct still serving Rome), and the earth's magnetic field. Often they took long walks together around the city, and during these outings Persico came to realize and marvel at his friend's extraordinary intelligence: "... in these adolescent talks Enrico brought a precision of ideas, a self-assurance, and an originality which continually surprised me. Furthermore in mathematics and in physics he showed that he was familiar with many subjects well beyond what was taught at school. He knew these topics not in a scholastic fashion, but in a way that enabled him to use them with the greatest skill and familiarity. Even then, to know a theorem or a scientific law meant to him, above all, knowing how to use it" [2].

During one of these strolls around the city, young Fermi discovered at a bookstall in Piazza Campo dei Fiori a 900-page treatise on mathematical physics —*Elementorum Physicae Mathematicae, Volumen Primum et Secundum* (Elements of Physical Mathematics, Volumes 1 and 2)— written in Latin in 1840 by Andrea Caraffa, a Jesuit priest who taught physics at the Collegio Romano, the secondary school run by his order in Rome.

Caraffa's book dealt with fluid mechanics, acoustics, optics and astronomy, and Fermi studied it thoroughly. The numerous notes he made on the margins, and the calculations he did on sheets of paper stuck between the pages, constitute an initial clear record of Enrico Fermi's natural tendency to reformulate in his own way problems already solved by other people using different methods. Fermi's study of this text was part of his cultural self-education.

During his years in secondary school, Fermi became acquainted with the engineer, Adolfo Amidei, one of his father's colleagues, who very soon "became convinced that Enrico was truly a prodigy" [3]. Amidei, a lover of mathematics and physics, was astonished to learn that the boy used second hand books for studying mathematics and physics "hoping to find one treatise that would scientifically explain the motion of tops and gyroscopes, but he could never find an explanation, and so, mulling the problem over and over again in his mind, he succeeded in reaching an explanation of the various characteristics of the mysterious movements by himself." Amidei suggested him that a rigorous explanation could be obtained only by mastering "a science known as 'Theoretical mechanics'; but in order to learn it he would have to study trigonometry, algebra, analytical geometry and calculus ...", and he supplied him with the books that he thought were "most suitable for giving him clear ideas and a solid mathematical base." In fact, Amidei contributed to Fermi's scientific education from his 13th to the end of his 17th year, by lending him many university textbooks, such as Bianchi's *Geometria analitica* (Analytical Geometry), Dini's *Analisi infinitesimale* (Infinitesimal Analysis) and Grassman's *Die lineale Ausdehnungslehre, ein neuer Zweig der Mathematik* (Geometric Analysis According to the *Ausdehnunglehre*), and, in particular, Poisson's *Traité de Mécanique* (Treatise on Mechanics), a classic treatise on rational mechanics, replete with discussions of mathematical methods for physics, that had a deep influence on young Fermi's scientific education. When he read a book, even once, he knew it perfectly and did not forget it; in fact, he had an exceptional memory together with a marvelous aptitude for the sciences. When Fermi returned the calculus book by Dini, Amidei told him that he could keep it for another year just in case he needed to refer to it again. He received this surprising reply: "Thank you, but that won't be necessary because I'm certain to remember it. As a matter of fact, after a few years I'll see the concepts in it even more clearly, and if I need a formula, I'll know how to derive it easily enough" [4]. This is exactly what he did during the rest of his scientific life.

1918

In July, Fermi got his secondary school diploma (having skipped the third year), and Amidei asked him "whether he preferred to dedicate himself to mathematics or to physics", Enrico replied: "I studied mathematics enthusiastically because I considered it necessary for the study of physics, *to which I want to dedicate myself exclusively.*" Then he asked him if his knowledge of physics was as vast and deep as his knowledge of mathematics and Fermi replied: "It is much wider and I think equally profound, because I've read all the best books of physics." For a year Amidei had advised him to study the German language so that he could read scientific publications without having to wait for the Italian and French translation, and now he thought that the proper moment had arrived to present to him his project for the future: Amidei suggested that Enrico take the entrance examination for the Scuola Normale Superiore in Pisa, and attend the University in that city. The many advantages would be a scholarship (including room and board), close contact with the faculty, an excellent library and extra-curricular teaching, plus a more serene atmosphere than Fermi's own home, which was still in mourning after Giulio's death.

To prepare for the examination, Fermi studied the French translation of the Chwolson's *Traité de Physique* (Treatise on Physics). At the end of July, he wrote his friend Persico that "I'm going through Chwolson fast, and I figure I'll have finished it in another month or month and a half, since I found about 1000 pages I could skip because I already knew them."

Chwolson's huge, nine-volume treatise dealt with all of the great turning points in physics; there he could find all the names that represent the transition from classical physics to the new physics: besides A. Einstein and M. Planck, there were W. Wien, H. Poincaré, H. A. Lorentz, M. Abraham, H. Minkowski, P. Ehrenfest, M. von Laue, J. Rayleigh and A. Sommerfeld. Also extensively cited were Vito Volterra and Tullio Levi-Civita, whom Fermi was soon to personally meet. At a 1909 conference in Salzburg, Einstein spoke admiringly of Chwolson's work, praising its "excellent" style and its exhaustive discussion of both theory and experiments. Fermi's study of Chwolson is what definitively attracted him to physics and its problems. On August 18, he wrote again to Persico, "The reading of Chwolson proceeds quickly and I expect to finish in the next three or four days. This is a study that I'm very happy to have done, because it has greatly deepened the notions of physics I already had and taught me many things I did not have the slightest idea. With these foundations I hope that I will be able to compete for Pisa with some probability of success."

As Amidei had expected, not only did Fermi win the competition, but he was also the first among the applicants. In fact, his essay for the entrance examination to the Scuola Normale, on "Caratteri distintivi del suono e le loro cause" (Distinctive properties of sounds and their causes), amazed the examiners for its depth and for his perfect mastery of mathematical tools. In the fall, Fermi enrolled at the University of Pisa as a Fellow of the Scuola Normale, where several of the most distinguished literary and scientific figures of modern Italy have studied.

1919

Fermi's intense extracurricular studies during his university years can be retraced through the letters he wrote to his friend Persico. In February he writes: "Since I have almost nothing to do for my class work and I have many books available, I try to enlarge my knowledge of mathematical physics and I shall try to do the same for pure mathematics." From the same correspondence we know that he read Poincaré's *Théorie des Tourbillons* and Appell's *Traité de Mécanique Rationelle* (Treatise on Rational Mechanics), which led him to explore the methods of analytical mechanics, and he studied chemistry textbooks such as Nernst's *Theoretische Chemie* (Theoretical Chemistry) and Ostwald's *Lehrbuch der Allgemeinen Chemie* (Manual on General Chemistry). He also carefully read Planck's *Thermodynamics*. Among Fermi's papers preserved in Chicago is a small leather-bound notebook where he wrote some notes between July and September 1919, in which he listed a series of subjects he had studied recently; as was his habit, it is written in pencil without almost any erasures. This sort of physics vademecum ranges from Hamilton and Jacobi's analytical mechanics to Lorentz's theory, special relativity, black body theory, diamagnetism and paramagnetism. There is also a bibliography listing many of the fundamental books, in particular Richardson's *Electron Theory of Matter* (the spring board for Fermi's first article on relativity), Rutherford's *Radioactive Substances and Their Radiations* (from which he compiled an extensive bibliography on "Radioactive Elements and Their Constants"), and Levi-Civita's notes on general relativity. Fermi's 1919 notebook also mentions Bohr's first works on the hydrogen spectrum, which at the time were certainly not well known and if they were not taken into consideration in Italy. There is also a brief but clear exposition of the kinetic theory together with Boltzmann's H-theorem. The booklet, totaling 102 pages, concludes with two bibliographies taken from Townsend's book on gas discharge, which deal with electrical properties of gases and photo electricity. Fermi's grasp of physics was far above the current teaching level, so that he was not only completely self-taught but was already moving confidently through all the fields of physics and mathematics, giving advice to his friend Persico and suggesting what books to read or what problems to tackle.

"This booklet shows many of the author's characteristics in an embryonic stage", first of all his capacity of choosing the material with "surprising discrimination". "Another characteristic", as Segrè remarks, "is that Fermi, although never repulsed or frightened by any mathematical difficulty, does not seek elegant mathematics for its own sake ... All told, it is surprising that after one year of university work a student should be able to put together such a booklet, which would be very creditable even for a teacher with a long educational career behind him", concluded Segrè [5].

1920

Initially Fermi chose mathematics as his major subject but soon changed to physics. He soon became friend with Franco Rasetti and Nello Carrara. The physicists with whom Fermi came in contact in Pisa were Luigi Puccianti, then professor of experimental physics and Director of the physics laboratory, and his assistant Giovanni Polvani, with whom Fermi became friend. Puccianti gave them free access to the research laboratories, and Fermi had very clear ideas about which experiments would constitute original research: "Carrara and I", recalls Rasetti [6], "who in the previous year had come to recognize Fermi's immense superiority in the knowledge of mathematics and physics, henceforth regarded him as their natural leader, looking to him rather than to the professors for instruction and guidance."

Besides studying physics Fermi also acquired a vast and thorough knowledge of mathematics, and whenever he needed an ingenious and powerful mathematical method, "Fermi always had it ready in some corner of his mind, even when it involved mathematical notions above and beyond the common knowledge of professional theoretical physicists" [7].

Fermi had already made a thorough study of what was then considered the Bible of atomic physics, Sommerfeld's *Atombau und Spektrallinien* (Structure of the Atom and Spectral Lines), and as a result of this he was able to master the quantum theory of the atom better than anyone else in Italy. The professors themselves considered him the indisputable authority on the Bohr-Sommerfeld theory, as Fermi wrote to his friend Enrico Persico on January 30, 1920, "At the physics department I am slowly becoming the most influential authority. In fact, one of these days I shall hold (in the presence of several magnates) a lecture on the quantum theory, for which I am always a great propagandist." He was then a third-year student, eighteen years of age.

1921

When the fourth German edition of Hermann Weyl's book *Raum, Zeit, Materie* (Space, Time, Matter) appeared, Fermi probably read it that same year (1). Fermi principally learned from Weyl the power of variational methods in mathematical physics. In fact, after studying quantum mechanics Fermi deepened and perfected his study of relativity, and during his third year at the Normal School he published his first works on problems in electromagnetism in the journal "Nuovo Cimento". The first article, "Sulla dinamica di un sistema rigido di cariche elettriche in moto traslatorio" (Dynamics of a Rigid System of Electrical Charges in Translational Motion – FP 1) (2), which dealt with the inert mass of a rigid system of electric charges, presents the interesting result that, in the most general cases, the mass is expressed by a tensor.

In the second article, "Sull'elettrostatica di un campo gravitazionale uniforme e sul peso delle masse elettromagnetiche" (Electrostatics of a Uniform Gravitational Field and the Weight of Electromagnetic Masses – FP 2), Fermi discussed, using general relativity methods, the effect of a uniform and static gravitational field on a system of electrical charges, and found that the charges have a *weight* equal to that of a material mass U/c^2 (where U is the electrostatic energy of the system), in perfect agreement with Einstein's principle of equivalence between mass and energy.

The formal apparatus present in these two Notes is the one typically used by mathematical physicists during those years, but Fermi was already focusing on concepts and findings closer to physics. According to Persico, the way Fermi worked was to "take the data on a given problem, process them, and then compare his results with the ones obtained by the published authors. In doing this type of work, he would sometimes pose new problems and solve them, or even correct mistaken solutions if they were universally accepted. This is how his first publications came into being" [8].

---

(1) The book, the outgrowth of lectures delivered by Weyl in Zürich in the summer of 1917 and appeared in 1918 (WEYL H., *Raum, Zeit, Materie* (Springer-Verlag, Berlin)) underwent a series of expansions and revisions culminating in the great classical work of 1923, the fifth edition. The fourth edition (1921) was translated also in French, and in English as "Space, Time, Matter" (Methuen, London 1922).

(2) Fermi's writings are chronologically ordered and numbered in "Note e Memorie" (Collected Papers). From now on, the abbreviated form FP followed by the corresponding number in the collection will be used.

1922

In January, Fermi wrote to Persico that he was "doing relativity". Relativity continued to be his great passion at the time when relativity was the center of attention, and he was developing his own research program. In the meantime, he had mastered the tensorial calculus methods developed by the Italian mathematicians Gregorio Ricci-Curbastro and Tullio Levi-Civita, which constituted the mathematical structure on which the theory of general relativity was based. His most notable result in the field of relativity appears in an article entitled "Sopra i fenomeni che avvengono in vicinanza di una linea oraria" (On the Phenomena Occurring Near a World Line – FP 3) where Fermi also demonstrated a theorem in absolute differential calculus whereby space near a world line behaves as if it were Euclidean. With the aim of treating problems involving gravitational and electromagnetic fields in a general and systematic way, in this study Fermi introduced a system of space-time coordinates (known as the Fermi coordinates) that was very effective in describing the temporal evolution of phenomena that occur in a small region of space (since this concept was significantly extended by the British mathematical physicist A. G. Walker in 1932, scientific literature generally refers to the Fermi-Walker coordinates). Even today, the Fermi coordinates are mentioned and used in many treatises on general relativity and it is still widely quoted in the most important treatises which deal with absolute differential calculus, first and foremost Levi-Civita's famous "Lezioni di calcolo differenziale assoluto" (Lectures on Absolute Differential Calculus), published in 1925.

On July 7th, Fermi graduated *cum laude* in physics. The dissertation, his first experimental work, was research on X-ray diffraction images. As to Fermi's reasons for choosing this topic, Rasetti recalled years later that "It must be explained that at the time in Italy theoretical physics was not recognized as a discipline to be taught in universities, and a dissertation in that field would have been shocking, at least to the older members of the faculty. Physicists were essentially experimentalists, and only an experimental dissertation would have passed as physics." Rational mechanics, the subject closest to theoretical physics, was taught by mathematicians, who considered it part of applied mathematics with complete disregard for its physical implications. "These circumstances explain why such topics as the quantum theory had gained no foothold in Italy; they represented a 'no-man's land' between physics and mathematics. Fermi was the first in the country to fill the gap." At any rate, Fermi had already developed a great desire to do experimental physics; in Rasetti's words: "He was from the first a complete physicist for whom theory and experiment possessed equal weights, even though for many years his fame was chiefly based on the theoretical contributions. But never has he been for a moment one of those theoreticians who, to use a joking expression later much in use among the Rome group, could not 'tell steel from aluminum'." The year before, Fermi had published a long theoretical paper, "I raggi Röntgen", (Röntgen Rays – FP 6) on the properties of X-rays in the journal "Il Nuovo Cimento", and in it he had shown to have perfectly mastered all the literature on the subject. At the time of his graduation, Fermi had already developed a well-defined scientific style and showed a clear tendency to follow a research program which was all his own, having already achieved an uncommon mastery of both theoretical and experimental physics. Three days later Fermi passed *cum laude* also the examination required for the "Diploma" of the Scuola Normale, with a thesis on probability which was found again only in 1959 in the Archive of the School. Fermi published a part of this work in 1926, entitled "Un teorema di calcolo delle probabilità ed alcune sue applicazioni" (A Theorem on Probability and Some of its Applications – FP 38*b*).

1923

By this time, at least in some circles, Fermi was already so well known, that he was asked to contribute to the Italian edition ("I fondamenti della relatività Einsteiniana" (Hoepli, Milano 1923)) of A. Kopff's book "Grundzüge der Einsteinschen Relativitätstheorie" (Leipzig, 1921) (Fundamentals of Einstein Relativity). Relativity was the main topic but most of the twelve essays in the appendix, written by highly reputed Italian physicists and mathematicians of the older generation, were very skeptical of and hostile to Einstein's theory. Fermi, writing one of the few essays favorable to Einstein, stood out for his ability to identify the most interesting developments in physics. In his short essay on "Le masse nella teoria della relatività" (Mass in Relativity Theory – FP 5), he underlined the fact that "The enormous conceptual importance of the theory of relativity" and the philosophical debate about its foundations "have perhaps diverted attention from another of its results, which, though less clamorous and (granted) less paradoxical, has no less noteworthy consequences for physics, and whose interest will very likely grow in the near future as the science develops. The result I am speaking of is the discovery of the relationship that binds a body's mass to its energy." After stressing the potential quantities of energy contained in a gram of matter according to the equation $E = mc^2$, Fermi continued: "It does not seem possible, at least in the near future, to find a way to release these dreadful amounts of energy —which is all to the good, because the first effect of the explosion of such a dreadful amount of energy would be to smash into smithereens the physicist who had the misfortune to find the way to do it."

Upon returning to Rome, Fermi became acquainted with Senator Orso Mario Corbino, Director of the Physics Institute at the University of Rome. Corbino was very influential both in politics —he had served as Minister of Education— and in Italy's scientific community; in Fermi's words he "was universally known as one of the most eminent scholars" ([3]). Gifted with a brilliant mind, Corbino immediately recognized Fermi's talent and decided to further the young man's career. Fermi now felt it was high time for him to go abroad, and enter a less provincial scientific community, so that he used his fellowship to study with Max Born in Göttingen. There he met Werner Heisenberg and Pascual Jordan, but he did not establish strong ties with this circle, remaining rather an outsider.

([3]) In E. Fermi, "Un maestro: Orso Mario Corbino" (FP 120) FNM I, p. 1017. (That same year, in 1923, Mussolini made Corbino Minister of National Economics, although he was not, and never became, a member of the Fascist Party.)

In 1923 Fermi published a series of works on analytical mechanics, in particular, the article entitled "Dimostrazione che in generale un sistema meccanico normale è quasi ergodico" (Demonstration that in General a Normal Mechanical System is Nearly Ergodic – FP 11*b*). This paper was translated into German, so that it could be evaluated and appreciated by Paul Ehrenfest, who, together with Einstein, was one of the greatest experts in statistical mechanics, and whose work at that time was of far-reaching importance for the foundations of mechanics ([4]). In the second of these 1923 publications, "Il principio delle adiabatiche ed i sistemi che non ammettono coordinate angolari" (The Adiabatic Principle and Systems that do not admit Angular Coordinates – FP 12), which Fermi wrote after barely a month in Göttingen, he defined the validity limits of Ehrenfest's principle of adiabatics, applicable to a mechanical system in which the forces or constraints are continuously modified over time, but very slowly compared with the periods proper to the system —or, as Ehrenfest termed it, "adiabatically". In the second article Fermi wrote at Göttingen, "Alcuni teoremi di meccanica analitica importanti per la teoria dei quanti" (Some Theorems in Analytical Mechanics of Importance for Quantum Theory – FP 13), he showed how Ehrenfest's principle could be used to quantize three-body atomic systems such as the molecule of ionized hydrogen (made up of two protons and an orbital electron) or the helium atom (nucleus plus two orbital electrons).

Much reference was later made to the above papers by Tullio Levi-Civita in his work "Sugli invarianti adiabatici" (On the Adiabatic Invariants, this volume, p. 85) presented in 1927 at the Como Conference, held during the celebration of the centenary of Alessandro Volta's death.

When George Eugene Uhlenbeck was about to go to Rome, in the autumn of 1923, Ehrenfest gave him a letter for Fermi, in which he asked the young physicist a series of questions. Fermi and Uhlenbeck struck up a great lifelong friendship. These circumstances were probably why Fermi decided to go to Leiden in September 1924 on the Rockefeller fellowship grant he had been awarded. Once back in Rome, he wrote a short essay, "Sulla probabilità degli stati quantici" (On the Probability of Quantum States – FP 17*a*), that can be considered for all intents and purposes his first major contribution to quantum mechanics.

1924

This year Enrico Fermi published "Über die Theorie des Stosses Zwischen Atomen und Elektrisch Geladenen Teilchen" (On the Theory of Collisions Between Atoms and Charged Particles – FP 23*b*), a semi-classical theory of *bremsstrahlung* that had some qualitative success, although many years would go by before it was rigorously justified by C. F. von Weizsäcker and E. J. Williams and widely employed for atomic and nuclear problems.

Fermi's version of the method was criticized by Bohr, who underlined the discrepan-

([4]) The German translation of this work appeared in two parts in "Physikalische Zeitung" of 1923 and 1924.

cies with experimental results on the distribution of electrons emitted in the collisions. This work can be considered as one of Fermi's most significant contributions before the discovery of quantum statistics. It is based on a simple, efficient idea that uses analogy as a scientific method: utilize consolidated results in a specific area of physics to solve an apparently remote problem. Fermi's article "Sopra la teoria di Stern della costante assoluta dell'entropia di un gas perfetto monoatomico" (On Stern's Theory of the Absolute Constant of the Entropy of a Perfect Monoatomic Gas – Fp 16), published the year before, is of historical interest because it was the first indication of Fermi's reflections on this subject, in particular, on the division of phase space into h^3-volume cells. But it was only in the work entitled "Considerazioni sulla quantizzazione dei sistemi che contengono degli elementi identici" (Considerations on the Quantization of Systems Containing Identical Elements – FP 19) that Fermi posed the problem of searching for a "missing principle" to explain the behavior of identical particles that obey the requirements of Bohr-Sommerfeld quantum mechanics, and more specifically to explain why entropy vanishes at absolute zero. Fermi's remarks bear some analogy to the hypothesis on the exchange of the coordinates of identical particles in the wave function of a system, as were later formulated in quantum mechanics. After considering the case of an atom with more than one electron, Fermi worked out the quantization of a gas of identical particles in a box, assuming that the box is divided into cells of equal volume. Only if the number of cells were equal to the number of particles, *i.e.* if there is one particle per cell, he would obtain the correct entropy. At least two years before writing his famous work on the statistics of the ideal gas, Fermi was unwittingly anticipating Pauli's principle, albeit for generic molecules, when he asserted, in concluding this paper on the quantization of systems containing identical elements, that "[The failure of Sommerfeld's rules for calculating the absolute value of the entropy of a gas is avoided by quantizing the motion of identical molecules] only in cases where each cell contains a single molecule, whereas if the gas is a mixture of two kinds of molecules and is quantized by enclosing the molecules in cells, so that each cell contains two molecules of different kinds, the correct result is obtained once again." So Fermi knew where the problem was —the quantization conditions— and he was close to the correct result —one particle per cell.

From September to December of 1924, Fermi was in Leiden on a fellowship from the Rockefeller Foundation, which he had been awarded because of the influence of the great mathematician, Vito Volterra, professor of mathematical physics in Rome and during those years president of the Accademia dei Lincei. Ehrenfest was a great mentor of Fermi's, thanks to his deep knowledge of physics and his great openness. Besides being in contact with Ehrenfest, Fermi met scientists like Hendrick Antoon Lorentz and Albert Einstein (who said he was "very fond" of Fermi), and became friends with other young physicists like Samuel Goudsmit and Jan Tinbergen (Tinbergen later went into economics). Aside from his lifelong friend, Enrico Persico, in Italy there were no physicists (in the strict sense of the term) with whom he could discuss his research on quantum mechanics and the theory of relativity, although important mathematicians like Tullio Levi-Civita (an absolute expert on relativity) and Federigo Enriques, Guido Castelnuovo, and to some extent Vito Volterra realized how brilliant Fermi was and were in touch with him.

1925

This was a crucial year for physics. In January, Pauli published his famous article on the exclusion principle, and in October, Uhlenbeck and Goudsmit announced their discovery of electron spin.

During the summer, Heisenberg, Born and Jordan laid the foundations of the new quantum mechanics (or matrix mechanics), which Fermi did not like because he found the formulation too abstract. Because of the type of mathematical training he had had, matrix algebra and operator mathematics, in general, were not particularly congenial to him, but partial differential equations were. This may explain why Fermi quickly accepted Erwin Schrödinger's version of wave mechanics but was diffident about Heisenberg's formulation, that is of course until Born, Jordan and Dirac demonstrated their complete equivalence. In any case, the unfamiliarity of matrix mechanics conceptual background seems also to be at the origin of his reluctance to accept the unaccustomed mathematics involved, as Emilio Segrè has reported: "Heisenberg's great paper on matrix mechanics of 1925 did not appear sufficiently clear to Fermi, who reached a full understanding of quantum mechanics only later through Schrödinger's wave mechanics. I want to emphasize that this attitude of Fermi was certainly not due to the mathematical difficulties and novelty of matrix algebra; for him, such difficulties were minor obstacles; it was rather the physical ideas underlying these papers which were alien to him" [9].

From the fall of 1924 to early 1926, Fermi taught theoretical mechanics and mathematical physics at the University of Florence, where his friend Rasetti had been an assistant professor since November 1922. The Physics Institute, directed by the physicist and Senator Antonio Garbasso, was located in Galileo's villa in Arcetri. Fermi and Rasetti performed a series of original experiments, the first of which was published in "Nature" and entitled "Effect of an Alternating magnetic field on the Polarization of the Resonance Radiation of Mercury Vapor" (FP 26). Rasetti recalls that this second foray of Fermi's into the experimental field after years of theoretical work "show his ingenuity in dealing with unfamiliar techniques; and constitute the first instance of an investigation of atomic spectra by means of radiofrequency fields, a technique that was to receive numerous applications many years later" [10].

At that time Fermi gave several lectures for the Mathematics Seminar at the University of Rome, and this gave him a good chance to write articles for the general public, such as "Sui principi della teoria dei quanti" (On the Principle of Quantum Theory – FP 22), "Sopra la teoria dei corpi solidi" (On the Theory of Solid Bodies – FP 29). These were published in "Periodico di Matematiche", a review devoted to the secondary school teachers. The periodical was directed by the mathematician Federigo Enriques, who hoped that through it his readers would be able to understand the basics of modern physics as well as mathematics.

E. Fermi at the age of 4 with his brother Giulio and his sister Maria.

E. Fermi student in Pisa (1920) with G. Pacella, G. Gotti and L. Fantappiè.

With N. Carrara and F. Rasetti during a trip in Alpi Apuane Mountains.

With E. Persico and M. Fermi (1923).

In Leiden (1924) with G. Dicke, S. Goudsmit, J. Timbergen, P. Ehrenfest and R. Kronig.

With N. Carrara, F. Rasetti and R. Brunetti (1925).

1926

This year Fermi published his first major contribution to theoretical physics entitled "Sulla quantizzazione del gas perfetto monoatomico" (Quantization of the Perfect Monoatomic Gas – FP 30), in which he formulated the theory of an ideal gas of particles obeying Pauli's exclusion principle.

The "preparatory role" of Fermi's works on Stern's method for the calculation of the entropy constant of a perfect gas and on the quantization of systems containing identical particles, is unanimously recognized. Unfortunately, little is known of the circumstances that led Fermi to the discovery of the statistical laws which govern such particles nor is much known about some of his most significant contributions in the theoretical field. "As a theoretician, Fermi was entirely self-sufficient; some of his work was done at home in the early hours of the morning, and occasionally even his closest associates had little information on the problem that had occupied his mind until he presented to them, in an informal lecture, the finished product of his meditations" [11]. In later years Fermi told Segrè that the division of phase space into finite cells was a problem he had thought a lot about, and that if Pauli had not announced the exclusion principle, he might have come up with it himself, starting from the entropy constant, which he had been working on since January of 1924, when he wrote his article on the quantization of systems containing identical particles.

When Pauli announced his "exclusion principle" in January 1925, Fermi immediately understood the underlying reasons that justified his statistics, which he was ready to develop on a basis and from a point of view relatively independent of the contemporary development of quantum mechanics. In fact, on July 2nd of 1924, the Indian physicist Satyendra Nath Bose had introduced a new kind of statistics for light quanta, producing the blackbody formula. Just over a week later, Einstein applied Bose's statistics to a gas of free particles. In considering the system as defined by the number of identical particles occupying the single-particle quantum states, Bose and Einstein had introduced a different way of counting microscopic states which applies to all particles of integral spin (such as photons, pions, He^4, and so forth). In fact, these occupation numbers are all it takes to define the state of the system. Fermi does not seem to have been greatly influenced by Einstein's theory, although in a footnote he points out the analogy between the two forms of statistics.

In introducing a new method of counting for particles of half integral spin, (like electrons, protons, neutrons, and so forth) which allowed only one particle per cell, Fermi arrived at a radically different form of the distribution function in which all levels are filled up to a transition region of energy where the fraction of cells filled drops continuously from one to zero over a region of median width kT. The median transition point is now called the Fermi energy or Fermi level. Upon reading Pauli's article, Fermi realized that he now possessed all the elements needed to formulate a theory of the ideal gas which would satisfy Nernst's principle at the absolute zero, give the correct Sackur-Tetrode formula for the absolute entropy in the low density and high temperature limit, and free of the various arbitrary assumptions necessarily introduced in order to derive a correct entropy value. He probably developed within a very short time the detailed and definitive form of the theory, which was published also in a German version ("Zur Quantelung des idealen einatomigen Gases" – FP 31) [11, 12]. Actually Fermi had come so close to the exclusion principle that, as Pontecorvo later recalled in his biography, "he was very disgruntled at not having managed to formulate Pauli's principle himself."

It is well known that Paul Adrien Maurice Dirac developed the same type of statistics independently of Fermi; but he was also the first to show that the two types of statistics, now usually designated as Bose-Einstein and Fermi-Dirac statistics, are related to the two possibilities of eigenfunctions of a system being either symmetric or antisymmetric with respect to the exchange of the coordinates of two identical particles. Dirac's work "On the Theory of Quantum Mechanics" was presented to the Royal Society on August 26th of that same year, whereas Fermi's first short note published in "Rendiconti della Reale Accademia dei Lincei" had been presented at the meeting of that Academy on February 7th. But Fermi's longer paper, which appeared in "Zeitschrift für Physik", did not seem to produce any immediate effect on the international scientific community, even if both publications antedate Dirac's by an appreciable time. The reason was that his formulation was based on procedures from the old quantum physics. Dirac himself recalled many years later that in reading Fermi's work, he had not realized how important it was because it seemed unrelated to the basic problems of the time.

According to the dates on which the articles reached the various scientific journals, Fermi-Dirac statistics were first applied to the field of astrophysics. On December 10, 1926, R.H. Fowler presented to the Royal Society a work entitled "Dense Matter", in which he demonstrated that an electron gas contained in a white dwarf must be a degenerate "Fermi gas". The importance of Fermi statistics for the electrons in a metal surfaced in an article by Pauli that appeared two months later in "Zeitschrift für Physik" (February 10, 1927), "Über Gasenartung und Paramagnetismus" (On the Degeneration of Gas and Paramagnetism). In this work, for the first time, Pauli applies Fermi's statistics to conduction electrons within a metal considered as a perfect degenerate gas, and uses it to explain the weak, temperature-independent paramagnetism of the alkalis. In honor of Fermi all particles that obey this type of statistics (*e.g.*, electrons, protons and neutrons) are now called fermions.

In the early months of 1926, Schrödinger's first work on wave mechanics appeared in "Annalen der Physik". On his return from Florence to Rome for summer vacation, Fermi met with Aldo Pontremoli and Enrico Persico who were studying and discussing this new formulation of quantum mechanics that used partial differential equations that thus seemed to most physicists to be much more practical than the "odd" matrix mathematics used by Heisenberg. These discussions led Fermi and Persico to write a joint paper entitled "Il principio delle adiabatiche e la nozione di forza viva nella nuova meccanica ondulatoria" (The Principle of Adiabatics and Kinetic Energy in the New Wave Mechanics – FP 37). Fermi's attitude towards Quantum Mechanics was "an entirely pragmatic one: Quantum Mechanics is acceptable *because* its predictions agree with experiment. (He once said 'the Schrödinger equation has no business agreeing so well ...'). Nothing else counted. He devoted no time to such topics as 'The Quantum Theory of Measurement'... He was immune to the 'Copenhagen spirit', both by temperament and by educational background. He was completely self-taught in quantum mechanics, an outsider of the Göttingen-Zürich-Copenhagen spirit circle of its founders ... It may be supposed that Fermi always needed to draw a firm line between physics and 'philosophy'" [13].

Soon after the discovery of the statistics Corbino succeeded in having a chair in theoretical physics, the first in Italy, established in Rome. The other two successful candidates were his childhood friend Enrico Persico (who replaced Fermi at Florence), and Aldo Pontremoli, who later was to disappear tragically in the Nobile polar expedition of 1928.

1927

When Fermi got his chair in theoretical physics at the University of Rome, no research group was active there. Persico was in Florence and Corbino was absorbed by his teaching and other commitments. But soon Rasetti was transferred from Florence to Rome and became Corbino's assistant leader. Emilio Segrè, who had met Rasetti through his friend Giovanni Enriques, the son of the mathematician Federigo, got to know Fermi during the summer and, in the fall of the same year, decided to follow his "old desire of studying physics". He has remarked how "without anybody taking special notice of it, I had become Fermi's first pupil, at least in the formal sense. The school in Rome had been started" [14].

In June of the same year Corbino appealed to his best engineering students to transfer to the physics Institute: Edoardo Amaldi, the son of the mathematician Ugo Amaldi, took up his invitation request. After a few months Segrè talked to his friend and schoolmate Ettore Majorana, and he also joined the group. The first core of an Italian school of modern physics took shape around Fermi. Emilio Segrè, Edoardo Amaldi and Ettore Majorana were still students, but, as Segrè recalled, the "speed at which it was possible to form a young physicist at that school was incredible" [15]. Bruno Pontecorvo joined the group later on. "We never had a regular course", recalled Segrè, in remembering the peculiar training they had had in via Panisperna in the beginning. "Fermi's seminar was always improvised and informal. In the late afternoon we would meet in his office, and our conversations might give rise to a lecture —for example, if we asked what was known about capillarity, Fermi would improvise a beautiful lecture on its theory. One had the impression that he had been studying capillarity up to that moment and had carefully prepared the lecture. I find in one of my notebooks on the discussions of those years the following topics: blackbody theory, viscosity of gases, wave mechanics (the establishment of Schrödinger equation), tensor analysis, optical dispersion theory, gaussian error curve, more quantum mechanics, and Dirac's theory of the spin" [16]. His course on mathematical physics was a sort of encyclopedia containing some electrodynamics, relativity theory, heat-conduction theory, elasticity theory and diffusion theory. The instruction was chiefly in theoretical physics, commented Segrè, and no distinction was made between future theoreticians or experimentalists.

By this time Fermi had become quite well known within the international scientific community; Einstein, writing to Lorentz in June of 1926, suggested that Fermi and Langevin would be better speakers on quantum statistics than he, at the fifth Solvay Conference in October. In September, an International Physics Conference was held in Como, one of the events arranged to celebrate the centenary of Alessandro Volta's death. All the great figures in the world of physics were there —Hendrik A. Lorentz, Ernest Rutherford, Max Planck, Niels Bohr. On this occasion the latter presented for the first time his principle of complementarity as an elaboration of the content of the uncertainty relations derived by Heisenberg in March of that same year. In some way, all this marked the end of the new quantum mechanics revolutionary period.

The event was rich in satisfactions and recognition for Fermi. In fact, Arnold Sommerfeld described a series of remarkable results that showed the importance of Fermi's new statistics for interpreting the behavior of electrons in metals, which could not be explained at all by the classical theories.

Fermi was deeply concerned with the application of the new statistics to the electrons in metals. His concern was a foreshadowing of his train of thought which within a short time would mature into the idea of applying the same reasoning to the completely degenerate state of the electrons in an atom. After developing the statistics of a gas of particles obeying Pauli's exclusion principle, he used the same method to calculate the effective potential that acts on the electrons of an atom, considering them as a gas of fermions at absolute zero maintained around the nucleus by Coulombian attraction. The cloud of degenerate free electron gas does not collapse into the nucleus as it would in a classical distribution because the exclusion principle limits the occupation of states. This idea was the basis of the work "Un metodo statistico per la determinazione di alcune proprietà dell'atomo" (A Statistical Method for the Determination of Some Atomic Properties – FP 43), known today as the Thomas-Fermi model of the atom. Fermi was unaware that L.H. Thomas had reached essentially identical conclusions a year earlier; the journal in which Thomas had published them was not widely circulated. The virtue of the Thomas-Fermi theory is its great simplicity in comparison to the full many-body quantum mechanical description of atoms or molecules. Fermi himself (see papers from 43 to 48) and his associates (described in several papers during the following years solved it numerically using a small mechanical hand calculator such as those available in the late 1920s. In 1930 Chandrasekhar realized that a model similar to the Thomas-Fermi model could explain why certain cold stars known as white dwarfs did not collapse under the influence of gravity. In fact, the Thomas-Fermi model giving an approximate description of the atomic density, can be used to provide a simple qualitative explanation of how the size of everything around us depends on a delicate balance between the electrostatic forces and Fermi pressure. The approximate expression for the shield of the Coulomb potential due to the electrons as a whole, introduced by the effective potential of the Thomas-Fermi model, was applied in solid-state physics throughout the second half of the 20th century.

1928

During Fermi's first years in Rome his work was devoted to atomic physics and spectroscopy. He was engaged in both theoretical and experimental work, although he often left the latter to Rasetti. In 1927-1928 Fermi worked out some of the main applications of the model but other members of the group also took part in it, such as Giovanni Gentile jr. and Ettore Majorana, who was still a student. Fermi published a series of articles in which he used the Thomas-Fermi model to calculate atomic properties that vary regularly with variations in atomic number. Over the years he was to apply his statistical method to many problems in very different fields; much later, he used it to analyze high-energy nuclear problems or questions.

In 1928, Fermi published his brilliant textbook "Introduzione alla fisica atomica" (Introduction to Atomic Physics), which filled a major Italian academic textbook gap. Very little of modern physics was known or taught in Italy at the time. Relativity had flourished in the hands of mathematicians such as Tullio Levi-Civita and Guido Castelnuovo, but quantum theory and atomic physics were essentially absent. Fermi was impatient with older peoples'resistance to the new developments of quantum mechanics; he spoke on the subject to professional mathematicians less familiar with the experimental background of physics. With the help of Corbino, Rasetti and Persico, he devoted a lot of time also to semi-popular lectures on these subjects, which were then published in technical periodicals for engineers and teachers. His commitment is evidenced by several articles, that appeared mainly in 1930, such as "I fondamenti sperimentali delle nuove teorie fisiche" (Experimental Foundations of the New Physical Theories – FP 56), "Atomi e stelle" (Atoms and Stars – FP 60), "La fisica moderna" (Modern Physics – FP 62).

1929

In 1927 Dirac published the first works in which he systematically extended the rules for quantizing mechanical systems to electromagnetic fields. During the winter of 1928-29 Fermi started studying the quantum theory of radiation, and his first step was to look at Dirac's works. But Amaldi tells us "The method used by Dirac did not appeal to Fermi, who preferred, as he did very often, to recast the theory in a form mathematically more familiar to him" [17]. In turning his attention to quantum electrodynamics, the new frontier of theoretical physics, Fermi came up against all the problems involved in trying to explain the interactions between the electromagnetic field and the charged particles of matter by the principles of quantum mechanics, and the consequent problem of photon emission and absorption. "He tried very hard to find some solution to the known difficulties of the divergencies of quantum electrodynamics and he filled several notebooks with attempts to find some form of Hamiltonian which would be satisfactory in this respects." The attempts did not bear fruit, but some interesting results are to be found in a series of papers from published from 1929 to 1932 where he formulated a relativistically invariant description of the interaction between charged particles and the electromagnetic field which treated both particles and electromagnetic field quantum mechanically. According to Amaldi, Fermi, at least in the beginning, did not study the papers on quantum electrodynamics by Pauli and Heisenberg, nor was he very familiar at the time with the formalism of second quantization. In all his QED papers from 1929 to 1932 Fermi described matter as particles obeying either Schrödinger or Dirac equations rather than quantized fields — *à la* Jordan-Klein, Jordan-Wigner or Heisenberg and Pauli.

In his first QED paper "Sopra l'elettrodinamica quantistica" (On Quantum Electrodynamics – FP 64) published in 1929, Fermi stressed that he wanted to formulate the equations of motion of classical electrodynamics in such a way that "they can readily be translated into a quantum form."

"While doing this work, Fermi taught his results to several of his pupils and friends including Amaldi, Majorana, Racah, Rasetti and Segrè. Every day when work was over he gathered the various people mentioned above around his table and started to elaborate before them first the basic formulation of quantum electrodynamics and then one after the other, a long series of applications of the general principles to particular physical problems. A striking feature of Fermi's method of working on a theoretical problem in public (so to speak) and of teaching at the same time, was the way in which he could say out loud what he was thinking, proceeding at a steady, unhesitating pace, never going extremely fast, but never failing to make progress" [17].

All this work on electrodynamics was summed up by Fermi in a lecture series delivered in April 1929 at the Institut Henri Poincaré in Paris "Sur la théorie de la radiation" (On the Radiation Theory – FP 66) and, in a more complete form, at the School of Theoretical Physics at the University of Michigan, in Ann Arbor, during the summer of 1930 ("Quantum Theory of Radiation" – FP 67). The latter was published in "Review of Modern Physics" in 1932, and contained the first notions of field theory. Through this monograph generations of researchers learned electrodynamics, as much later Hans Bethe recalled: "Many of you probably, like myself, have learned their first field theory from Fermi's wonderful article in the "Review of Modern Physics" of 1932. It is an example of simplicity in a difficult field, which I think is unsurpassed. It came after a number of quite complicated papers and before another set of quite complicated papers on the subject, and without Fermi's enlightening simplicity I think many of us would never have been able to follow into the depths of field theory. I think I am one of them" [18].

Fermi's QED work and his exposition of the theory in "Review of Modern Physics" had a great influence on an entire generation of physicists during the 1930s, and it was the point of departure for Richard Feynman's formulation of QED.

1930

As could already be seen in his works on relativity, Fermi usually avoided purely epistemological reflections and adopted a pragmatic approach whereby he was satisfied when a mathematical formalism was able to explain experimental data. An exception was the debate on the correct interpretation of the uncertainty principle. In an article entitled "L'interpretazione del principio di causalità nella meccanica quantistica" (Interpretation of the Causal Principle in Quantum Mechanics – FP 59), he tried to explain what it means to say that future events cannot be determined in quantum mechanics, and showed clearly that the new theory is characterized not so much by the indetermination of the development of events over time, as by the uncertainty necessarily implicit in descriptions of physical states. This work also stemmed from heated discussions with the mathematicians Guido Castelnuovo, Tullio Levi-Civita and Federigo Enriques after a series of lectures on quantum mechanics that Fermi delivered at the Mathematics Seminar organized at the University of Rome.

The German physicist Hans Bethe was in Rome during this period as a Rockefeller Foundation fellow, and he was struck by Fermi's working style: "My greatest impression of Fermi's method in theoretical physics was of its simplicity. He was able to analyze into its essentials every problem, however complicated it seemed to be. He stripped it of mathematical complications and of unnecessary formalism ... This method was particularly impressive to me because I had come from the school of Sommerfeld in Munich who proceeded in all his work by complete mathematical solution. Having grown up in Sommerfeld's school, I thought that the method to follow was to set up the differential equation for the problem (usually the Schrödinger equation), to use your mathematical skill in finding a solution as accurate and elegant as possible, and then to discuss this solution. In the discussion finally, you would find out the qualitative features of the solution, and hence understand the physics of the problem. Sommerfeld's way was a good one for many problems where the fundamental physics was already understood, but it was extremely laborious. It would usually take several months before you knew the answer to the question. It was extremely impressive to see that Fermi did not need all this labor. The physics became clear by an analysis of the essentials, and a few order of magnitude estimates. His approach was pragmatic ... Fermi was a good mathematician. Whenever it was required, he was able to do elaborate mathematics; however, he first wanted to make sure that this was worth doing" [19]. In 1932 Fermi and Bethe published an article they had co-authored, "Über die Wechselwirkung von Zwei Elektronen" (Interaction Between Two Electrons – FP 70), where they describe the interaction between charged particles in terms of photon exchanges, thus showing that quantum field theory brings to physics a new way of understanding forces.

Gathered around Fermi and Franco Rasetti were now the very young physicists Emilio Segrè, Edoardo Amaldi and Ettore Majorana. Around 1930 the leaders of the Rome group, Fermi Corbino and Rasetti, became involved in discussions on the future development of physics. There had always been a strong spectroscopic tradition in Pisa, Florence and Rome, and most experimental work had been in that field. For some time, however, Fermi had felt that physicists would be ready in the near future to attack the problems of nuclear structure and that efforts should be directed to that end. A very lucid presentation of these arguments had been given in a speech presented by Corbino at the Italian Association for the Advancement of Science in 1929. His conclusion had been that "the only possibility of great new discoveries in physics is offered by the chance that somebody will succeed in modifying the atomic nucleus." The move to nuclear physics was gradual but determined. This represented a radical change in the research project of the group, but, in fact, according to Rasetti [20], Fermi had set his mind on learning the techniques of nuclear physics many years before his name became famous in that field. He put that decision to work in 1930, when members of the Rome group began to go to the major foreign research centers to learn about experimental techniques unknown in Italy at the time. Rasetti went to study with Robert Millikan at Pasadena, where he did important work on the Raman effect, and then spent a year in Berlin-Dahlem with Lise Meitner, learning nuclear techniques; Segrè went to Otto Stern in Hamburg and Pieter Zeeman in Amsterdam, to study forbidden spectral lines; Amaldi went to Peter Debye's laboratory in Leipzig, where he worked on X-ray diffraction of liquids. Fermi himself began to devote a great amount of time to these problems, although his theoretical work in that period was along totally different lines. As a first step, in the winter of 1930-31 he worked on the construction and operation of a cloud chamber, with Amaldi's assistance. The weakness of Rome University's physics Institute was its lack of apparatuses and the inefficiency of the machineshop where only very simple instruments could be built, and this took a great deal of time and effort. These circumstances persuaded Fermi to use do-it-yourself methods that typified his way of doing both experimental and theoretical work. The cloud chamber was built and used, but the tracks of alpha particles observed in it were of poor quality because the expansion mechanism was inefficient and it was hard to eliminate the old tracks. Photographs were not even attempted.

1931

In the spring of 1931, the difficulties encountered at the practical level persuaded Fermi to abandon the cloud chamber project and once again devote himself entirely to theoretical work, though he continued to participate in the experimental work done in Via Panisperna, which until that year was almost entirely in the field of atomic and molecular spectroscopy, partly because they knew the technique well and had suitable instruments. The standard Fermi tool, the statistical model of the atom, was applied to the study of hyperfine structure and the nuclear magnetic moments.

At the end of the 1920's the model of the nucleus then commonly accepted, *i.e.* a system composed of A protons and $A-Z$ electrons came up against a series of difficulties. Most of them originated from the application of the "then new" quantum mechanics to the interpretation of nuclear phenomena. One of the difficulties, and by far the most dramatic one, was which of the two quantum statistics is obeyed by nuclei. As Rasetti's experiments on the Raman effect of the rotational spectra of biatomic molecules with equal nuclei of 1929/1930 on Raman spectra had proved, ^{14}N has spin $I = 1$ and obeys Bose-Einstein statistics and not Fermi-Dirac statistics ([5]). Since ^{14}N was supposed to consist

([5]) The Raman effect was one of the main topics discussed at the Leipzig conference on molecular structure that year. Rasetti participated reporting about an apparently anomalous Raman effect in CO_2, which was correctly explained by Fermi in his "Über den Ramaneffekt des Kohlendioxids" (The Raman Effect in Carbondioxides). This was followed by a joint article with Rasetti, "Über den Ramaneffekt des Steinsalzes" (The Raman Effect in Rocksalt), and by a review article, "L'effetto Raman nelle molecole e nei cristalli" (The Raman Effect in Molecules and of 21 particles obeying Fermi-Dirac statistics it itself had to obey the same statistical law. This conclusion and its comparison with the experimental results of Rasetti gave rise to different opinions. The more popular view was expressed by stating that "when inside a nucleus, the electrons loose some of the properties which they have outside" [21].

Taking an extreme position Bohr pointed out "the failure of the fundamental quantum mechanical rules of statistics when applied to nuclei ..." and that "according to experimental evidence, the statistics of an ensemble of identical nuclei is determined solely by the number of protons ... while the intranuclear electrons show in this respect a remarkable passivity" [22]. Experimental results had also definitively proved that the spectrum of the electrons emitted in beta decay is continuous, so that only two theoretical possibilities were open for its interpretation: a) the principle of energy conservation is valid only statistically in the processes of beta decay; b) the conservation of energy is strictly valid in all single beta decay processes, but simultaneously with the electron, also another radiation is emitted which escapes observation.

The apparent non-conservation of energy, was a hypothesis that Bohr took into serious consideration. This was in fact the most astonishing aspect of beta nuclear decay; the emitted electron seemed not to possess all the energy that should have been released in the transition. But Pauli thought, "Bohr is on a completely wrong track" and suggested that the second hypothesis should be valid: one or more neutral particles are emitted together with the electron. In beta disintegration, the energy released in the process is split between the electron and the neutrino, so that the electron's energy can be any value from zero up to a certain maximum. Pauli had informally proposed this in a famous letter written on December 4, 1930. To take stock of the unresolved questions in nuclear physics, Fermi decided to organize an international conference on the subject. The first international gathering in the field, was held under the aegis of the Accademia d'Italia and lavishly sponsored by the Fondazione Volta, where Corbino was very influential. It was held in Rome from October 11 to 17, 1931, and the presence of the leading brains engaged in the nuclear field gave the event enormous importance. In private conversations during the conference, Wolfgang Pauli again advanced the hypothesis of the existence of a new particle, "neutral, light and very penetrating, to save the principle of the conservation of energy in beta decay", as he told Franco Rasetti years later. To avoid its being confused with the neutron, the "Pauli particle" was renamed "neutrino", which in Italian means "a small neutral object"; it seems the term was jokingly suggested by Amaldi to denote the phantom particle emitted together with the electron in beta decay —a particle whose existence would be confirmed only by the C. Cowans and F. Reines experiment in 1956.

The nuclear physics conference —followed by one in London in 1934— put the seal on the new disciplinary identity, and helped the group to familiarize with current problems. Times were getting ripe for Fermi's group, likely the first modern research team in physics, to enter upon the international stage [23].

Crystals), presented at a meeting of the Accademia d'Italia in 1932, where Fermi summarized results on the theory.

1932

A series of significant discoveries and events helped focus the attention of an important part of the physics community on nuclear phenomena and the new research possibilities in this field. These possibilities were enormously increased by the development, availability and productive use of new particle accelerators. It all started in January of 1932. Harold Urey announced the discovery of a heavy isotope of hydrogen, which he dubbed deuterium. In February, James Chadwick demonstrated the existence of a new nuclear particle, the neutron. In April, John Cockcroft and Ernest Thomas Walton obtained the first nuclear disintegration by bombarding light nuclei with artificially accelerated protons. In August, Carl Anderson discovered a new particle —the positively charged electron, immediately named "positron"— in cosmic ray tracks. This was experimental confirmation of the existence of the antielectron foreseen by Dirac's relativistic electron theory. Patrick Blackett and Giuseppe Occhialini immediately sought and detected not only the tracks of positrons in exposed emulsions, but also events in which the creation of the electron-positron pair under the effect of very-high-frequency gamma radiation is more evident. That same summer, Ernest Lawrence, Stanley Livingston and Milton White used the cyclotron devised by Lawrence —a device that a few months later would be able to generate around 5 MeV— to disintegrate nuclei.

In July the Fifth International Conference on Electricity was held in Paris. Fermi was invited to give a report and he chose the nucleus, reading the paper published as "Lo stato attuale della fisica del nucleo atomico" (The Present State of Atomic Nucleus – FP 72), his first on a nuclear subject. He stressed the difficulties inherent in a nuclear model based on electrons and protons as constituents of the nucleus, and mentioned Pauli's hypothesis of the neutrino.

In 1931 and again in 1932, Franco Rasetti had visited Lise Meitner's lab at the Kaiser Wilhelm Institute for Chemistry to learn nuclear techniques. He was there soon after the discovery of the neutron, wrote some papers, and returned to Rome after having worked with neutron sources, counters, cloud chambers, and radioactive substances. Upon Rasetti's return from Berlin-Dahlem in the fall of 1932, Fermi and his colleagues definitively decided to start a program on nuclear physics research. They designed the necessary devices, which were then built by outside firms. One was a cloud chamber similar to the ones L. Meitner was using in Berlin; it worked perfectly right from the start. Another was a crystal spectrometer for gamma rays; a special technique was used to prepare large bismuth crystals for this instrument, described by Fermi and Rasetti in the article "Uno spettrografo per raggi 'gamma' a cristallo di bismuto" (A Gamma-Ray Crystal Spectrometer – FP 78), published in 1933. Various types of Geiger-Müller counters were built. Rasetti, who had learned the techniques for preparing radioactive sources in Berlin, devised a polonium-beryllium neutron source. These developments were made possible because Italy's National Research Council amply funded it. By the end of 1933, the nuclear instrumentation would have been adequate to do research along different lines.

E. Fermi in the family Villa in Caorso (1926).

Como Conference (1927).

With Heisenberg and Pauli in Como (1927).

The marriage of Enrico Fermi and Laura Capon (1928).

Assembly of the Italian Academy of Sciences (1931).

Nuclear Physics Conference in Rome (1931) with O. M. Corbino, M. Cantone and A. Sommerfeld.

With J. Franck in Hamburg (1931).

With A. Rostagni, G. Wataghin, E. Persico and M. Rostagni in Gressoney (1932).

"I Ragazzi di Via Panisperna", 1934 (O. D'Agostino, E. Segrè, E. Amaldi, F. Rasetti and E. Fermi).

Berkeley 1937, J. R. Oppenheimer, E. Fermi and E. O. Lawrence.

With W. Zinn.

1933

At the end of October Fermi attended the seventh Solvay Conference on Physics held in Brussels. About forty scientists attended it; its overall subject was the "Structure and Properties of Atomic Nuclei". Nuclear physics had taken giant steps on the experimental level, and the model of the nucleus made up of protons and neutrons had been fairly well established through the work of Werner Heisenberg, Dmitri Iwanenko, Ettore Majorana and Eugene P. Wigner. Nearly all the most important nuclear physicists were there, including James Chadwick, Patrick Blackett, Ernest Rutherford, Lise Meitner, Frédéric Joliot and Irène Curie, Ernest Lawrence (the only American invited), Ernst Rutherford, and theoreticians like Niels Bohr, Werner Heisenberg, Wolfgang Pauli, Paul Adrien Maurice Dirac and George Gamow. Pauli presented once again his idea of the neutrino, and it appeared for the first time in the conference proceedings. However, the hypothesis was still fairly vague and at any rate no one had yet come up with a formal theory of the beta decay process. Of the three types of radioactive emission, two, the α and γ rays, did not pose substantial conceptual problems. Many details still remained to be clarified, but in 1933 the nature of alpha emission was well understood, and the essentials were well interpreted by Gamow's successful 1928 theory. In γ radioactivity photons are emitted. Although many details were missing in 1933, this phenomenon was clearly a close analogue of photon emission in atoms, *i.e.* a transition between two different quantum states of the same nucleus. It was also well known that β rays are electrons, but what was their origin? Before Chadwick's discovery of the neutron, the current hypothesis was that nuclei were composed of protons and electrons. A helium atom, for instance, would have contained four protons, providing most of the mass, and two electrons. In the proton-electron model of the nucleus β decay would have been similar to α decay, the emission of an electron already present in the nucleus, a true disintegration. In fact, the proton-electron model had failed an important test with Rasetti's determination, in 1929, of the statistics of nitrogen's atoms, a result which would be fully appreciated later. After the discovery of the neutron the transition to the modern proton-neutron model was very rapid. But if electrons are not present in the nucleus, what happens in β decay? There was in fact a more serious problem posed by β decay: α and γ rays are emitted with an energy equal to the difference between the energies of the initial and final nucleus, so as to guarantee the overall conservation of energy; in β decay, on the contrary, electrons are emitted, in any given transition, with a continuous energy spectrum. Niels Bohr had suggested that electrons in the nucleus could not be described within the framework of quantum mechanics, and was led to propose that energy is not exactly conserved in nuclear processes like β decay. On the other hand, Wolfang Pauli was unwilling to abandon energy conservation, and proposed a solution for this second puzzle: in β decay a second particle was emitted together with the electron, so that the two would share in different ways the available energy. This would certainly explain why the electron appears with a range of energies. The "second particle" —a new particle, never observed before— would have been neutral, and available data excluded that it could be a photon. Pauli was very prudent about his idea, which he probably considered too extreme, and he did not publish it. He discussed it in the corridors of physics conferences, but never officially, until the Solvay meeting, when he finally communicated his ideas about the neutrino in the discussion following Heisenberg's lecture.

Fermi liked the idea, and two months after the Solvay conference he completed his celebrated work on the explanation of beta decay, in which he applied quantum field theory to beta radioactivity. The emission of an electron is similar, he said, to the emission of light by an excited atom —neither the beta particle nor the light quantum is contained in the atom before its emission— but the emission of the beta particle is due not to electromagnetic interaction, but to a new class of forces (much later on it would be known as weak interaction). At the core of Fermi's theory is the idea that electrons and neutrinos can be created and destroyed, and in the disintegration process, every transition from neutron to proton must be accompanied by the creation of an electron and a neutrino. These processes are well understood in the quantum theory of the electromagnetic field, developed by Dirac soon after Heisenberg's quantum mechanics came into being. That the formalism of quantized fields could be applied to the creation and absorption of any particle —which must necessarily obey either the Bose-Einstein statistics, as photons do, or, as in the case of an electron, the Fermi-Dirac statistics— had been shown since 1927 by Dirac, Klein, Jordan and Wigner. "Apparently he had some difficulty with the Dirac-Jordan-Klein method of the second quantization of fields, but eventually also mastered that technique and considered a beta decay theory as a good exercise on the use of creation and destruction operators ... The theory that he built on these foundations is remarkable for its ability to withstand almost unchanged two and a half decades of revolutionary advances in nuclear physics. One might say that seldom was a physical theory born in such a definitive form" [24].

In this way, Fermi formally struck electrons off the list of nucleus components and opened up a new field in elementary particle physics: the physics of weak interactions.

His first article appeared in the Italian National Research Council's journal "La Ricerca Scientifica", and had the title "Tentativo di una teoria dell'emissione dei raggi β" (Tentative Theory of Beta Rays – FP 76), but, as a matter of fact, Fermi had intended to announce the results of his theory in a letter to "Nature". The submission was rejected with a note explaining that it contained too many abstract speculations "too remote from physical reality to be of interest to the readers." He then sent a longer paper ("Tentativo di una teoria dei raggi β" – FP 80*a*) to "Il Nuovo Cimento" and its German translation to "Zeitschrift für Physik" ("Versuch einer Theorie der β-Strahlen. I". – FP 80*b*); both appeared early in 1934.

The impact of Fermi's 1933 and 1934 papers actually goes well beyond the study of weak interactions. Fermi's paper was the first in which quantum field theory was used in the modern sense, and it must thus be considered the first modern paper on the physics of elementary particles. In using the language of quantum fields to describe entirely new phenomena, he greatly influenced, for example, Yukawa's meson theory.

At the moment Fermi's results opened up new perspectives for research already under way on nuclear forces, in fact, Emilio Segrè recalled that "Fermi was fully aware of his accomplishment and he said he would be remembered for this paper, which he thought it would turn out to be his masterpiece, remembered by posterity" [25].

During those same years, a group of young researchers was gathering together in Florence with Antonio Garbasso, head of the University's Physics Department and, like Corbino, a man with a great deal of influence on the institutional level. In 1928 Gilberto Bernardini had become Enrico Persico's assistant; like Fermi, Persico had won a chair in theoretical physics two years earlier and at that time was teaching the new quantum mechanics. Bruno Rossi had moved to Florence in 1928, at the age of 23, just after graduating at the University of Bologna; thanks to his intuition, a promising long-term research program on the physics of cosmic radiation was set up in 1929. The young Rossi built his famous multiple-coincidence electronic circuit, which made it possible to observe simultaneous pulses ("coincidences") between more than two Geiger-Müller counters and detect the trajectories of individual corpuscles of penetrating radiation. At the time, physicists were still debating whether these corpuscles represented the primary phenomenon, coming from cosmic space, or were generated in the upper atmosphere as a result of another kind of radiation. If the intensity of the penetrating radiation were found to depend on geomagnetic latitude (the latitude effect), the discovery would bear out the primary-phenomenon hypothesis. Rossi had foreseen that in this case the earth's magnetic field would have had to cause an east-west effect too; that is, an asymmetry in cosmic ray intensity relative to the plane of the planet's magnetic meridian. In 1930 and 1931 Rossi performed a series of experiments to discover whether an east-west effect exists; Rossi thought their negative outcome might be ascribed to atmospheric absorption. Before reaching any definite conclusion, it was necessary to improve the theory of geomagnetic effects. During one of his frequent visits to Rome, he discussed the problem of absorption with Fermi, and in 1933 they published a jointly written article entitled "Azione del campo magnetico terrestre sulla radiazione penetrante" (Effect of the Earth's Magnetic Field on Penetrating Radiation – FP 74). They concluded that in the vicinity of the equator the east-west effect should have been clearly observable. "On the strength of this prediction" recalled Rossi, "I decided to organize a cosmic-ray expedition to Africa, which was carried out shortly thereafter and provided valuable data on the east-west effect whose existence had been meanwhile announced by Johnson and by Alvarez and Compton" [26]. In fact, because of what Rossi called "logistical problems", his team was nosed out by T. H. Johnson and L. Alvarez (a young student of Arthur Compton's), who moreover made the discovery using Rossi's "telescope", a particular set of counters assembled in a multiple-coincidence electronic circuit.

1934

In January Irène Curie and Frédéric Joliot announced the discovery of new radioisotopes they had obtained artificially by bombarding the nuclei of light elements with alpha particles.

Fifty years later Edoardo Amaldi described what happened in Rome: "After the papers of Joliot and Curie were read in Rome, Fermi, at the beginning of March 1934, suggested to Rasetti that they should try to observe similar effects with neutrons by using the Po_α + Be source prepared by Rasetti. About two weeks later several elements were irradiated and tested for activity by means of a thin-walled Geiger-Müller counter but the results were negative due to lack of intensity."

"Then Rasetti left for Morocco for a vacation while Fermi continued the experiments. The idea then occurred to Fermi that in order to observe a neutron induced activity it was not necessary to use a Po_α + Be source. A much stronger Rn_α + Be source could be employed, since its beta and gamma radiations (absent in Po_α + Be sources) were no objection to the observation of a delayed effect ... All one had to do was to prepare a similar source consisting of a glass bulb filled with beryllium powder and radon. When Fermi had his stronger neutron source (about 30 millicurie of Rn) he systematically bombarded the elements in order of increasing atomic number, starting from hydrogen and following with lithium, beryllium, boron, carbon, nitrogen and oxygen, all with negative results. Finally, he was successful in obtaining a few counts on his Geiger-Müller counter when he bombarded fluorine and aluminium. These results and their interpretation in terms of (n, γ) reaction were announced in a letter to "La Ricerca Scientifica" on March 25, 1934. The title: "Radioattività provocata da bombardamento di neutroni – I" (Radioactivity Produced by Neutron Bombardment – FP 84*a*) indicated his intention to start a systematic study of the phenomenon which would have brought to the publication of a series of similar papers" [27] (FP 84-94).

In fact, Fermi and his group, pioneered a new line of research whose outcomes no one could have foreseen at the time. To speed things up, Fermi asked Amaldi and Segrè to work with him, summoned Rasetti back from Morocco, and invited the chemist Oscar D'Agostino, who was just back for Easter vacation from the Joliot-Curie laboratory, where he had learned the techniques of radiochemistry, to join the group. Some sixty elements were irradiated with neutrons over a short period of time, and new radioactive elements were discovered, and often identified, in at least forty of them. The results obtained by the "Via Panisperna boys" demonstrated all the advantages of teamwork in science, which Fermi had introduced for the first time in Rome. The great importance of these results was immediately clear. On April 23rd, Ernest Rutherford —the father of nuclear physics— wrote to compliment Fermi on the success of his experiments. "I congratulate you", he said, "on your successful escape from the sphere of theoretical physics! You seem to have struck a good line to start with."

Proceeding with their systematic bombardment, Fermi and his group reached thorium (atomic number 90) and uranium (atomic number 92), but the natural activity of these elements made it hard to identify the new artificial radionuclides obtained. "We attempted since the spring of 1934 to isolate chemically the carriers of these activities, with the result that the carriers of some of the activities of uranium are neither isotopes of uranium itself, nor of the elements lighter than uranium down to the atomic number 86. We concluded that the carrier was one or more elements of atomic number larger than 92" [28].

The group was fairly certain it had produced and identified two transuranian elements, which were named hesperium and ausonium, even if they were very cautious about expressing such a possibility when writing the paper "Possible production of elements of atomic number higher than 92" (FP 99), appeared in "Nature" [29]. An idea they did not go into was the possibility that the uranium nucleus could split into "many large pieces", each of which could be an isotope of a known element located far from uranium and thorium on the periodic table. This hypothesis had been explicitly advanced by the German radio chemist Ida Noddack in an article published in 1934 and sent to Fermi, but Noddack's idea was quickly set aside at the time. In fact it took four more years of research at the major laboratories of the time to solve the enigma of the transuranians.

During the summer Amaldi and Segrè went to Cambridge, in England, where they brought the manuscript of a paper summarizing all their results which was immediately published in the "Proceedings of the Royal Society" ("Artificial Radioactivity produced by Neutron Bombardment" – FP 98). They helped T. Bjerge and H. C. Westcott to tackle an important question regarding the production of radioactive isotopes of the target. In fact there was great deal of doubt whether this would be the result of (n, γ) or (n, 2n) reaction, as it is presently called. They found what they thought was a very clear cut of an (n, γ), so that back in Rome Amaldi and Segrè irradiated other substances in order to find more examples. They thought they had found one in aluminum, but soon thereafter they found a different decay period for irradiated aluminum which showed that the so-called (n, γ) did not occur. They were quite sure of their results, even if it seemed that they were unable to reproduce them consistently. In the fall of 1934, Fermi assigned Amaldi and Bruno Pontecorvo (who had received his degree the year before) the task of establishing a quantitative scale of the radioactivity induced in the bombarded elements. As Amaldi recalled, "We immediately found, however, some difficulty because it became apparent that the activation depended on the conditions of irradiation. In particular in the dark room, where usually we carried out the neutron irradiation, there were certain wooden tables near a spectroscope that had miraculous properties. As Pontecorvo noticed accidentally silver irradiated on those tables gained more activity than when it was irradiated on the usual marble table in the same room."

The results baffled them for several days, until they decided to try to filter the radiation that produced the artificial radioactivity. To solve the mystery, on October 18th they started to make systematic observations, as appears from notebook B1 (now in the archives of the Domus Galileana, in Pisa), where the data from this period were recorded. In particular, Amaldi made a series of measurements inside and outside a sort of small box ("castelletto") with 5-centimeter-thick lead walls which served as a radiation shield. The results showed clearly that outside the lead chamber, activation capacity decreases rapidly as the distance between source and irradiated element increases, but inside the chamber the decrease is much slower. These measurements were recorded on page 3 of notebook B1. To measure the lead's absorption, a wedge of this substance was prepared with the idea of placing it between the neutron source and the detector, and then comparing its absorption at different thicknesses with that of a large block of lead with the same thickness. On the morning of October 22nd, the group members were involved with examinations and Fermi decided to proceed on his own to solve the "enigma of lead". Enrico Persico, visiting from Florence, was the only person with him, and it was Persico who wrote the measurements down on pages 8 and 9 of notebook B1 as Fermi read them out. As Fermi told Subrahmanyan Chandrasekhar years later: "One day, as I came to the laboratory, it occurred to me that I should examine the effect of placing a piece of lead before the incident neutrons. And instead of my usual custom, I took great pains to have the piece of lead precisely machined. I was clearly dissatisfied with something: I tried every "excuse" to postpone putting the piece of lead in its place. When finally, with some reluctance, I was going to put it in its place, I said to myself 'No! I do not want this piece of lead here; what I want is a piece of paraffin.' It was just like that: with no advanced warning, no conscious, prior, reasoning. I immediately took some odd piece of paraffin I could put my hands on and placed it where the piece of lead was to have been" [30].

It appears that Fermi, well acquainted with the results of the experiments carried out by the Joliots carried out in France and by Chadwick in Cambridge, already at the end of 1933, was aware both of the increase of the scattering cross section when the energy decreases and of the larger efficiency of paraffin with respect to lead in the slowing-down and in the absorption of neutrons. In this respect his decision would be the result of a subconscious elaboration of what was already known to him [31].

The paraffin-filtrated neutrons coming from the Rn-Be source had the miraculous effect of powerfully activating the irradiated substances, being far more effective than those that fell directly on the target. "The interpretation of these results was the following. The neutron and the proton having approximately the same mass, any elastic impact of a fast neturon against a proton initially at rest, gives rise to a partition of the available kinetic energy between neutron and proton; it can be shown that a neutron having an initial energy of 106 volts, after about 20 impacts against hydrogen atoms has its energy already reduced to a value close to that corresponding to thermal agitation. It follows that, when neutrons of high energy are shot by a source inside a large mass of paraffin or water, they very rapidly lose most of their energy and are transformed into 'slow neutrons'" [32].

Neutrons slowed down to the energy of the molecules' thermal excitement by elastic collisions with the hydrogen nuclei contained in paraffin spend more time close to the target nuclei and thus become more effective in inducing artificial radioactivity. This explained the "miraculous" properties of wooden tables, which, in fact, yielded slow neutrons whereas the source on the marble did not. In about half an hour they could explain the disagreement between Amaldi's and Segrè's results. Both were vindicated.

That same evening, October 22nd, the group wrote a letter to "La Ricerca Scientifica" —"Azione di sostanze idrogenate sulla radioattività provocata dai neutroni. I" (Effect of Hydrogenated Substances on Radioactivity Induced by Neutrons. I – FP 105a))— in which they announced that "A layer of paraffin a few centimeters thick inserted between the neutron source and the silver increases the activity rather than diminishing it." All this contradicted the spread knowledge that the more energetic the particles, the greater would be their effectiveness in producing reactions. This article was soon followed by a second one, with the same title (FP 106*b*).

The discovery had immediate practical applications; artificial radioactive isotopes could be used, for instance, as tracers in physics, chemistry and biology. Over the next six weeks, by December 6th, Orso Mario Corbino persuaded Fermi and his collaborators to take steps to share in the profit of any future industrial exploitation of neutron activation and patent the process together with some of the products. This was done and resulted in Italian patent No. 324458, of October 26, 1935.

The results of this work can be found in about 20 papers published by Fermi and collaborators during 1934, and most of them were translated into English. They are the culmination of the experimental activity of Fermi in Italy and cover one of the most fruitful periods of his career.

1935

Unsuccessful attempt were made in January-February to explain the great number of new activities induced in thorium and uranium, which had been isolated by the group in Rome as well as by other groups. Since, at that time, they tried to interpret all the activities observed as due to transuranic elements and their possible decay products, they explored the possibility that some of them were alpha emitters with a very short lifetime, and therefore (according to the Geiger-Nuttal law) had high energy. They always carried out the experiments using thorium and uranium covered with a layer of aluminum in order to suppress the low energy alpha emission due to decay of these elements. Obviously they failed to see alpha particles emitted by transuranic elements, moreover, the aluminum layer stopped all the fission fragments and thus, prevented them from getting inside the ionization chamber. In Segrè's words: "It was this aluminum layer which prevented us from seeing the big ionization pulses characteristic of fission" [33]. But, as recalled also by Rasetti, "In retrospect, it is difficult to say whether the fission explanation would have been found even if the phenomenon had been seen" [34].

The discovery of slow neutrons had, in fact, multiplied the strength of their sources, and in this sense constituted also an improvement in the study of old and new phenomena. The main results of all their work were summarized in a paper for the Royal Society, "Artificial Radioactivity produced by Neutron Bombardment. Part II" (FP 107), received on February 15th. By this time Fermi had made very substantial progress in the study of the theory of the slowing-down process, and further significant work was done during the following months.

The need to move from natural radioactive sources to accelerators was felt by Fermi and his team as early as 1935. By 1939 cyclotrons were working, or under construction, in several American laboratories and in Cambridge, Copenhagen, Liverpool, Paris and Stockholm in Europe; towards the end of the decade, "a laboratory without a cyclotron could no longer compete in the interdisciplinary nuclear science invented in Berkeley" [35].

Thanks to money provided by the Fondazione Volta, the Rome physicists traveled to the States during the summers of 1935 and 1936 to learn about the different kinds of accelerators being developed. That summer Rasetti was in California, visiting Robert Millikan's laboratory in Pasadena, where he studied a high-voltage linear accelerator, and the Radiation Laboratory at Berkeley, which had the new accelerator invented by Ernest Lawrence. Rasetti then spent the 1935-36 academic year at Columbia. Around this time, the group began to break up. Segrè moved to Palermo, where he had won the chair of experimental physics; Pontecorvo worked with Gian Carlo Wick, and then went to Paris in the spring of 1936 to work with the Joliot-Curies; D'Agostino began to work at the National Research Council's Chemistry Institute.

"When the fall of 1935 came", recalled Rasetti, "and we all should have returned to Italy, the Ethiopian crisis was at its height ... The group was virtually disbanded. Only Amaldi and Fermi were left in Rome with neutron sources. From there on the investigation concentrated mainly on the study of the slowing-down process and on the 'groups of neutrons'... Fermi and Amaldi worked extremely hard." They were reacting against an atmosphere that was growing bleaker and bleaker. In Germany, Nazi persecution of political opponents and ethnic minorities was in full swing, and Hitler's expansionist policy foreshadowed another full-scale war in Europe. All year long Italy had been preparing to fight a colonial war in Ethiopia; the invasion began on October 3rd and on November 18th the League of Nations voted to apply sanctions against Italy.

1936

They worked very hard on what has become the famous study of the motion of the neutrons in hydrogenated substances. As Amaldi recalls, they worked with "incredible stubbornness", "We would begin at eight o'clock in the morning and take measurements almost without a break until six or seven in the evening, and often later. The measurements were taken on a chronometric schedule ... They were repeated every three or four minutes, according to need, for hours and hours and for as many successive days ... Having solved one problem, we immediately attacked another without a break or feeling of uncertainty" [36].

The result that the cross-section for the capture process of a neutron by a nucleus should be inversely proportional to the velocity of the neutron was in qualitative agreement with the high efficiency of the slow neutron bombardment, observed experimentally. On the other hand it failed to account for several features of the absorption process, which, instead, appeared to obey more complicated laws, as other authors had shown: the absorption of slow neutrons by a given element appeared, as a rule, to be larger when the slow neutrons were detected by means of the activity induced in the same element.

Fermi approached the problem by developing a diffusion equation in which significant values are assumed for the distance a neutron travels from the point where it was created until it has been completely thermalized. This distance is measured by a parameter that has been known ever since as "Fermi age" or "Fermi lifetime". From the systematic study of the absorption properties of the various neutron groups an interesting new phenomenon emerged: each absorber of the slow neutrons has one or more characteristic absorption bands, and independently of this or these absorption bands, the absorption coefficient is always large also for thermal neutrons. Some elements, especially cadmium, absorb thermal neutrons very strongly, while being almost transparent to neutrons of higher energies. These experiments established that neutron groups could be explained as energy differences of resonant absorption lines. Bohr, as well as Breit and Wigner independently explained the above anomalies, as due to resonance with a virtual energy level of the "compound nucleus" (*i.e.* the nucleus composed by the bombarded nucleus and the neutron). It appears that Fermi's experiments directly influenced Bohr's thought leading him to his most significant contribution to nuclear physics during that period.

In a few months' time, from October 1935 to May 1936, Fermi and Amaldi published a series of works in "La Ricerca Scientifica" (FP 112-119), culminating in a long article "Sopra l'assorbimento e la diffusione dei neutroni lenti" (On the Absorption and the Diffusion of Slow Neutrons – FP 118*a*), in which they described a series of conclusive results from their systematic study of the absorption and diffusion of slow neutrons (6).

During the summer Fermi gave a course at Columbia University. Amaldi followed him, staying in New York for a month, where he translated this paper into English for publication in "Physical Review" ("On the Absorption and the Diffusion of Slow Neutrons" – FP 118*b*), and then he used his grant from the Volta Foundation to work on the project for the construction of a 1 MeV proton accelerator in Washington D.C.

1937

If the Ethiopian war marked the beginning of the decline of the work at the Institute, the sudden death, on January 23rd, of Corbino left Fermi without his political and scientific backing. Fermi wrote a brief eulogy, "Un maestro: Orso Mario Corbino" (FP 120) mainly emphasizing Corbino's scientific work, but also mention of his strong personality and generosity was made. Corbino's chair was given to Amaldi, who had won a competition for a chair of experimental physics in Sardinia. He was only twenty-nine years old, all of his excellent qualities had been enhanced under the guidance of Fermi and Corbino, and now he was ready to face up to the heavy and unexpected responsibilities resulting from impending dramatic events.

In fact, the person appointed to succeed Corbino as director of the Institute was not Fermi, as might have been expected, but Antonio Lo Surdo. This change at the top, plus the rapidly worsening political situation, foreshadowed imminent catastrophe.

The considerable effort in the experimental work with neutrons had left the Institute with few artificial neutron sources. By that time all the major laboratories were using linear accelerators and cyclotrons that provided much more intense artificial neutron sources than those that could be obtained with a maximum of 800 millicuries of radon plus beryllium contained in little glass cylinders. In June, Fermi and his group built a small-scale, 200 kV prototype of the accelerator in the Physics Institute, which had moved from its building on Via Panisperna to the University's new campus on the east side of Rome. The facility for producing neutrons by means of deuterium ions accelerated to 200 keV was described in an article published in "La Ricerca Scientifica", "Un generatore artificale di neutroni" (An Artificial Neutron Generator – FP 121).

In the meantime Fermi and Domenico Marotta, Director of the Istituto Superiore di Sanità (Public Health Institute), had also managed to secure approval for a proposal to build a 1 MV Cockcroft-Walton type accelerator capable of accelerating particles up to

(6) For a description of the work done at that time see Amaldi E., ref. [36] pp. 808-811.

energy of one million volts. The 1 MV accelerator, whose construction began soon after completion of the 200 kV prototype, was completed two years later, by that time Fermi had already left Italy.

At this time Fermi came to realize that an absolute priority for the development of physical research in Italy was the creation of a national laboratory for physics, independent of any single university institute and modeled on the most advanced research institutions abroad. On the basis of information received from Cockcroft, Joliot and Scherrer, in January of 1937 he submitted to the National Research Council a proposal for the creation of a national laboratory modeled on the most advanced research institutions abroad. During a visit to Ernest Lawrence in the summer of 1937, Fermi looked also into the possibility of building a "cheap" cyclotron in Italy.

Then Guglielmo Marconi died suddenly, in July. As president of the National Research Council and of the Italian Academy, Marconi had been a valid supporter of Fermi's group.

This was also the year of Ernest Rutherford's death; Fermi wrote an eulogy that appeared in "Nature" ("Tribute to Lord Rutherford" – FP 123).

1938

The death of Guglielmo Marconi in July 1937, following that of Corbino deprived Fermi of most of the support he could count on in the National Research Council. A final decision on Fermi's proposal for the creation of a national laboratory for physics was taken only in June 1938: considering that "for the creation of an Institute of Radioactivity ... much larger and more conspicuous means would be required that those approximately estimated by Professor Fermi", the Presidency of CNR resolved not to take into consideration the creation of such an institute, leaving to the Directory the decision whether to grant Fermi an annual budget "to organize researches in the field of radioactivity". By now it was clear that Italy's most important scientific institution was unable to give Fermi and his collaborators the means to pursue research that could compete with what was being done in the world's most advanced laboratories. Meanwhile, the political situation was precipitating. In March, Nazi Germany annexed Austria; in July, the Italian government fired the opening guns in its own anti-Semitic campaign officially announced in the *Manifesto della Razza.* Over the next few months, racial laws were promulgated and Jews were expelled from all public schools, universities and academies, and from all government jobs. Fermi's wife, Laura Capon, was Jewish. At the beginning of September, the couple decided to emigrate.

On November 10th, Fermi received the official announcement that he had won the Nobel Prize, and decided that after the award ceremony in Stockholm, he and his family would go straight on to New York. The political situation had forced them to make a decision, which he had told only a few intimate friends. During his previous visits to the United States, Fermi had developed a deep liking for that country. As Emilio Segrè recalled, "He was attracted by the well-equipped laboratories, the abundant funds for research, the enthusiasm he sensed in the new generation of physicists." Segrè emphasized that Fermi's decision to emigrate was "more the execution of a long-meditated plan than a sudden decision determined by circumstances." On December 10th, Fermi received the Nobel Prize for Physics "for his demonstrations of the existence of new radioactive elements produced by neutron irradiation, and for his related discovery of nuclear reactions brought about by slow neutrons." In his Nobel lecture, Fermi mentioned the hypothesis of the transuranians esperium and ausonium, and cited research done by Otto Hahn and Lise Meitner, who had apparently identified elements with atomic numbers up to 96. But an incredible chain of events had been set off in the meantime. Otto Hahn, an able radio chemist, and Fritz Strassmann, an expert in chemical analysis, had detected the presence of radioactive barium in elements produced by bombarding uranium with neutrons, and on December 22nd they sent an article to the journal "Naturwissenschaften" announcing their discovery.

Fermi's Nobel lecture, "Artificial Radioactivity Produced by Nuclear Bombardment" (FP 128) —a review of the work on the slow neutrons— was published the next year, so that in the proofs he could add a footnote: "The discovery by Hahn and Strassmann of barium among the disintegration products of bombarded uranium, as a consequence of a process in which uranium splits into two approximately equal parts, makes it necessary to re-examine all the problems of the transuranic elements, as many of them might be found to be products of a splitting of uranium" [37].

1939

Fermi had barely arrived in New York, on January 2nd, when he learned of the discovery of uranium fission. Otto Hahn and Fritz Strassmann's article was published in January but as Segrè recalled, "the news of these sensational discoveries spread by word of mouth, letter and telegram as the work was going on, before any result had been published." Hahn kept in touch by mail with his erstwhile colleague, L. Meitner, who was Jewish and had just escaped from Germany to Sweden because of the Nazi persecution. The letter Hahn sent to Meitner announcing the amazing news that barium was present in the products of uranium bombardment by neutrons reached her near Göteborg, where she was vacationing with her nephew, Otto Frisch, one of Niels Bohr's collaborators. Barium could only be a fission product. Together they devised the first theoretical interpretation of the fission process, calculated the energy released, understood that the transuranic elements were fission fragments, and figured that a large amount of energy would be released in the fission process.

When Frisch got back to Copenhagen, he found Bohr about to leave for the United States and told him the news. Bohr arrived in New York in mid-January, and in no time the news spread, reaching Fermi among others. The cyclotron had just started operating at Columbia and Herbert Anderson, a graduate student at the time, was setting up an ionization chamber-linear amplifier combination for his doctoral research on neutron scattering. Anderson proposed a simple modification of his ionization chamber, which would make it possible to observe the intense ionization caused by the fragments from the fission of uranium. This is his account of it: "We saw the pulses generated by the fission of uranium on the screen of my cathode-ray oscilloscope on January 25, 1939" [38]. The next day, in Washington, at the Fifth Theoretical Physics Conference, Fermi was able to speak about the fission process with the conviction of personal experience. He advanced the hypothesis that in such a violent reaction, neutrons might be released too. If the arrangements were such that the emitted neutrons could produce further fissions, the process might become multiplicative. If the circumstances were favorable enough a chain reaction might be developed and large amounts of energy released. A demonstration of the fission process was organized for the attendees. "By the time he returned to Columbia, Fermi knew what questions he wanted to answer. Were neutrons emitted in the fission of uranium? If so, in what numbers? How could these neutrons be brought to produce further fissions? What competitive processes were there?" And above all, added Anderson, "Could a chain reaction be developed?" Nuclear fission was becoming more than a scientific curiosity.

Fermi insisted on the need to make quantitative measurements. A month after his arrival at Columbia, Fermi, together with a working group that included Anderson and his thesis adviser J. R. Dunning, signed his first "American" article, "The Fission of Uranium" (FP 129), where the value of the fission cross-sections for slow neutrons as well as for fast neutrons was reported. In addition, evidence was found which indicated that the fission cross-section had $1/v$ dependence at low energies, and implication was made that it was the rare uranium isotope U-235, which was involved. In this period Fermi concentrated on the problem of neutron emission rather than on the aspects of the fission process which seemed less directly related to the possibility of producing a chain reaction ("Production of Neutrons in Uranium Bombarded by Neutrons" – FP 130). These experiment were parallel to those carried out independently by H. von Halban, F. Joliot, and L. Kowarski in Europe.

The question of the production of neutrons in uranium bombarded by neutrons was the subject of several reports published in "Physical Review" in 1939. After that, it became impossible to follow Fermi's work through public periodical literature; his reports were designated top secret. In "Simple Capture of Neutrons by Uranium" (FP 131), Anderson and Fermi analyzed the simple capture of slow neutrons by U-238, which, according to Meitner Hahn, and Strassmann's discovery of 1937, transmutes into the radioactive isotope U-239 which decays in 23 minutes into an element that would have atomic number 93 and mass 239. U-238's capture of a single neutron is thus a process that leads through neptunium to the production of plutonium, and that later proved to be of great importance. The process of production of U-239 competes with fission in taking up the neutrons which are needed to sustain a chain reaction, that is why Fermi and Anderson wanted to know how much of the absorption was due to this process, which in creating a chain reaction has to be avoided.

In the spring, Fermi, Anderson and Leo Szilard published "Neutron Production and Absorption in Uranium" (FP 132), where they reported that the number of neutrons emitted by uranium under the effect of slow neutrons is greater than the number absorbed. This is the necessary condition for setting off a chain reaction. This article also stressed the importance of resonance absorption and provided the clue to how, by lumping the uranium, the losses due to the neutron losses caused by this effect could be reduced. These experiments also highlighted a fundamental problem: thermal neutrons absorption by hydrogen was too large thus water was not the preferred medium for slowing down neutrons in a chain reaction. This was the first and last experiment Fermi did with Szilard, whose style of working on an experiment was not at all congenial to Fermi; from then on they lead two different groups.

Fermi was the first to inform U.S. military leaders about the possible military implications of atomic energy. In March he gave a lecture at the Navy Department, after which he was granted a small budget for this research at Columbia. In early summer, Szilard and Paul Eugen Wigner, both from Hungary, persuaded Albert Einstein to sign a letter addressed to President Roosevelt that described Fermi's and Szilard's chain reaction research in the United States and Joliot's in France, and said it was nearly certain that this result would be achieved "in the immediate future". After outlining the possibility of building what he called "a new type of bomb", Einstein ended by stressing the need to act speedily and create "a permanent liaison between the government and the group of physicists who are working on the chain reaction in America", because similar research might be under way in Germany. On September 1 the Second World War broke out. In October Roosevelt set up an Advisory Committee on Uranium to coordinate fission research in the various U.S. laboratories. Fermi and Szilard were regularly invited to the meetings. The Committee succeeded in obtaining a funding of $ 6,000 for the American armed forces to pursue chain reaction research with graphite.

In the summer, Fermi went to the University of Michigan at Ann Arbor (where he had lectured in 1929) for the Summer School of Theoretical Physics. During this period he kept in touch with Szilard by mail. In July, both of them reached the conclusion, independently, that water was not suitable as a moderator because hydrogen's absorption of thermal neutrons is too high. Since water could not be used to slow down neutrons in a chain reaction, it would be better to try graphite.

Meanwhile, Fermi's attention had been captured by the very lively debate on the discovery, two years earlier, of unstable particles in cosmic rays. These particles were thought to have a mass one or two hundred times the electron's and a mean lifetime of about 2 microseconds this was immediately measured by Bruno Rossi and collaborators. The pioneer of cosmic-ray research in Italy had fled to the U.S. after the Fascist regime's promulgation of racial laws. The hypothesis was that the particle in question might be the "heavy quantum" of the nuclear force field, postulated by the Japanese physicist Hideki Yukawa as a mediator of nuclear forces. For this reason the particle detected by Carl Anderson and Seth Neddermeyer at the California Institute of Technology was given the name "mesotron". Upon his return to Columbia, Fermi wrote a short note on the subject, "The Absorption of Mesotrons in Air and in Condensed Materials" (FP 133) a period then followed (1938-1943) in which physicists measured the properties of the mesotron (mean lifetime and decay properties), trying to fit it into the framework of Yukawa's theory, albeit with increasing difficulties.

While Fermi was on the verge of becoming involved in events which nobody could have anticipated a few years earlier, Rasetti had emigrated to Canada, Segrè had been dismissed from his Palermo post because of racial laws, and settled in Berkeley. Amaldi had gone to the United States to look for a job. When Germany invaded Poland, and his family was refused a passport, he returned to his post in Rome and in the following years he succeeded admirably in facing the arduous task of maintaining the international position that Italian physics had acquired during the decade of Fermi's preminence.

1940

The spring of this year brought experimental confirmation that the fission process produced by slow neutrons concerns only the rare isotope U-235 as Bohr had understood. (Natural uranium contains only tiny quantities (0.7%) of U-235, together with the more abundant U-238). In February Fermi was in Berkeley to give the Hitchcock Lectures, and there he met up with Segrè, whom he had not seen for two years. Together they used the 60-inch cyclotron recently activated to demonstrate the splitting of uranium by bombardment with alpha particles ("Fission of Uranium by Alpha-Particles" – FP 135). But on that occasion they did not discuss problems related to the chain reaction of uranium "because of the secrecy involved" [39].

When he returned to Columbia, Fermi went back to his experiments on neutron absorption and diffusion by graphite, and to his theoretical analysis of the chain reaction, an extension of his research in Italy on the slowing-down of neutrons in hydrogenated substances. The results of this work not only confirmed that graphite was the best choice for slowing down neutrons, but also were of the greatest importance in providing an initial theoretical basis for describing the behavior of neutrons contained in these substances. The slowing-down process was described by a diffusion equation according to a theory, which later became known as "Fermi age theory". Fermi and Anderson wrote the joint work "Production and Absorption of Slow Neutrons by Carbon" (FP 136), where also the absorption of thermal neutrons was investigated. The latter recalled how "Fermi returned with enthusiasm to the quest for the way to make the chain reaction. This was the kind of physics he liked best" [40].

Scientists at other universities were working on similar problems, and Fermi, considered the world's greatest expert on neutrons, was asked for advice on both theoretical and experimental problems. Meanwhile, as Hitler's troops swept through Europe, Roosevelt set up a National Defense Research Committee (NDRC) to coordinate military research and organize the mobilization of the scientific community for military purposes. The Uranium Committee was placed under its jurisdiction and foreign-born scientists who had not been nationalized like Fermi and Szilard were not allowed to be on it.

1941

After the success of the graphite measurements, Fermi and Anderson got interested in studying the fission process itself in detail. In particular, they wanted to know what was the probability, called the branching ratio, that when fission occurred in uranium, a given radioactive series would appear. From this work stemmed the paper "Branching Ratios in the Fission of Uranium (235)" (FP 137).

The advantages of graphite as a means of slowing down neutrons became apparent after the experiment, which used a pile of graphite to measure the absorption of carbon. The neutrons were slowed down more than in water, but once they reached thermal energies, they diffused more and reached greater distances from the source. The pile could be used to separate the thermal neutrons from higher energy neutrons and Fermi saw many ways to advantageously use this, like determining the average number of neutrons produced by uranium upon the capture of a thermal neutron ("Production of Neutrons by Uranium" – FP 138).

The idea of using the uranium in lumps, in order to reduce the resonance absorption, had by now taken hold, but Fermi wanted to know how big to make the uranium oxide lumps and how to space them in the graphite. ("Capture of Resonance Neutrons by a Sphere embedded in Graphite" – FP 139). Until the summer of 1941, research focused on the possibility of using the chain reaction for production of power rather than for explosion in a bomb. Fermi, because of this, wrote a reportin which he mainly discussed some general points of view as to the methods that could be employed for using a chain reaction in uranium as a source of energy, and submitted it to the Uranium Committee on June 30, 1941 ("Some Remarks on the Production of Energy by a Chain Reaction in Uranium" – FP 141). The separation of U-235 from U-238 and the production of plutonium were by then being studied, and some results had been achieved on a very small scale, but both enterprises seemed impossible from the technological standpoint. The other route to fission was based on Hahn's and Meitner's demonstration that U-238 transmutes into U-239 by capturing a neutron then decays in 23 minutes into an element that should have atomic number 93 and mass 239. The first transuranic element, eventually called neptunium, had been identified at Berkeley in the summer of 1940; as it decays, neptunium emits electrons with a period around two days long, giving rise to an isotope with mass 239 of the element with atomic number 94, namely plutonium (Pu-239). The study of this new element's properties showed it could be a potential nuclear explosive.

By the spring several members of the Uranium Committee had mainly taken into consideration a controlled chain reaction and doubted that atomic energy would be ready in time to affect the war. Scientists like Szilard were, on the contrary, convinced that atomic bombs were feasible. By that time, at the request of the National Defense Committee, the National Academy of Sciences appointed a special committee to review the military importance of the uranium work. The opinion of the committee evolved rapidly: after a first cautious report in May, the results of work on plutonium were revealed in July, and the possibility of a plutonium bomb was mentioned. On November 6th the National Academy committee submitted a third and most encouraging report, which discussed also the feasibility and critical size of a U-235 bomb.

The year ended with the Japanese attack on Pearl Harbor, on December 7th, and the United States' entry into the war against Japan, Germany and Italy. Research efforts to build a nuclear bomb were accelerated to top speed financially, scientifically and technically. If atomic weapons were feasible the United States had to have them before the Nazis.

Receiving the Nobel Prize from Gustav V, King of Sweden (1938).

With his wife Laura and their children, Giulio and Nella when arriving in New York (1939).

With N. Bohr at Carnegie Institution, Washington D.C. (1939).

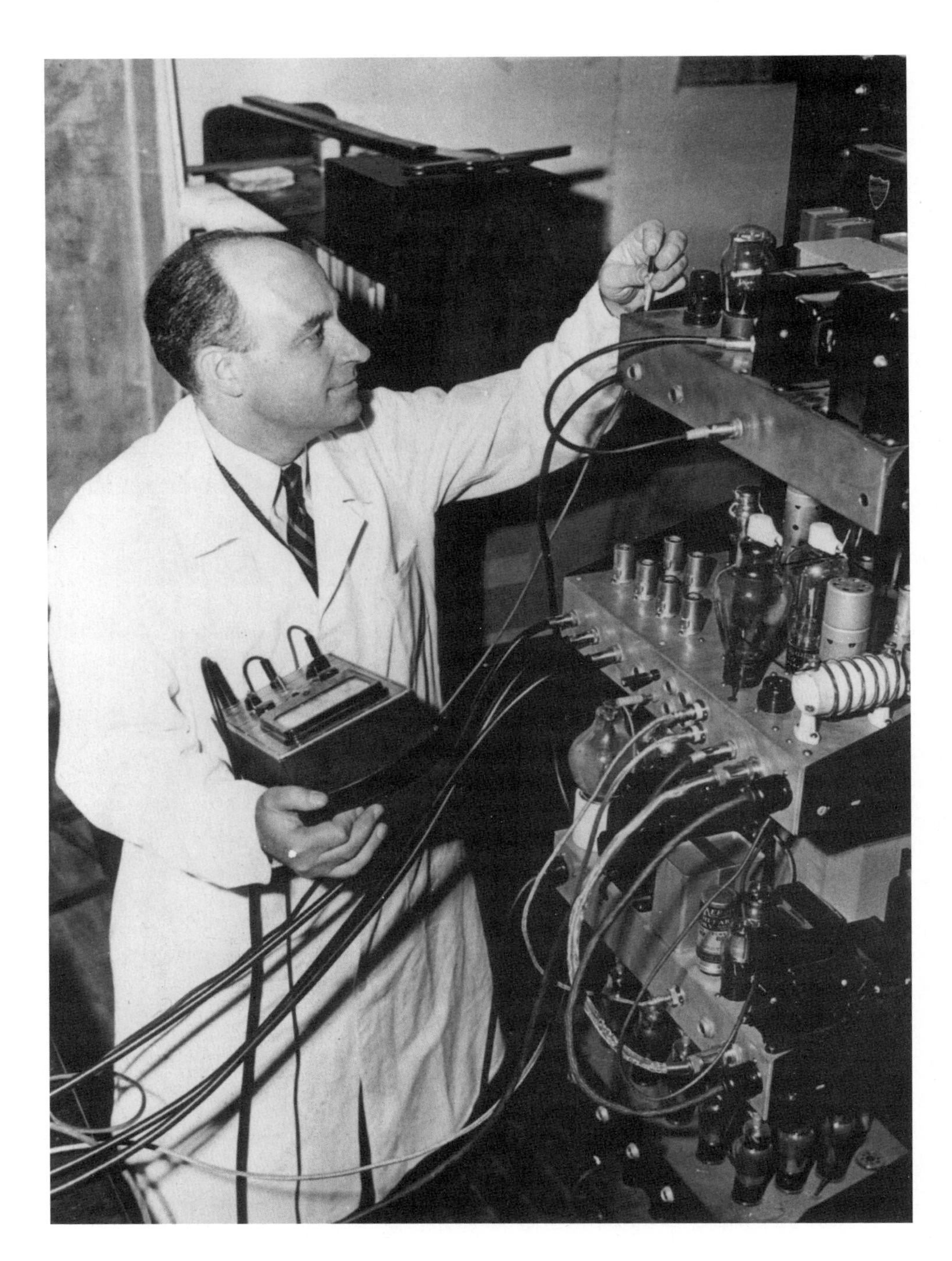

E. Fermi at the cyclotron (1940).

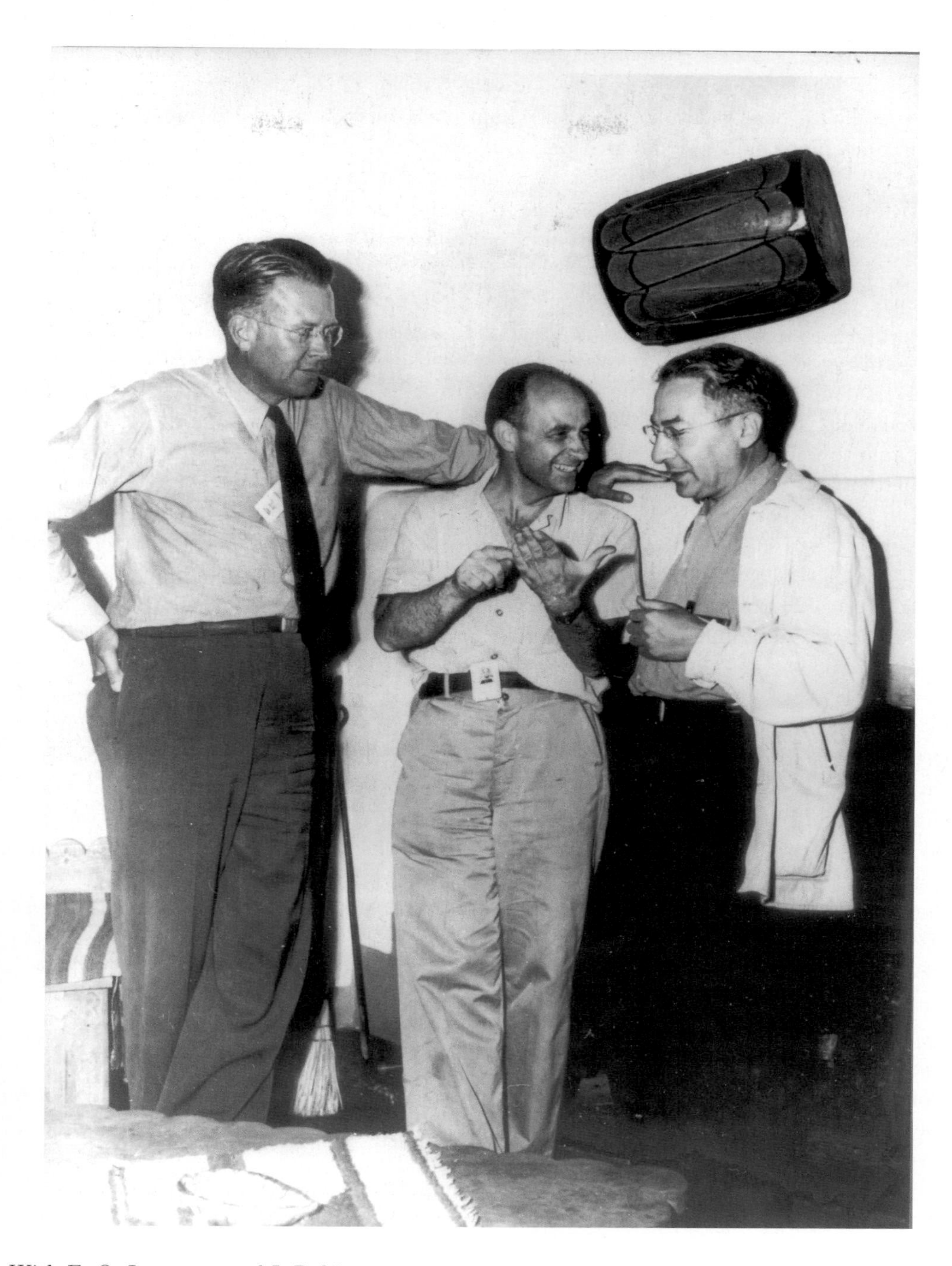

With E. O. Lawrence and I. Rabi.

1942

It was these circumstances which lead to the creation of the Metallurgical Laboratory.; its primary purpose was to develop the chain reaction with natural uranium, using this to produce plutonium. Arthur H. Compton, of the University of Chicago, was chosen as director. Fermi, being an enemy alien, was not considered eligible to hold such a position.

Before the move to Chicago, the center of the work, experiments were performed to test the neutron loss by leakage through the confines of the graphite-uranium structure. For a divergent chain reaction to occur, great care must be taken to avoid neutron leakage, especially along the boundary areas. To reduce such undesirable losses, it is necessary to use a sufficiently large structure. To perform the test on a smaller scale, Fermi invented the exponential experiment, which measures the exponential decrease in neutron density in a square-based uranium-graphite column. That is why these piles were called "exponential piles". Based on his "age theory", dating back to 1936, Fermi was able to know point by point the properties of the neutron flux inside the uranium-graphite system. The fundamental quantity in this theory is the k reproduction factor, and in fact these experiments measured the "reproduction factor" k, namely the mean number of neutrons produced in fission processes due to a given "generation" of neutrons. A chain reaction can occur only if k is greater than 1.

The most important of these experiments was performed to ascertain "whether a given lattice of uranium oxide lumps embedded in graphite would produce a divergent chain reaction if its dimensions were sufficiently large." A column of base 8 feet by 8 feet and 11 feet high gave $k = 0.87$. This was appreciably under 1, but it was possible to think of enough improvements in the purity, geometry, and density of uranium so that the prospect for a k greater than 1 looked fairly promising. The paper "Neutron Production in a Lattice of Uranium and Graphite" (FP 149) is the basic paper on the chain reaction, and makes clear how the results of the exponential experiment were to be interpreted. Experiments were set up to find the source of the impurities which caused the difference between the results observed and what was expected from the cross sections. The paper "Neutron Production in a Lattice of Uranium Oxide and Graphite [Exponential Experiment]" (FP 150) was a report on the first of that long series of tests conducted on different lattice dimensions and materials of different kinds and shapes, which eventually led to the first successful chain reaction.

In the last exponential experiments performed at Columbia University the results were encouraging, and gave $k = 0.918$ ("Preliminary Report on the Exponential Experiment at Columbia University" – FP 151). The committee now saw in the chain reaction not only the possibility of producing power but also its application as a producer of plutonium a likely competitor of U-235 for making atomic bombs

In the spring Fermi, Szilard and the other Columbia University physicists permanently settled in Chicago and started working at the Metallurgical Laboratory, the top-secret wartime project, where a nuclear reactor was to be built using natural uranium and graphite. Fermi's role was greatly changed, he was the scientific head of the project, besides tactfully directing the engineers, who had to deal with totally new problems, he had to attend meetings, write reports and give advice on technical matters. Instead of doing experiments himself, he had to put everything in the hands of trusted collaborators except the data analysis, which he reserved for himself. He was much sought after for advice: as he himself stated he seemed to be "doing physics by telephone".

A series of the so-called "exponential" experiments was now repeated on a much larger scale. The work headed by Fermi is described in over sixty declassified reports.

In June, President Roosevelt made the decision to go ahead with a large-scale program aimed at building nuclear bombs, and put the Army in charge of what was to be called the Manhattan Project.

In March, June and September Fermi organized a series of seminars for the staff of physicists who were working in Chicago on neutron physics and the chain reaction, and this was a good opportunity to review these ideas and check the soundness of his plans. These seminars were extraordinary examples of Fermi's greatness as a teacher, which stemmed from the fact "that he made little, if any, distinction between teaching and research" [41].

The exponential pile experiments at Chicago continued to improve in the purity of the graphite and uranium oxide, and in August multiplication factors greater than 1 were obtained. This meant that a large enough pile of natural uranium and graphite could sustain a chain reaction. By November the rate at which materials arrived had increased sufficiently so that it became apparent that there would be enough for the chain reaction by the end of the month. The original project envisaged building the atomic pile in the Argonne Forest, near Palos Park, but a strike slowed construction to the point that Fermi suggested using instead the space under the West Stands of Chicago's Stagg Field stadium, in the center of a densely populated district. Fermi was so sure of himself that he managed to persuade Arthur Compton, the head of chain reactor research, and General Leslie Groves, director of the Manhattan Project.

Fermi had decided to build the chain-reacting pile in a spherical shape in order to minimize the losses of neutrons from the surface. Construction was begun within a wooden framework, and the first layer of CP-1 was installed on 16 November. Graphite layers containing uranium oxide formed onto pseudo spheres in the pile alternated with layers of pure graphite blocks. Slots were machined in the graphite so that wooden rods wrapped in thin sheets of cadmium (a powerful neutron absorber, as Fermi and his group had discovered years earlier) could be inserted into the pile. Every evening cadmium rods were removed to check the neutron intensity. One special control rod, called Zip, operated by gravity. It was to be pulled out before the pile went into operation and was fastened with a rope. Another safety rod was attached to a hook held by a solenoid magnet; if the power failed it would automatically be pulled into the pile. The solenoid was also connected to an ionization chamber through a relay; if the ionization current exceeded a specified value, it would automatically fall in place inside the pile.

On the night between December 1st and 2nd, the last layer of uranium and graphite was stacked on top of the pile, during Anderson's shift. This was layer 57, the one that, according to the measurement results and Fermi's calculations, would have made the pile critical, triggering the chain reaction.

The next morning some forty people came to witness the event, nearly all of them scientists working at the Metallurgical Laboratory. Fermi had prepared a routine for an approach to criticality under perfect control. After a series of proofs Fermi said: "I'm hungry. Let's go to lunch".

After lunch the procedure started again with the rod 21 set at one of its early morning positions. "The last cadmium rod was pulled out step by step. At each step a measurement was made of the increase in the neutron activity, and Fermi checked the result with his prediction, based on the previous step ... The process converged rapidly, and he could make predictions with increased confidence of being accurate. So it was then when he arrived at the last step, Fermi was quite sure that criticality would be attained then" [42]. Indeed, once the cadmium rod was pulled out entirely, the pile went critical. At this point Fermi broke into a big cheerful smile. He put away his slide rule and announced, "The reaction is self-sustaining".

The pile was allowed to function for twenty-eight minutes, at a maximum power of 0.5 watt. Eugene Wigner produced a bottle of Chianti, and Leo Szilard, who had done so much to foster the use of nuclear energy, commented: "This is an unfortunate day for the history of man". Fermi, in his monthly report for December, "Experimental production of a Divergent Chain Reaction" (FP 180) —declassified only ten years later— wrote simply "The activity of the Physics Division in the past month has been devoted primarily to the experimental production of a divergent chain reaction. "The chain-reacting structure has been completed on December 2 and has been in operation since then in a satisfactory way."

Fermi's own description of the work on the pile can be found in the paper "The Development of the First Chain Reacting Pile" (FP 223) presented at the *Symposium on Atomic Energy and its Implications* which took place in Philadelphia on November 16 and 17, 1945, as well as in "Physics at Columbia University" (FP 269) a verbatim transcript of his last address delivered informally before the American Physical Society on January 30, 1954.

1943

"What thrilled Fermi most about the chain reacting pile was not so much its obvious promise for atomic energy and atomic bombs, which many others were now prepared to pursue, but an entirely new and unsuspected feature. It was a marvelous experimental tool", wrote Anderson. Three months later, the atomic pile's potentialities had been explored far and wide; by then people knew how to build it, with improvements. By the middle of March, the CP-2 was rebuilt at the Argonne Laboratory and reached criticality. Fermi insisted on calibrating the pile because "He wasn't happy unless he had control of and could account for, in a quantitative way, the behavior of all the elements that entered into his experiments. Fermi's role in this work is recorded in the report of Anderson's group "Standardization of the Argonne Pile" (FP 188).

According to Leona Woods, who had joined Fermi's group of collaborators in the spring of 1942 and married John Marshall, another member, in the summer of 1943, every day in the late afternoon they would all go for a swim in Lake Michigan. On Sunday afternoons they would all go on a bike ride or take a walk in the woods. "On these occasions, Enrico liked to show superendurance, to swim farther, to walk farther, to climb farther with less fatigue, and he usually could. In the same way he liked to win at throwing the jacknife, pitching pennies, or playing tennis, and usually he did. These qualities of gaiety and informality of his character made it easy for the young members of the laboratory to become acquainted with him ... On hikes and swims we talked about wind and waves, geophysics, origins of the Solar System, novas and supernovas, and the physical world ... He sketched his ideas qualitatively and briefly so that one caught the image and felt a desire for deeper understanding. The same quality of brevity extended as well to his serious teaching, in that he strongly and colorfully described the main principles, but rarely filled in the details. In this way the student felt both excited by Fermi's broad view-point but also tantalized by a desire to understand a great many newly glimpsed problems ... In data analysis he was especially impressive. For example, he insisted that integrals could be evaluated numerically in less time than it takes to look them up in a table, and he drove his colleagues to lightning slide rule manipulations to feed numbers to his speed integration on a desk computer" [43].

At Argonne, 20 miles outside Chicago, Fermi for a while could go back to basic research, the work he enjoyed the most. Pile engineering problems were being studied in other places. In fact, after the success of Enrico Fermi's atomic pile, two parallel projects for the production of fissile materials were undertaken: a plant for separating U-235, at Oak Ridge, and the construction of nuclear reactors for the production of plutonium. In the laboratories at Los Alamos, New Mexico, known as "site Y", research on problems more strictly connected to the bomb was underway.

In 1943 and 1944 Fermi, Anderson, Leona and John Marshall, Walter Zinn, and others did many experiments in theoretical physics with the Argonne pile. Though the results remained secret until after the war, these experiments opened up a new field of research in which neutrons were used for studying a wide variety of problems. Once again, Fermi foresaw the potentialities offered by neutrons and began studying the solid state. In this research he introduced a new facility he called the "thermal column", on which he set great stock. "Neutrons of energy much lower than thermal were produced by filtering a beam of thermal neutrons through a block of graphite 23 cm long. Neutrons having a de Broglie wavelength comparable with the interatomic distances of crystalline substances filtered through microcrystalline graphite virtually unimpeded by scattering." The discovery was described in the joint papers with Anderson and L. Marshall "Production of Low Energy Neutrons by Filtering Through Graphite" (FP 191), and in "Slowing Down of Fission Neutrons in Graphite" (FP 197) written with Leona and John Marshall, who, together with Herbert Anderson, were his principal collaborators in this work.

1944

With some help from Segrè, during a visit at Los Alamos, Fermi also designed a mechanical velocity selector which he used to observe neutron reflection and refraction ("A Thermal Neutron Velocity Selector and Its Application to the Measurement of the Cross-Section of Boron" – FP 200). Now Fermi could use the highly collimated, high-intensity neutron beams to study crystal lattices, and induce interference effects in beryllium, bismuth and sulfur. He thus began systematic research into the optical properties of neutrons in collaboration with Walter Zinn ("A Thermal Neutron Velocity Selector and Its Application to the Measurement of the Cross-Section of Boron", "Collimation of Neutron Beam From Thermal Column of CP-3 and the Index of Refraction for Thermal Neutrons", "Reflection of Neutrons on Mirrors"). The construction of the heavy water pile CP-3 had been completed in June, and immediately had been used for experimental purposes. In the immediate post-war period Fermi continued to conduct experiments with the neutron fluxes obtained from the CP-3 reactor —key experiments leading to a new chapter in solid-state physics.

Samples of plutonium became available in the spring, and reports for April show that Fermi carried out some preliminary experiments. At the time Fermi was once again in the best conditions for doing research, the thing he liked best; he was relatively free of the organizational problems he had had to deal with before. Nonetheless, his competence in neutron physics and his knowledge of the atomic pile were absolutely unique, and, as Arthur Compton related, he was often summoned to serve as a "safety anchor" at the various laboratories where fissile materials were being produced.

During the summer J. Robert Oppenheimer, the scientific director of "Project Y2 at Los Alamos, went to Chicago several times and persuaded Fermi to move to Los Alamos. He had already visited Los Alamos laboratories, the principal center of the whole Manhattan Project, where research on problems closely related to the construction of the atom bomb was being carried out. In September the Fermis moved to Los Alamos, in New Mexico.

1945

In the spring of 1945, more than two thousand people were working at Los Alamos. Under the direction of Robert Oppenheimer, a brilliant theoretical physicist who had introduced quantum mechanics to the United States, the top theoretical and experimental physicists were trying to solve a series of fundamental problems for the construction of a fission bomb. Oppenheimer felt that Fermi should have an official position, and so he was named Associate Director of the project.

At that time, at the age of 44, Enrico Fermi was the greatest neutron physicist in the world, he became the leader of the especially organized special Division F (the F stood for his own last name), whose general responsibility was to investigate problems that did not fit in the work of other divisions. Fermi also served as a super-consultant, as Segrè recalled, "Fermi was a sort of oracle to whom any physicist in trouble could appeal and more often than not come away with substantial help. There was no limit to the variety of problems that were brought to him ..." [44].

At Los Alamos, Fermi struck up a friendship with John von Neumann, who played a fundamental role in the development of the first electronic computers. They shared an interest in computers. Fermi, always the brilliant manipulator of numerical calculations, immediately took a particular interest in the newly installed IBM electromechanical machines, which Nicholas Metropolis, Richard Feynman and other computationally inclined physicists, had assembled at Los Alamos to replace teams of human "computers", in order to perform the necessary calculations required to predict the behavior of implosion designs of fission weapons. According to Metropolis, Fermi spent hours upon hours in the computer room to study them and test them.

In March it was decided that no amount of experimental work would yield as much information on how to determine the critical size of the U-235 bomb, as an actual explosion. Plans were made for an atomic test, under the code name "Project Trinity".

The first atomic bomb was exploded at Alamogordo, in the desert land of southern New Mexico, on July 16th.

At the moment of the explosion he stood with the others some 10000 meters away from the steel tower supporting the atomic device. In his characteristic style Fermi had foreseen a simple way to estimate the energy released by the bomb: he dropped a few small pieces of paper on the floor and measured the displacement produced when the front of the shock wave reached his observation point. His estimate well agreed with that obtained by much more elaborate operations.

In May, President Truman, Roosevelt's successor, had created an Interim Committee, chaired by the Secretary of War, to address the issue of the possible use of the atom bomb. The Interim Committee was assisted by a scientific subcommittee made up of Oppenheimer, Fermi, Lawrence and Compton, the scientific leaders of the project and had the delicate task of providing technical advice on the use of the nuclear weapon. After Germany's surrender, on May 8th, the scientists working on the Manhattan Project were forced to ask themselves whether it made sense at this point to use the nuclear weapons being built at Los Alamos. During the days just after the end of the war with Germany, Arthur Compton gave the responsibility of addressing these issues to a committee made up of scientists from the Metallurgical Laboratory in Chicago and chaired by James Franck, an able German physicist who had escaped Nazi persecution. On June 11th, the committee submitted a long memorandum advising against "the use of nuclear bombs for an early attack against Japan", and insisting that the power of the atom bomb should be demonstrated to Japan's leaders in an uninhabited area, before it was used for military purposes. The report met with no response. Szilard, who had been on the committee and was firmly convinced that the use of the atom bomb against Japan might trigger an arms race, had made several fruitless attempts to bring the highest government officials and the other scientists around to this view, but by then the decision-making process was unalterably oriented toward using the new nuclear weapons. The Interim Committee's scientific subcommittee (Fermi, Oppenheimer, Lawrence and Compton) did not think a demonstration would guarantee the end of the war in the Pacific. At the end of June, the members wrote in their official report to the Interim Committee: "We were asked to give an opinion on the first use of the new weapon ... The opinions of our scientific colleagues are not unanimous, and range from the proposal of a purely technical application to that of a military application more likely to persuade the Japanese to surrender ... We are unable to propose any technical demonstration that would end the war; we see no acceptable alternative to direct military use." A uranium bomb was dropped on Hiroshima on August 6th, a plutonium bomb was dropped on Nagasaki on the 9th, and hostilities ceased on the 15th.

After the Trinity test and the end of the war Fermi's attention turned to teaching, including an interesting series of lectures on thermonuclear reactions. Shortly after his arrival at Los Alamos, Fermi had begun giving lectures on a variety of topics and in the fall of 1945 he taught a regular course in neutron physics where several students took practically verbatim notes ("A Course in Neutron Physics" – FP 222).

1946

"I am going to talk on peaceful applications of atomic energy", with these words Fermi began his lecture on atomic energy at a public meeting, where he was invited to speak in May ("Atomic Energy for Power" – FP 224). Since 1944, when the development of reactors was in its initial stages, he had been very interested in the possibility of using them to generate power. In particular, he had immediately considered the idea that a reactor could produce more fissile material than it consumes —that is, the possibility of breeder reactors ("Discussion on Breeding" – FP 211, "Relation of Breeding to Nuclear Properties" – FP 221). In other words, the amount of plutonium produced in an atomic pile fueled by natural uranium would be larger than the quantity of U-235 employed.

Fermi had left Los Alamos with his family on December 31, 1945. He had been appointed Charles H. Swift Distinguished Service Professor of Physics at the Institute of Nuclear Studies of the University of Chicago. Compton wanted him to run the Institute, recently founded together with one for radiobiology and another for metals studies, but he turned down Compton's offer, not feeling suited for administrative tasks, and asked Samuel K. Allison, his closest collaborator, to become the first director.

After focusing on applied physics during wartime years, he wanted to go back to pure science, and was ready for a major change, comparable to the one he had made in the 1930s, when he had left atomic and molecular physics, by then well established, to devote himself and his group to the more promising field of neutron physics.

The war years had contributed to the advent of *Big Science*, in financial, scientific and engineering terms, with the construction of large machines such as nuclear reactors and particle accelerators, and also on the organizational level. It was a totally new way of doing research. In January of 1946, Fermi wrote to Edoardo Amaldi and Gian Carlo Wick (who had taken over Fermi's chair of theoretical physics at the University of Rome in 1939): "Since January I've been in Chicago, more or less definitively ... It seems we'll be getting rather unlimited funds and we've started to use them by ordering a 100-MeV betatron ... now that people are convinced that with physics you can make atom bombs, everybody is talking casually about figures in the millions of dollars. The impression one gets is that the biggest difficulty from the financial standpoint will be imagining enough things to spend on. Also, we expect that the number of students will grow considerably ..." [45]. Fermi would be one of the key actors in this new field of physics, having deeply influenced the birth and rapid growth of high-energy physics during the period 1946-1954, after his return to Chicago from Los Alamos till his death in 1954.

One of Fermi's outstanding contributions to XX Century's Physics is his role as a founder of the so-called "Chicago School", where many important particle physicists were trained. As Segrè put it, "The word soon spread, and an extraordinary constellation of students was formed at Chicago, attracted at least in part by Fermi's reputation" [46]. Between 1946 and 1953, Fermi gave twenty-three courses, teaching many first-class students. According to Jack Steinberger "It was a great privilege to be a student in that department at that time ... Fermi's courses, in particular, were models of transparent and simple organization of the most important concepts. He went to a great length to show those of us who had finished the courses and were working on our Ph.D., theses how to attack a variety of simple, general problems in different branches of physics, by gathering us together one or two evenings a week [...] proposing a problem, and then, perhaps later, going through the solution" [47]. Among these students there was Chen Ning Yang who later recalled: "Fermi gave extremely lucid lectures. In a fashion that is characteristic of him, for each topic he always started from the beginning, treated simple examples and avoided as much as possible 'formalisms'. (He used to joke that complicated formalism was for the 'high priests'). The very simplicity of his reasoning conveyed the impression of effortlessness. But this impression is false: The simplicity was the result of careful preparation and of deliberate weighing of different alternatives of presentation." Now Fermi resumed his old habit of giving informal unprepared lectures. The small group of graduate students gathered in his office and someone, either Fermi himself or one of the students, would propose a specific topic for discussion. As Yang has amimately described, "Fermi would search through his carefully indexed notebooks to find his notes on the topic and would then present it to us ... The fact that Fermi had kept over the years detailed notes on diverse subjects in physics, ranging from the purely theoretical to the purely experimental, from such simple problems as the best coordinates to use for the three-body problem to such deep subjects as general relativity, was an important lesson to all of us. We learned that *that* was physics. We learned that physics should not be a specialist's subject; physics is to be built from the ground up, brick by brick, layer by layer. We learned that abstractions come *after* detailed foundation work, not before" [48].

Through the influence of his students, Fermi effectively revolutionized the training of physicists in the United States and, hopefully, in the whole Western world" [49]. And, indeed, he taught physics to a generation of young scientists, many of whom played a major role in the revolution that marked the development of physics in the second half of the XX century.

In Los Alamos Desert (1944).

With E. Segrè at Los Alamos (1944).

On the boat (1946).

The atomic pile group (1947).

1947

"At the end of the war, the physicists who returned to the University of Chicago to form the Institute for Nuclear Studies found a Physics department with bare shelves. Almost every usable meter, tool, and accessory of particle physics had long since been absorbed into wartime research ... It was reasonable, therefore, that we turned to the excellent heavy water reactor facility of the Argonne Laboratory with its high thermal neutron flux, to investigate aspects of neutron physics which had been bypassed in the drive to the wartime objectives" [50].

The field of neutron optics, which Fermi had begun to explore before the war, now received strong impetus. He renewed his interest in making use of the intense neutron flux from the CP-3 reactor doing a series of experiments with Leona Marshall ("Phase on Neutron Scattering" – FP 227, "Interference Phenomena of Slow Neutrons" – FP 228). The research fields that Fermi opened up during this period grew enormously as the years went by, and now constitute entirely new chapters in the physics of solids and neutrons.

This year Fermi was appointed to the General Advisory Committee (GAC), made up of eight scientists and chaired by Oppenheimer, whose task was to provide the Atomic Energy Commission (AEC) with scientific and technical opinions on civilian and military programs set up for the development of atomic energy. Fermi remained on this committee until August of 1950. His presence in Chicago helped persuade the AEC to choose a nearby location for the permanent headquarters of the Argonne National Laboratory.

In the meantime, however, Fermi's attention was captured once again by the problem of mesons. At the end of 1946 Edoardo Amaldi, who was then in Washington, for a three-month trip to the US, wrote him about the latest results of an important experiment done in Rome that year by Marcello Conversi, Ettore Pancini and Oreste Piccioni, which was going to be published [51]. Starting in 1943 the Italian physicists had carried out a series of experiments finding that the processes of mesotron decay and absorption are very different from what they had expected. Tomonaga and Araki, starting from a previous work of Yukawa and Okayama, had pointed out that because of the nuclear Coulomb field "the competition between nuclear capture and spontaneous decay must be different for mesons of different sign." Slow positively charged mesotrons traversing the matter should prefer to decay rather than be absorbed by a nucleus, since Coulomb repulsion would prevent them from reaching the nucleus; on the other hand, negative Yukawa particles should strongly prefer absorption to decay, because negative mesotrons after being slowed down can approach the nuclei and disappear by nuclear interactions. In fact they had observed a different behavior in positive and negative mesotrons (as they were then called) that ending up in iron or graphite. But, in iron the negative mesotrons were captured before decaying, if a light element like carbon was used to stop mesotrons, both the positive and negative mesotrons decayed and emitted (respectively) electrons or positrons, while the prediction indicated that negative mesons should undergo nuclear capture "in any material". All of that pointed to a fundamental discrepancy between the cosmic-ray mesotrons they had observed and the expected behavior of Yukawa's scheme. Fermi, who in 1939 had already studied the anomalous absorption of cosmic rays in air, immediately realized the importance of the results of this experiment, which today is considered the experiment that initiated the era of high-energy physics. In a few days, working with Edward Teller and Victor Weisskopf, he completed a detailed analysis of the phenomenon demonstrating that the time of capture from the lowest orbit of carbon was not less than the time of natural decay, *i.e.* 10^{-6} s, in complete disagreement with the estimate of the theory. The interaction between the so-called μ-mesons and nucleons was, therefore, much weaker than demanded by Yukawa's theory: mesotrons in cosmic rays could not be the same as Yukawa's particle ("The Decay of Negative Mesotrons in Matter" – FP 232, "The Capture of Negative Mesotrons in Matter" – FP 233).

In early June, the results of the Rome experiment and its theoretical implications were discussed at the Shelter Island Conference, where important conjectures were advanced. The enigma was solved definitively a few months later at Bristol. Cesare Lattes, Giuseppe Occhialini and Cecil Frank Powell, using the technique of exposing photographic emulsions to cosmic rays at high altitudes, discovered that the "mesotron" observed at sea level was none other than the μ-meson (or muon) produced by the decay of a new particle, the π-meson (or pion), which is precisely the meson hypothesized by Yukawa [52]. The discovery confirmed the so-called two-meson hypothesis put forward by Bethe and Marshak at the Shelter Island Conference. It was the beginning of high-energy nuclear physics.

1948

Fermi was now gradually giving up experimental work at Argonne. By that time the Swedish physicist Hannes Alfvén, who had always been interested in cosmic-scale electromagnetic phenomena, was in Chicago too. He had been invited by Edward Teller: "At that time I was playing with the idea that cosmic rays might be accelerated in the neighborhood of the Sun ... During this visit Fermi learned from Alfvén about the probable existence of greatly extended magnetic fields in our galactic system. Since this field would necessarily be dragged along by the moving and ionized interstellar material, Fermi realized that here was an excellent way to obtain the acceleration mechanism for which he was looking" [53]. In an article published during the following year ("On the Origin of Cosmic Radiation" – FP 237) Fermi used this phenomenon to propose a theory of the origin of cosmic radiation according to which cosmic rays originate and accelerate primarily in the interstellar space of the galaxy due to collisions against moving magnetic fields." The article stirred up heated discussions with Teller, and Fermi succeeded in finding a model which explained most of the observed properties of cosmic rays, but the chief difficulty is that it fails to straightforwardly explain the heavy nuclei observed in the radiation, like protons.

By that time new machines were opening new horizons in elementary particle physics. Occhialini and Powell's discovery of pion's tracks in photographic emulsions was quickly followed by the production of "artificial" pions (positive, negative and neutral) at the Berkeley accelerators. The era of High Energy Physics had begun. Fermi and the other physicists at Chicago were anxious to obtain a large accelerator themselves. The construction of a 170-inch synchrocyclotron in Chicago was slated to begin in 1949, under the direction of Herbert Anderson and John Marshall.

Fermi kept a close watch on this work, and did the calculations for the radius of the emergent beam's orbit, using a graphic computer which he had invented. Even before the new machine was working, Fermi had decided to devote himself to theoretical studies, starting with the new theories of elementary particles.

With G. Racah (1949).

With B. Pontecorvo and G. Enriques at the "Olivetti Factory" in Ivrea (1949).

1949

When the news came that the USSR had exploded an atom bomb, the upper echelons of the US military began to discuss the development of a hydrogen bomb. Fermi and Isidor Rabi, another member of the GAC, expressed their view: "The fact that the destructive power of this weapon is unlimited makes its existence, and the ability to build it, a danger for all mankind." They invoked both moral and technical reasons for their advice: since a thermonuclear bomb could be made as large as its maker pleased, there was no natural limit to the devastation it could wreak.

In the summer Fermi returned to Italy, after an absence of nearly eleven years, to attend the International Conference on the Physics of Cosmic Rays, in Como where he presented the paper "An Hypothesis on the Origin of Cosmic Radiation" (FP 238). He also gave two series of lectures on atomic physics, in Rome and Milan, organized by the Accademia dei Lincei and the Donegani Foundation. This was his chance to see old friends and meet the young physicists of the new generation, for whom Fermi was a legend. Some of the topics Fermi dealt with in his lectures —particle physics, nuclear orbits, and new developments in quantum electrodynamics— were very influential in orienting the interests of Italian physicists toward particle physics.

That same summer, Fermi and Chen Ning Yang co-authored an article entitled "Are Mesons Elementary Particles?" (FP 239). In this article Fermi and Yang suggested that π-mesons, or pions, might be composite particles formed by the association of a nucleon and an antinucleon. The model was generalized in 1956 by Shoichi Sakata with some success, but eventually superseded by the quark model. At Chicago Fermi took a very active part in all seminars and in many discussions, and often his acute remarks sowed the seeds of further developments: "At that time Enrico Fermi had become interested in the magic numbers. I had the great privilege of working with him, not only at the beginning, but also later. One day as Fermi was leaving my office he asked: 'Is there any indication of spin-orbit coupling?' Only if one had lived with the data as long as I, could immediately answer: 'Yes, of course, and that will explain everything'". And indeed Maria Mayer thanks Enrico Fermi for his remark that was the origin of her classical paper on the shell model [54].

1950

Since the times of the famous experiments of Conversi, Pancini and Piccioni in Rome, Fermi's interests had gravitated towards high-energy physics; already at Los Alamos Fermi had the feeling that his next phase of activity would not be in neutrons but in something new. He reminded his former student Emilio Segrè "the switch from atomic to nuclear physics about fifteen years earlier and quoted (with his ironic smile) one of the slogans of Mussolini that the Fascists used to paint on Italian buildings: 'Rinnovarsi o perire' (Renew ourselves or perish)" [55].

The nature of the mesons that carried the nuclear force was an important focus of theoretical and experimental physics in the postwar era, as Fermi pointed out in his *Silliman Lectures* at Yale University in 1950. Part of his talks on the subject appeared also in the *Donegani Lectures* he had given in Italy; a more complete version of his lectures was later published in a book entitled "Elementary Particles". One example of this preparatory work was the article "High Energy Nuclear Events" (FP 241), which Fermi wrote for a Japanese journal on the occasion of the eighteenth anniversary of Yukawa's formulation of his theory. The idea came from an attempt to figure out what happens during high-energy nucleon-nucleon collisions with multiple production of particles. With this forage into high energy physics, his old deep inclination for statistical methods was now again at work, even if applied to an extremely simplified model, from which one could get orders of magnitudes, and begin to understand what was going to happen.

In his second work on these problems ("Angular Distribution of the Pions Produced in High Energy Nuclear Collisions" – FP 242), which appeared the following year, Fermi compared the experimental results with the theoretical model. To do so, he once more used a statistical method. "Fermi needed a framework in which to set the information which came to him in this way ...", wrote Anderson, "For this he developed simplified methods for calculating the orders of magnitude of the pertinent quantities, the cross-sections of the processes of interest. His position was that the meson theories were not correct anyway, so why take the trouble to calculate anything with them exactly. Fermi's methods were a boon to the experimentalists, who had difficulty in following the sophisticated way in which the theorists liked to put forth their theories" [56]. In reality, Fermi's model was oversimplified, just as he wanted it to be, and though it did not give detailed results —they would be found in the future— it did serve as a standard for measuring experimental results of multiple production of mesons and reveal non-statistical processes.

On January 27, 1950, Klaus Fuchs confessed that he had passed secret information on atomic weapons to the Soviets from 1942 to 1949. Three days later, Fermi and the other members of the GAC met and reached the conclusion that Fuchs might also have given the Soviets very important information about the hydrogen bomb, a project under way at Los Alamos from the time the atom bomb was being built. President Truman was informed of the GAC's conclusion, and on January 31st he announced the decision to give top priority to the development of the super bomb, Fermi, who as a member of the GAC had fought against the development of the H bomb, was among the first to get back to work on it. He spent the summer at Los Alamos doing H-bomb research. In November 1951, the United States exploded the first hydrogen bomb, which released nearly a thousand times the energy of the Hiroshima bomb. In August 1953, the Soviets exploded their own super bomb; the first portable H-bomb followed it in 1955.

1951

In the spring Chicago's new synchrocyclotron finally started operating, and Fermi went back to experimental physics. It was, for a few years, the highest energy accelerator in the world, as Anderson recalls: "it could accelerate protons at 450 MeV and a copious number of pions could be produced with these. The machine had been built with the idea that Fermi would be its principal user, and when it was finally complete he spent a great deal of time familiarizing himself with its operation, laying out the pion beams and measuring their intensity and energy" [57].

Right from the start Fermi spent a great deal of his time getting familiar with its aligning the pion beams and measuring their intensity and energy. He was very proud of at least one of his contributions to its construction, namely the target-carrier that he had built himself in his little shop. The "Fermi trolley" was a wheeled device with which the target could be readily moved around the periphery of the cyclotron, in a region of high magnetic field and was controlled from the outside by the action of the magnetic field on currents sent through coils to which the wheels were connected; it worked perfectly for years. In the last three years of his life he was to concentrate on interaction between pions and nucleons (that is, protons and neutrons), the fundamental process which in Yukawa's theory was known as strong interactions. The debut of Chicago's new cyclotron was the occasion for organizing an "International Conference on Nuclear Physics and the Fundamental Particles", held at the Institute for Nuclear Studies, from September 17th to 22nd. Some two hundred scientists attended, forty of them from foreign countries. Interesting results were reported on the first experiments, carried out the previous summer, on the measurement of cross sections of pions on liquid hydrogen. Fermi delivered the opening talk on elementary particles ("Fundamental Particles" – FP 246) and took part in all the discussions.

The conference came just a few days before his 50th birthday, on September 29th, and he celebrated it informally with some of his old friends. Fermi delivered the first paper of the conference listing some 21 "fundamental" particles, expressing the conviction that "philosophically, at least some of these 21 particles must be far from elementary. The requirement for a particle to be elementary is that it be structureless. Probably some of these 21 particles are not structureless objects. They may even have some geometrical structure, if geometry has any meaning in such a small domain" [58].

In this period Fermi was also working on a theoretical problem involving the instability of the surface dividing two fluids when the lighter of them accelerates the heavier one (Taylor Instability). Principally John von Neumann had investigated this topic at Los Alamos, because it was important for the implosion method of producing atomic bombs. Fermi's interest in the subject is evidenced by three works, two of them published in 1951 ("Excerpt from a Lecture on Tayor Instability Given During the Fall of 1951 at Los Alamos Scientific Laboratory" – FP 243, and "Taylor Instability of an Incompressible Liquid" – FP 244) and the third ("Taylor Instability at the Boundary of Two Incompressible Liquids" – FP 245) in 1953. The last two were written with von Neumann.

1952

The main work performed with Chicago's 450-MeV synchrocyclotron was the systematic study of pion-nucleon interactions. The initial experiments measured the transmission of first negative and then positive pions through liquid hydrogen targets. These were followed by a detailed study of the angular distribution of pion scattering in hydrogen. This work took a great deal of time, and Fermi devoted himself to it at length, helped by a team whose permanent members were D. E. Nagle and H. L. Anderson. The first article that appeared on this subject, "Total Cross-Section of Negative Pions in Hydrogen" (FP 248) showed that the cross-section grows rapidly with energy from the low value (85 MeV) found earlier by others. These results were indicative of the great strength of the pion-nucleon interaction. The article entitled "Ordinary and Exchange Scattering of Negative Pions" (FP 249) reported a surprising result: scattering with charge exchange occurs about twice as frequently as ordinary scattering. The next article, "Total Cross-Section of Positive Pions in Hydrogen" (FP 250), held a still greater surprise. The cross-section for positive pions is far larger than the maximum found for negative pions. All the values of the total cross-section for $\pi^+\pi^-$ in hydrogen were contained in three letters published in the 1952 March issue of "Physical Review"; all of them were received on the same day, January 21st. In fact the group began to carry on π^+ measurements on the last days of December. Anderson remembered quite well how "This anomaly puzzled Fermi very much ... He kept shaking his head because it kept coming out so high ... On this day there was a preprint of a paper by Keith Bruckner on meson nuclear scattering. 'Enrico,' I said after glancing at one of the curves, 'here's a guy who seems to think the π^+ cross-section should be higher than the π^-.'... 'Let me have a look at that paper.' Then, 'Will you take over for 20 minutes while I go up to my office?' I suppose he consulted his 'Artificial Memory' for he was back in 20 minutes with a broad grin. 'The cross-sections will be in the ratio $9:2:1$ for $\pi^+ : \pi^0 : \pi^-$ scattering,' he announced" [59]. Fermi was referring to the π^+ elastic, the π^- charge exchange, and the π^- elastic processes in that order, he had argued that the isotopic spin $3/2$ interaction was very strong. In reading Brueckner's paper he had grasped the importance of isotopic spin (or "isospin"), and succeded in linking the results to this essential element ([7]).

A few months later, during a meeting of the American Physical Society in New York, Fermi made an announcement regarding pion-meson interaction, for which he had a series of results and had found an explanation that implied an important principle. In strong pion-nucleon interaction, isotopic spin is conserved. An old idea, rather neglected up to that time, thus assumed new importance. Meanwhile, there had appeared significant indications of the existence of the first pion-nucleon resonance —the first excited state of the nucleon— which would lead scientists to believe that nucleons have a structure. Definitive confirmation came only after Fermi's death. The community of theoretical physicists followed these experiments with great interest because they seemed able to provide the key to understanding nuclear forces.

The problem had captured the attention of the brilliant young American theoretician Richard Feynman, whom Fermi had known well since their Los Alamos days. Feynman wrote a letter to Fermi containing predictions based on different theories for mesons and for pion-nucleon cross-sections. Fermi's reply showed his way of analyzing experimental results and stressed the importance of experimental evidence of the conservation of isospin.

To complete the study of the scattering of pions by protons, it was necessary to observe the angular distribution of the scattered particles, which yielded a much more complete analysis of the process. Fermi used the method of the phase shifts in analyzing these experiments, and was able to select out in a quantitative way which states were important in the scattering. A more critical test of the principle of isotopic spin conservation was provided by these results. Between March and April three new letters were received by the "Physical Review": "Deuterium Total Cross Sections for Positive and Negative Pions" (FP 252), "Angular Distribution of Pions Scattered by Hydrogen" (FP 253) and "Scattering and Capture of Pions by Hydrogen" (FP 254). The first two were written with Anderson, Nagle and Yodh, the last was signed by Anderson and Fermi.

They carried out so many angular distribution measurements in the next six months that Fermi began to think that the phase shift problem might best be handled with a computer. Fermi had a longstanding interest in numerical analysis, and promptly recognized the potential of computers as natural tools for studying problems in physics and astrophysics, particularly to process the big amount of data coming out of accelerators.

During this period, Fermi liked to spend the summer in Los Alamos. Nicholas Metropolis, an American physicist of Greek descent, recalled his first discussion with Fermi in the computer room at Los Alamos, where Fermi had spent the summer of 1945:

([7]) See H. L. Anderson (ref. [59] pp. 135-153) for a discussion on Fermi's notebooks, and his idea that this might be a resonance.

"We started to discuss the features of some of the electro-mechanical accounting machines that were being used for scientific calculations, but before very long he casually reached into his shirt pocket for a piece of paper on which he 'happened' to have written an equation, and simply asked: 'How would we do this on the machines?' It was semi-empirical formula for atomic masses that he had derived. The question had immediate effects. The calculation was planned in terms of elementary steps... At each stage he would listen to the minimum explanation and then proceed with the task at hand, doing every detail himself" [60]. The computers in question were the IBMs used at the time of the Manhattan Project to perform calculations concerning fission bomb implosion problems.

The interdisciplinary use of computers at Los Alamos to solve weapons related problems had opened the way to the development of techniques useful in nuclear physics studies. In 1952, a computer called MANIAC (acronym for Mathematical Analyzer, Numerical Integrator and Computer) was completed at Los Alamos, and for some years it was to remain one of the most powerful computers around. Fermi was obviously fully aware that computers should now be considered a most fundamental tool of nuclear physics, together with accelerators, reactors and detectors. In the spring of 1952 Fermi and his collaborators at Chicago had measured the scattering of negative pions by proton and had made a preliminary analysis of the data in terms of phase shifts using a Marchant table calculator. The phase-shift analysis, however, required the solution of nine equations. These preliminary attempts convinced him that the computer could accelerate the solution. At about the same time, the computer MANIAC was completed, so Fermi immediately suggested using the machine to perform a more complete analysis of the huge amount of experimental data amassed by his group, including experimental results obtained by the Columbia and Carnegie University groups. He brought the problem to Los Alamos in the summer of 1952 where he ran it on Metropolis's machine. The result of these calculations, together with a general explanation of the techniques used, was reported in an article entitled "Numerical Solution of a Minimum Problem" (FP 256), the fruit of Fermi's collaboration with Nicholas Metropolis. From then on he became a strenuous defender of computers as scientific instruments, always advertising numerical techniques during his lectures. This is when he developed a passion for the Monte Carlo technique that his colleagues and friends John von Neumann, Stanislaw Ulam and Nicholas Metropolis had invented. Originally developed to study the neutron diffusion in critical assemblies, it was quickly extended to other applications, such as nuclear weapons design and statistical calculations.

By the time he returned to Chicago, Fermi was an expert in computer analysis and, full of enthusiasm for these machines, held a series of classes on their use and programming. Fermi presented the results of MANIAC's calculations at the Rochester Conference at the end of the year: "With the use of an electronic computer the phase shifts can be computed in five minutes, since there is one code for all calculations. With each calculation only taking about five minutes, one can learn something of the mathematics of the problem by varying the conditions a little... the phase shifts are then used to calculate the cross-section. The results invariably want the cross-section to look as they do experimentally. In this calculation on the S and P phase shifts are used" [61]. Metropolis and his associates later completed the work on phase shift analysis that Fermi had begun at Chicago; his pioneering work, was largely carried forward in the 1950s, through the building of new computers, which were more and more applied to the problems of data analysis and to control the accelerator and other detectors in high energy physics.

At the third Rochester Conference on high-energy nuclear physics (December 18-20) the experiments on the scattering of pions by protons commanded the center of the stage, and the audience was anxious to hear Fermi's report ("Report on Pion Scattering" – FP 255). The research he did in the last years of his life opened a new chapter in theoretical and experimental physics. It is impossible to get an idea of the vast bulk of his theoretical work solely from his publications. Fermi selected his works by the most demanding standards, publishing only a very small part, but he transcribed the unpublished results in summary form in a great many notebooks —what he called his "artificial memory".

1953

Anderson described in detail the organization of Fermi's group experimenting with the cyclotron. They started with a preliminary informal session in his office discussing developments in the technique, improvements in counters, in electronics, and in handling the liquid hydrogen. Fermi usually contributed in the preparatory stage by building some needed accessory in his own little shop, and sometimes helped construct electronic devices or checked the functioning of the new scintillation counters. He took active part in all the stages of the experiment, and when it ended "With all aspects of the measurement under his control in the counting room, Fermi seemed happiest, most relaxed ... Once the next measurement was started he would tap the Marchant calculator on the desk in front of him, and when the counts had been reduced to a cross section and neatly listed with all the rest he might sit back and say, 'You see, these pions like to scatter backwards'." Anderson remarked how this "was not merely the result that finally appeared after reducing all the data, it was something he had been noticing all along." ("Angular Distribution of Pions Scattered by Hydrogen" and "Nucleon Polarization in Pion Proton Scattering" – FP 257) [62].

More detailed information about the scattering process could be obtained through angular distribution measurements, but since the relation between the phase shifts and the cross sections is not a simple one, this looked like an ideal problem for the electronic computer. Needing more data on pion scattering for the analysis Fermi intended to make during the summer at Los Alamos, he did one last experiment on the pion scattering and wrote what was to be his last article on an experimental topic, "Scattering of Negative Pions by Hydrogen" (FP 259). During the summer, Fermi, Metropolis and E. F. Alei co-authored "Phase Shift Analysis of the Scattering of Negative Pions by Hydrogen" (FP 260), which was published the following year. This analytic method was to become the standard one used to process experimental data for all problems of this kind.

In the summer of 1952, during one of his frequent visits to Los Alamos, Fermi and the mathematician Stanislaw Ulam discussed the kind of future problem that developments in computer science would make it possible to study. "We decided to try a selection of problems for heuristic work where in absence of closed analytic solutions experimental work on a computing machine would perhaps contribute to the understanding of properties of solutions ... This could be particularly fruitful for problems involving the asymptotic-long time or 'in the large' behavior of non-linear physical systems ... The plan was then to start with the possibly simplest such physical model and to study the results of the calculation of its long-time behavior. Then one would gradually increase the generality and the complexity of the problem calculated on the machine" [63]. The paper "Studies of Non Linear Problems" (FP 266) presents the results of their very first attempt. The work had been planned in the summer of 1952, but they performed the calculations the following summer, in 1953, and the paper was published after Fermi's death, in 1955. The physicist John Pasta, a recent arrival at Los Alamos, helped Fermi and Ulam work out a flow chart, write a program and run it on the MANIAC computer. At the time there were no sets of instructions, ready-made programs or automated procedures —what we now call *software*— and the enterprise was enormously more difficult than it would be today.

Their problem turned out to be very well chosen. The results were totally different from what even Fermi, with his great knowledge of wave motion, had expected. In fact it turned out quite clearly that in the presence of a small nonlinear perturbation things evolved in such a way that during the transition from a state far out of equilibrium to a state of thermal equilibrium at a given temperature the energy initially concentrated on a single normal mode of the linear system did not become shared among all the modes of the system. "The results show very little, if any, tendency toward equipartition of energy among the degrees of freedom", the authors remarked in the abstract at the beginning of the paper.

In the discussions preceding the setting-up and running of the problem on the machine Fermi and Ulam envisaged a two-dimensional version as a next step, and Fermi had suggested studying something purely cinematic, such as the motion of a chain of points subject only to constraints but no external forces, moving on a smooth plane convoluting and knotting itself indefinitely. These preliminary studies served to build models for the motion of systems in which "mixing" and "turbulence" might occur. This research on the evolution of nonlinear systems was a path breaking enterprise —the first digital experiment in statistical mechanics— and opened the way to research on the statistical mechanics of equilibrium and nonequilibrium, also marking the return of the problems of nonlinear mechanics among the interests of Physicists after they had been for several decades a subject left to mathematicians.

This was not the first time Fermi and Ulam had worked together. In the summer of 1950 they had made a study of the behavior of a thermonuclear reaction in a mass of deuterium. The aim was to obtain (via a schematization, but a fairly elaborate one) a description of the evolution of a series of physical processes involved in the propagation of this type of reaction. The mathematics was enormously complex, and all the calculations were done with desk computers and slide rules. The long and demanding project Fermi and Ulam undertook next was organized and performed on the electronic computer machine under von Neumann's guidance. In both qualitative and quantitative terms, it largely confirmed the behavior of the system they had estimated and predicted in their final report, a combination of intuitive evaluations, model equations and hand-made calculations.

During the spring of 1953 Herbert Anderson, Fermi's major collaborator, got very ill. In that period Fermi gradually changed his role and spent more and more time helping others by discussion and lending a hand in the experiment, but never to the extent that would allow him to admit that the work was his own. Free from experimental work, he now considered the possibility of working with Chandrasekhar on astrophysics problems related to his earlier interest in the origin of cosmic radiation.

In 1952 and 1953 they talked regularly about astrophysics; the outcome of their discussions appeared in the papers "Magnetic Fields in Spiral Arms" (FP 261) and "Problems of Gravitational Stability in the Presence of a Magnetic Field" (FP 262) submitted on March 23 to "Astrophysical Journal". In the first they estimated the magnetic field in the arms of a spiral galaxy, while the second is a more extensive study of the gravitational stability in the presence of a magnetic field, and it is, the best demonstration, perhaps, of Fermi's willingness to solve any problem of physics. In the second the authors consider a number of problems relating to the dynamical and gravitational stability of cosmical masses in which there is a prevalent magnetic field. In the discussion of these problems, the assumption was made that the medium is effectively of infinite electrical conductivity.

Interest in magneto-hydrodynamics grew enormously in the postwar years; the difficulties in this field of research were due primarily to the problem of clearly visualizing the often conflicting tendencies to which the motions of an electrically conducting fluid are subject in the presence of a magnetic field. It was only natural that this type of phenomenon fascinated Fermi, who was always interested in problems which challenge a physical understanding. He had been the first to grasp the importance of magnetic fields for the structure and evolution of a galaxy.

By now Fermi had enormous prestige in the American physics community, and at the beginning of the year he had been elected president of the American Physical Society. Moreover, astrophysicists warmly welcomed his interests in their field, and he was asked to deliver an important lecture to the American Astronomical Society; he was the first "non-astronomer" to deserve this honor and he was especially proud of it. He took the opportunity to review his earlier ideas on the origin of cosmic radiation in the light of subsequent developments in knowledge of the intensity and behavior of magnetic fields. The concluding sentence of his Lecture "Galactic Magnetic Fields and the Origin of Cosmic Radiation" delivered on August 28 1953 and published on "Astrophysical Journal" in 1954 —was a prophetic utterance, as Chandrasekhar remarked: "In conclusion, I should like to stress that, regardless of the details of the acceleration mechanism, cosmic radiation and magnetic fields in the galaxy must be counted as very important factors in the equilibrium of interstellar gas" [64].

In Varenna (1954). Lectures on mesons.

Visit to the "Moto Guzzi" Factory near Varenna (1954).

At leisure playing football table with E. Amaldi in Varenna (1954).

The last time in Italy on Como Lake (1954).

1954

The American scientific community was thrown into turmoil by the Oppenheimer trial. On November 7th of 1953, William L. Borden, executive director of the joint congressional committee on atomic energy, had written to F.B.I. chief J. Edgar Hoover that "in all probability" the physicist J. Robert Oppenheimer had been and still was a Soviet agent. President Eisenhower immediately revoked Oppenheimer's clearance for access to atomic secrets and ordered the Atomic Energy Commission to investigate. The Oppenheimer trial began on April 13th, and ended with a guilty verdict that was never reversed, though many years later, in 1963, President John Kennedy awarded Oppenheimer the Fermi Prize (actually delivered to Oppenheimer by Lyndon Johnson). At the time Oppenheimer was the head of the Institute for Advanced Studies in Princeton, and a very sick man. He died on February 18, 1967, having refused a new trial. Fermi testified on his behalf on April 20, 1954. In a particularly significant part of his testimony, Fermi had this to say about the nuclear arms race: "My opinion at the time was that the super bomb should be banned before it was born. I thought it would be easier, through some international agreement, to ban something that didn't exist."

On Saturday morning, January 30, Fermi delivered his address before the American Physical Society. The speech, transcribed from a tape recording, was left deliberately in an unpolished and unedited form and published on "Physics Today", in November 1955 ("Physics at Columbia University" – FP 269).

In February, Emilio Segrè paid Fermi a visit and told him about the results of some recent experiments his group at Berkeley had done on the polarization of scattered protons. Similar experiments had been attempted in Chicago without success, and Fermi and Segrè had discussed the negative results already in November 1953. Fermi was very interested in Segrè's results and wanted to check right away whether the spin-orbit coupling, which plays a fundamental role in the shell model, could also account for the polarization in high-energy scattering. Fermi did the calculation later reported in "Polarization of High Energy Protons Scattered by Nuclei" (FP 267) on his blackboard from ten in the morning to about noon. He started off on the wrong foot by using Born's approximation, which gave a null result, but quickly corrected himself, and then proceeded rapidly while Segrè took notes for the paper, which was completed within very few days. After completing this paper Fermi undertook an exact calculation to prove the validity of the Born approximation and presented the results of this calculation in a private communication, which was not published at the time ("Polarization in the Elastic Scattering of High Energy Protons by Nuclei" – FP 268). Fermi loved the simplicity of the method and the results, and lectured on it not long afterward in Varenna ("Lectures on Pions and Nucleons" – FP 270). This was the last time Segrè saw Fermi at work solving a problem in his own style, so familiar to Segrè since the old days in Rome.

In 1954 Fermi also went back to his theoretical study of pion-nucleon and nucleon-nucleon collisions, applying the statistical methods he had used earlier to the case of multiple pion production at the energies (up to 2.5 BeV) of the Brookhaven Cosmotron, the proton-synchrotron that started operating fully at Brookhaven in the spring of 1953.

He showed how to take into account the consequences of charge independence, and how purely statistical effects enter in the analysis of experiments of this type ("Multiple Production of Pions in Pion-Nucleon Collisions" – FP 263, "Multiple Production of Pions in Nucleon-Nucleon Collisions at Cosmotron Energies" – FP 264).

In the summer, Fermi returned to Italy for the second time after the end of the war. At Villa Monastero in Varenna, a town near Como, he delivered a memorable series of lectures on the physics of pions and nucleons. As Bernard Feld recalled, "Here was Fermi at the height of his powers, bringing order and simplicity out of confusion, finding connections between seemingly unrelated phenomena; wit and wisdom emerging from lips —white, as usual, from contact with chalk— in that clear resonant voice of his that had never lost the soft Italian vowel endings, on a perfectly colloquial American delivery" [65].

During his stay in Italy, Fermi's health deteriorated seriously. Back in the United States, he was diagnosed as having stomach cancer. He died in Chicago on November 28th, at the age of 53. Fermi had devoted his last days to revising his notes for a nuclear physics course. After his death, they were edited by three of his students and published in book form under the title "Nuclear Physics". During his stay in Varenna, two researchers of the University of Pisa —Marcello Conversi and Giorgio Salvini— had asked Fermi for his opinion about the several possibilities on how to spend a substantial amount of money at that time available to the University. He answered that the idea of building an electronic computer in Pisa would be by far the best. It was his last contribute to the progress and the development of Italian research.

* * *

The author wishes to gratefully thank G. HOLTON for his careful reading of the manuscript and his precious suggestions. The photographs reproduced in this paper are in part collected in the SIF Archive and in part courtesy of the "Domus Galileiana", Pisa.

REFERENCES

[1] PERSICO E., "Souvenir de Enrico Fermi", *Scientia 90* (1955), pp. 1-9, on p. 1.
[2] SEGRÈ E., *Enrico Fermi Physicist* (the University of Chicago Press, Chicago and London) 1970, p. 5.
[3] See AMIDEI's letter quoted in SEGRÈ E., "Biographical Introduction", FNM I, pp. XLVI-IL.
[4] Quoted in SEGRÈ E., "Biographical Introduction", FNM I, pp. XLVIII-IL.
[5] SEGRÈ E., "Biographical Introduction", FNM I, p. LII.
[6] RASETTI F., Introductory note to paper 7, FNM I, p. 55.
[7] SEGRÈ E., "Biographical Introduction", FNM I, p. LIII.
[8] PERSICO E., ref. [1] p. 2.
[9] SEGRÈ E., "Biographical Introduction", FNM I, p. LV.
[10] RASETTI F., Introductory note to papers 26, 27, 28, FNM I, p. 159.
[11] RASETTI F., Introductory note to papers 30, 31, FNM I, pp. 178-181.

[12] SEITZ F., "Fermi Statistics and Its Applications" in *Memorial Symposium Held in Honor of Enrico Fermi at the Washington Meeting of the American Physical Society, April 29, 1955*, *Rev. Mod. Phys.*, **27**, no. 3 (1955) pp. 249-275, on p. 249; SEBASTIANI F. and CORDELLA F., "Fermi Towards Quantum Statistics 1923-1925", in *Proceedings of the International Conference "Enrico Fermi and the Universe of Physics" Rome, September 29 – October 2, 2001* (Enea, Rome) 2003. For a thorough comment about Fermi's fundamental contribution see ref. [11].
[13] TELEGDI V. L., "Enrico Fermi in America", in *Symposium Dedicated to Enrico Fermi on the Occasion of the 50th Anniversary of the First Reactor (Rome, 10 December 1992)* (Accademia Nazionale dei Lincei, Rome) 1993, pp. 71-90, on p. 83.
[14] SEGRÈ E., "Biographical Introduction", FNM I, p. LVIII.
[15] SEGRÈ E., "Biographical Introduction", FNM I, p. LX.
[16] SEGRÈ E., ref. [2] p. 52.
[17] AMALDI E., Introductory note to papers 50, 52, 64, 65, 66, 67, 70, FNM I, p. 305.
[18] BETHE H. A., "Remarks at the Memorial Symposium in Honor of Enrico Fermi", *Rev. Mod. Phys.* **27**, no. 3 (1955) p. 253.
[19] Quoted in SEGRÈ E., ref. [2] p. 59.
[20] From RASETTI's Autobiographical Notes, in "Archivio Edoardo Amaldi", box 1E, folder 2, Museo di Fisica, Università "La Sapienza", Roma.
[21] CHADWICK J., "The Existence of a Neutron", *Proc. R. Soc. London, Ser. A*, **136** (1932) pp. 692-708.
[22] BOHR N., "Atomic Stability and Conservation Laws", *Proceedings of the "Convegno di fisica nucleare, ottobre 1931"*, Reale Accademia d'Italia, Fondazione Alessandro Volta (1932).
[23] HOLTON G., *The Scientific Imagination. Case studies* (Cambridge University Press, Cambridge) 1978.
[24] See RASETTI F., Long introduction to the beta decay papers on FNM I; AMALDI E., "Beta Decay Opens the Way to Weak Interactions", *J. Phys. (Paris), Colloq.*, C**8**, suppl. Vol. **43**, no. 12, (1982); KONOPINSKI E. J., "Fermi's Theory of Beta-Decay", *Rev. Mod. Phys.*, **27**, no. 3 (1955) pp. 254-257. An excellent treatment of the subject is also in PAIS A., *Inward Bound* (Oxford University Press) 1986.
[25] SEGRÈ E., ref. [2] p. 72.
[26] ROSSI B., Introductory note to paper FNM I, 74, p. 509.
[27] AMALDI E., "From the Discovery of the Neutron to the Discovery of Nuclear Fission", *Phys. Rep.*, **111** (1-4) (1984) pp. 1-331, on p. 124.
[28] FERMI E., "Artificial Radioactivity Produced by Neutron Bombardment", Nobel lecture, paper 128, FNM I, pp. 1039-1040.
[29] See paper 94, FNM I, pp. 704-705.
[30] CHANDRASEKHAR S., Introductory note to papers 261, 262, FNM II, pp. 926-927.
[31] DE GREGORIO A., "Chance and Necessity in Fermi's Discovery of the Properties of the Slow Neutrons", *Giornale di Fisica,* **42**, 4 (2001) pp. 195-208.
[32] FERMI E., see paper 128, FNM I, pp. 1040-1041.
[33] SEGRÈ E., ref. [2] p. 86.
[34] RASETTI F., Introductory note to papers 84*a* to 110, FNM I, pp. 639-644, on p. 643.
[35] HEILBRON J. L., "The First European Cyclotrons", *Rivista di Storia della Scienza,* **3** (1986) pp. 1-44, on p. 7.
[36] AMALDI E., Introductory note to papers 112-119, FNM I, p. 811.
[37] FERMI E., see paper 128, FNM I, pp. 1037-1043, on p. 1040.
[38] ANDERSON H. L., Introductory note to paper 129, FNM II, p. 1.
[39] SEGRÈ E., ref. [2] p. 116.

[40] ANDERSON H. and ANDERSON H. L., Introductory note to paper 136, FNM II, p. 31.
[41] FELD B. T., Introductory note to papers 147, 150, FNM II, p. 1004.
[42] ANDERSON H. L., Introductory note to paper 180, FNM II, p. 269.
[43] MARSHALL L., Introductory note to paper 188, FNM II, p. 328.
[44] SEGRÈ E., ref. [2] p. 140.
[45] From FERMI's letter to Amaldi E. and Wick G. C., January 24, 1946, in "Archivio Edoardo Amaldi", box E1, folder 2, Museo di Fisica, Università "La Sapienza", Roma.
[46] SEGRÈ E., ref. [2] p. 168.
[47] STEINBERGER J., "A Particular View of Particle Physics in the Fifties", in BROWN L. M., DRESDEN M., HODDESON L. (Editors), *Pions to Quarks: Particle Physics in the 1950s*, (Cambridge University Press, Cambridge) 1989, pp. 307-330, on pp. 307-308.
[48] YANG C. N., Introductory note to paper 239, FNM II, pp. 673-674.
[49] TELEGDI V. L., ref. [13] p. 71.
[50] MARSCHALL L., Introductory note to papers 227-231, 234, 235, FNM II, p. 578.
[51] CONVERSI M., PANCINI E. and PICCIONI O., "On the Disintegration of Negative Mesons", *Phys. Rev.*, **71** (1947) pp. 209-210. The paper was published on February 1, 1947.
[52] LATTES C. M. G., MUIRHEAD H., OCCHIALINI G. P. S. and POWELL C. F., "Process involving Charged Mesons", *Nature*, **159** (1947) pp. 694-697.
[53] TELLER E., Introductory note to papers 237, 238, 264, FNM II, p. 655.
[54] GOEPPERT MAYER M., "The Shell Model", `http://www.nobel.se.physics`; "On Closed Shells in Nuclei. II", *Phys. Rev.*, **75** (1949) pp. 1969-1970, on p. 1970.
[55] SEGRÈ E., ref. [2] p. 166.
[56] ANDERSON H. L., Introductory note to papers 241, 242, FNM II, p. 789.
[57] ANDERSON H. L., Introductory note to paper 246, FNM II, p. 825.
[58] FERMI E., "Fundamental Particles", paper 246, FNM II, p. 826.
[59] ANDERSON H. L., "Meson Experiments with Enrico Fermi", *Rev. Mod. Phys.*, **27**, no. 3 (1955) p. 270.
[60] METROPOLIS N., Introductory note to paper 256, FNM II, p. 861.
[61] FERMI E., "Reports on Pion Scattering", paper 257, FNM II, p. 857.
[62] ANDERSON H. L., Introductory note to papers 257, 258, FNM II, p. 872.
[63] ULAM S., Introductory note to paper 266, FNM II, p. 977.
[64] FERMI E., "Galactic Magnetic Fields and the Origin of Cosmic Radiation", paper 265, FNM II, p. 976.
[65] FELD B. T., Introductory note to paper 270, FNM II, p. 1004.

About the Author

LUISA BONOLIS deals with the history of physics and has collaborated to the scientific organization of the Centenary from the birth of E. Fermi under an INFN (Istituto Nazionale di Fisica Nucleare) contract.

Fermi's bibliography related to the papers presented in this book

The complete bibliography of the papers published by Enrico Fermi during all his life is collected in the two volumes "Note e Memorie" (Collected papers) jointly published by the Accademia Nazionale dei Lincei and by the University of Chicago, thanks to the efforts of his friends, colleagues and coworkers who either collaborated with him in the referred papers or who were direct witnesses of the circumstances in which they were written. From this original list we extracted those works that for their contents we considered as the most helpful to the reading of the papers presented in this book. In order to facilitate their retrieving we kept the original numbering given in the two volumes of "Note e Memorie".

VOLUME I

3. *Sopra i fenomeni che avvengono in vicinanza di una linea oraria.* "Rend. Lincei" **31** (1), 21-23, 51-52, 101-103 (1922).

5. *Le masse nella teoria della relatività.* In A. KOPFF, *I fondamenti della relatività Einsteiniana.* Italian edition edited by R. Contu and T. Bembo. Hoepli, Milano 1923, 342-344.

11*b*. *Dimostrazione che in generale un sistema meccanico normale è quasi ergodico.* "Nuovo Cimento", **25**, 267-269 (1923).

13. *Alcuni teoremi di meccanica analitica importanti per la teoria dei quanti.* "Nuovo Cimento", **25**, 271-285 (1923).

15. *Generalizzazione del teorema di Poincaré sopra la non esistenza di integrali uniformi di un sistema di equazioni canoniche normali.* "Nuovo Cimento" **26**, 105-115 (1923).

16. *Sopra la teoria di Stern della costante assoluta dell'entropia di un gas perfetto monoatomico.* "Rend. Lincei", **32** (2), 395-398 (1923).

17*a*. *Sulla probabilità degli stati quantici.* "Rend. Lincei", **32** (2), 493-495 (1923).

19. *Considerazioni sulla quantizzazione dei sistemi che contengono degli elementi identici.* "Nuovo Cimento", **1**, 145-152 (1924).

22. *Sui principi della teoria dei quanti.* "Rend. Seminario matematico Università di Roma", **8**, 7-12 (1925).

30. *Sulla quantizzazione del gas perfetto monoatomico.* "Rend. Lincei", **3**, 145-149 (1926).

31. *Zur Quantelung des idealen einatomigen Gases.* "Z. Physik", **36**, 902-912 (1926).

35. *Sopra l'elettrone rotante.* F. RASETTI and E. FERMI. "Nuovo Cimento", **3**, 226-235 (1926).

38*b*. *Un teorema di calcolo delle probabilità ed alcune sue applicazioni.* Diploma thesis of Scuola Normale Superiore. Pisa, 1922. *Unpublished.*

43. *Un metodo statistico per la determinazione di alcune proprietà dell'atomo.* "Rend. Lincei", **6**, 602-607 (1927).

44. *Sulla deduzione statistica di alcune proprietà dell'atomo. Applicazione alla teoria del sistema periodico degli elementi.* "Rend. Lincei", **7**, 342-346 (1928).

50. *Sopra l'elettrodinamica quantistica.* "Rend. Lincei", **9**, 881-887 (1929).

56. *I fondamenti sperimentali delle nuove teorie fisiche.* "Atti Soc. It. Progr. Sci.", 18th Meeting, vol. **1**, 365-371 (1929).

58. *Problemi attuali della fisica.* "Annali dell'Istruzione media", **5**, 424-428 (1929).

59. *L'interpretazione del principio di causalità nella meccanica quantistica.* "Rend. Lincei", **II** 980-985 (1930); "Nuovo Cimento", **7**, 361-366 (1930).

60. *Atomi e stelle.* "Atti Soc. It. Progr. Sci.", 19th Meeting, vol. **1**, 228-235 (1930).

62. *La fisica moderna.* "Nuova Antologia", **65**, 137-145 (1930).

64. *Sopra l'elettrodinamica quantistica.* "Rend. Lincei", **12**, 431-435 (1930).

67. *Quantum Theory of Radiation.* "Rev. Mod. Phys.", **4**, 87-132 (1932).

72*b*. *Lo stato attuale della fisica del nucleo atomico.* "Ric. Scientifica", **3** (2), 101-113 (1932).

74. *Azione del campo magnetico terrestre sulla radiazione penetrante.* E. FERMI and B. ROSSI. "Rend. Lincei", **17**, 346-350 (1933).

76. *Tentativo di una teoria dell'emissione dei raggi "beta".* "Ric. Scientifica", **4** (2), 491-495 (1933).

77*a*. *Sulla ricombinazione di elettroni e positroni.* E. FERMI and G. UHLENBECK, "Ric. Scientifica", **4** (2), 157-160 (1933).

79. *Le ultime particelle costitutive della materia.* "Atti Soc. It. Progr. Sci.", 22nd Meeting, vol. **2**, 7-14 (1933); "Scientia", **55**, 21-28 (1934).

80*a*. *Tentativo di una teoria dei raggi β.* "Nuovo Cimento", **II**, 1-19 (1934).

83. *Statistica, meccanica.* "Enciclopedia Italiana di Scienze, Lettere ed Arti", Istituto G. Treccani, Roma, vol. 32°, 518-523 (1936).

84*a*.-92*b*. *Radioattività provocata da bombardamento di neutroni.* Letters to "Ric. Scientifica":

84*a*. I. "Ric. Scientifica", **5** (1), 283 (1934).

85*a*. II. "Ric. Scientifica", **5** (1), 330-331 (1934).

86*a*. III. "Ric. Scientifica", **5** (I), 452-453 (1934). E. Amaldi, O. D'Agostino, E. Fermi, F. Rasetti, E. Segrè.

87*a*. IV. "Ric. Scientifica", **5** (1), 652-653 (1934), E. Amaldi, O. D'Agostino, E. Fermi, F. Rasetti, E. Segrè.

88*a*. V. "Ric. Scientifica", **5** (2), 21-22 (1934). E. Amaldi, O. D'Agostino, E. Fermi, F. Rasetti, E. Segrè.

89*a*. VII. "Ric. Scientifica", **5** (2), 467-470 (1934). E. Amaldi, O. D'Agostino, E. Fermi, B. Pontecorvo, F. Rasetti, E. Segrè.

90*a*. VIII. "Ric. Scientifica", **6** (1), 123-125 (1935). E. Amaldi, O. D'Agostino, E. Fermi, B. Pontecorvo, F. Rasetti, E. Segrè.

91*a*. IX. "Ric. Scientifica", **6** (1), 435-437 (1935). E. Amaldi, O. D'Agostino, E. Fermi, B. Pontecorvo, E. Segrè.

92*a*. X. "Ric. Scientifica", **6** (1), 581-584 (1935). E. Amaldi, O. D'Agostino, E. Fermi, B. Pontecorvo, E. Segrè.

97. *Nuovi radioelementi prodotti con bombardamento di neutroni.* E. Amaldi, E. Fermi, F. Rasetti and E. Segrè. "Nuovo Cimento", **II**, 442-447 (1934).

99. *Possible Production of Elements of Atomic Number Higher than 92.* "Nature" (London), **133**, 898-899 (1934).

104. *La radioattività artificiale.* "Atti Soc. It. Progr. Sci.", 23rd Meeting, vol. **I**, 34-39.

105*a*. *Azione di sostanze idrogenate sulla radioattività provocata da neutroni.* I. E. Fermi, E. Amaldi, B. Pontecorvo, F. Rasetti, E. Segrè. "Ric. Scientifica", **5** (2), 282-283 (1934).

106*a*. *Effetto di sostanze idrogenate sulla radioattività provocata da neutroni.* II. E. Fermi, B. Pontecorvo, F. Rasetti. "Ric. Scientifica", **5** (2), 380-381 (1934).

108. *Ricerche sui neutroni lenti.* E. Fermi and F. Rasetti. "Nuovo Cimento", **12**, 201-210 (1935).

120. *Un maestro: Orso Mario Corbino.* "Nuova Antologia", **72**, 313-316 (1937).

121. *Un generatore artificiale di neutroni.* E. Amaldi, E. Fermi, F. Rasetti. "Ric. Scientifica", **8** (2), 40-43 (1937).

126. *Prospettive di applicazioni della radioattività artificiale.* "Rendiconti dell'Istituto di Sanità Pubblica", vol. **1**, 421-432 (1938).

* * *

Books of the Italian period:

Introduzione alla fisica atomica, pp. 330 Zanichelli, Bologna, 1928.

Fisica ad uso dei Licei, Vol. I, pp. 239 and Vol. II, pp. 243, Zanichelli, Bologna, 1929.

Molecole e cristalli, pp. 303, Zanichelli, Bologna, 1934 – *Moleküle und Kristalle*, translated into German by M. Schön and K. Birus, pp. VII-234 Barth, Leipzig, 1938.

Thermodynamics, pp. VII-160, Prentice-Hall, New York, 1937 – *Termodinamica*, translated into Italian by A. Scotti, pp. 179, Boringhieri, Torino 1958.

Fisica per le Scuole Medie Superiori, Fermi-Persico, pp. 314, Zanichelli, Bologna, 1938.

VOLUME II

129. *The Fission of Uranium.* H. L. ANDERSON, E. T. BOOTH, J. R. DUNNING, E. FERMI, G. N. GLASOE and F. G. SLACK. "Phys. Rev.", **55**, 511-512 (1939). (Letter).
132. *Neutron Production and Absorption in Uranium.* H. L. ANDERSON, E. FERMI and L. SZILARD. "Phys. Rev.", **56**, 284-286 (1939).
136. *Production and Absorption of Slow Neutrons by Carbon.* H. L. ANDERSON and E. FERMI. Report A-21 (September 25, 1940).
179. *Feasibility of a Chain Reaction.* Report CP-383 (November 26, 1942).
181. *Experimental Production of a Divergent Chain Reaction.* "Am. J. Phys.", **20**, 536-558 (1952).
222. *A Course in Neutron Physics.* Part I, Document LADC-255 (February 5, 1946) (Notes by I. HALPERN). Part II (Declassified in 1962).
224. *Atomic Energy for Power.* The George Westinghouse Centennial Forum. *Science and Civilization-The Future of Atomic Energy*; also MDDC-I (May, 1946).
225. *Elementary Theory of the Chain-Reacting Pile.* "Science", **105**, 27-32 (1947).
232. *The Decay of Negative Mesotrons in Matter.* E. FERMI, E. TELLER and V. WEISSKOPF. "Phys. Rev.", **71**, 314-315 (1947).
233. *The Capture of Negative Mesotrons in Matter*, E. FERMI and E. TELLER. "Phys. Rev.", **72**, 399-408 (1947).
234. *On the Interaction Between Neutrons and Electrons*, E. FERMI and L. MARSHALL. "Phys. Rev.", **72**, 1139-1146 (1947).
237. *On the Origin of the Cosmic Radiation.* "Phys. Rev.", **75**, 1169-1174 (1949).
238. *An Hypothesis on the Origin of the Cosmic Radiation.* "Nuovo Cimento", **6**, Suppl. 317-323 (1949).
239. *Are Mesons Elementary Particles?* E. FERMI and C. N. YANG. "Phys. Rev.", **76**, 1739-1743 (1949).
240. *Conferenze di Fisica Atomica* (Fondazione Donegani). Accademia Nazionale dei Lincei (1950).
244. *Taylor Instability of an Incompressible Liquid.* Part 1 of Document AECU-2979 (September 4, 1951).
245. *Taylor Instability at the Boundary of Two Incompressible Liquids.* E. FERMI and J. VON NEUMANN. Part 2 of Document AECU-2979 (August 19, 1953).
246. *Fundamental Particles.* Proceedings of the International Conference on Nuclear Physics and the Physics of Fundamental Particles. The University of Chicago (September 17 to 22, 1951) (Lecture).
247. *The Nucleus.* "Physics Today", **5**, 6-9 (March 1952).
251. *Letter to Feynman* (1952).
255. *Report on Pion Scattering.* Excerpts from the Proceedings of the Third Annual Rochester Conference (December 18-20, 1952).
256. *Numerical Solution of a Minimum Problem.* E. FERMI and N. METROPOLIS. Document LA-1492, (November 19, 1952).

261. *Magnetic Fields in Spiral Arms.* S. CHANDRASEKHAR and E. FERMI. "Astrophysical Journal", **118**, 113-115 (1953).
262. *Problems of Gravitational Stability in the Presence of a Magnetic Field.* S. CHANDRASEKHAR and E. FERMI. "Astrophysical Journal", **118**, 116-141 (1953).
265. *Galactic Magnetic Fields and the Origin of Cosmic Radiation.* "Astrophysical Journal" **119**, 1-6 (1954).
266. *Studies of the Nonlinear Problems. I.* E. FERMI, J. PASTA, S. ULAM. Document LA-1940 (May 1955).
270. *Lectures on Pions and Nucleons.* "Nuovo Cimento", **2**, Supp., 17-95 (1955) edited by B. T. Feld.

* * *

BOOKS OF THE AMERICAN PERIOD:

Elementary Particles, pp. XII+110, Yale University Press, New Haven, 1951 – *Particelle elementari*, translated into Italian and revised by P. CALDIROLA, pp. 192, Einaudi, Milano (1952).

Nuclear Physics. A course given at the University of Chicago. Notes Compiled by J. OREAR, A. H. ROSENFELD, and R. H. SCHLUTER, pp. VII+246, The University of Chicago Press, Chicago, 1949.

Conferenze di Fisica Atomica (Fondazione Donegani), Accademia Nazionale dei Lincei, Roma 1950.

Notes on Quantum Mechanics, pp. VII+171, The University of Chicago Press, Chicago 1961.

* * *

RECOMMENDED FOR FURTHER READING ON THE WORK AND LIFE OF E. FERMI:

FERMI L., *Atoms in the Family* (University of Chicago Press) 1954.

SEGRÈ E., *Enrico Fermi Physicist* (University of Chicago Press) 1970.

DE LATIL P., *Fermi, la vita le ricerche le testimonianze* (Edizioni Accademia) 1974.

PONTECORVO B., *Enrico Fermi* (Edizioni Studio Tesi) 1993.

DE MARIA M., *Fermi, un fisico da via Panisperna all'America*, "Le Scienze" n. **8**, aprile (1999).

SEGRÈ E., *A Mind Always in Motion* (University of California Press) 1993.

AMALDI E., *From the Discovery of Neutron to the Discovery of Nuclear Fission*, "Physics Reports" **111** (1-4), 1979, pp. 1-331.

ANALYTICAL INDEX

A

B

C

G

H

I

J

K

L

M

P

Q

HIC
ANIMO TOT INTER RERVM MIRA PACATO
ARCANA NATVRAE PRIMORDIA
INTRA ATOMOS VOLVENTIA
DOCTORVM COETVI POSTREMVM APERVI
MEVM VNDE NOMEN IAM IMMORTALE

Finito di stampare nel mese
di giugno 2004 da
Compositori Industrie Grafiche